D1654313

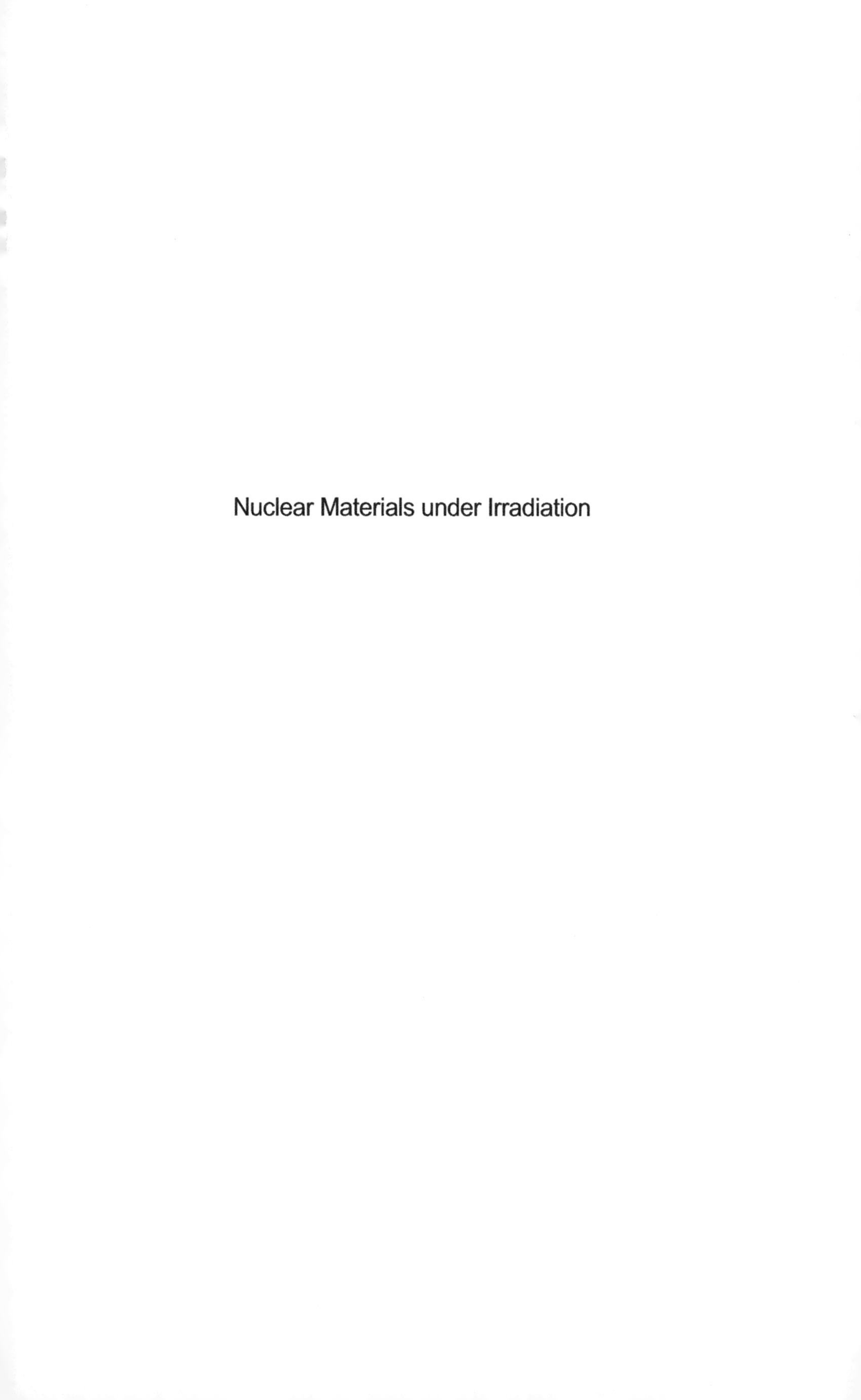

Nuclear Materials under Irradiation

SCIENCES

Energy, Field Directors – Alain Dollet, Pascal Brault

Nuclear Energy, Subject Head – Sylvain David

Nuclear Materials under Irradiation

Coordinated by
Serge Bouffard
Nathalie Moncoffre

WILEY

First published 2023 in Great Britain and the United States by ISTE Ltd and John Wiley & Sons, Inc.

ISTE Ltd
27-37 St George's Road
London SW19 4EU
UK

www.iste.co.uk

John Wiley & Sons, Inc.
111 River Street
Hoboken, NJ 07030
USA

www.wiley.com

Library of Congress Control Number: 2023938464

British Library Cataloguing-in-Publication Data
A CIP record for this book is available from the British Library
ISBN 978-1-78945-148-1

ERC code:
PE2 Fundamental Constituents of Matter
PE2_3 Nuclear physics
PE4 Physical and Analytical Chemical Sciences
PE4_14 Radiation and Nuclear chemistry

Contents

Preface

Serge BOUFFARD[1] and Nathalie MONCOFFRE[2]
[1] *CIMAP, CEA – CNRS – ENSICAEN, Université Caen Normandie, France*
[2] *Institut de Physique des 2 Infinis de Lyon,*
Université Claude Bernard Lyon 1, CNRS/IN2P3, Villeurbanne, France

The nuclear power industry is based on the major discoveries of the early 20th century: the discovery of natural and then artificial radioactivity, the neutron and lastly the fission of uranium. America's mobilization for the atomic weapon during World War II allowed the first nuclear reactor to be built very quickly. Thus, between 1942 and the end of the war, the Chicago pile 1 was followed by seven others. Their goal was to produce the plutonium that was necessary for the atomic bomb. Even though material considerations were not the focus of attention for these graphite-gas reactors of modular design, the effects of irradiation on materials were beginning to be known. Consequently, as early as 1942, Eugene Paul Wigner warned of the defects created in graphite and the energy stored in it.

However, it was with the first nuclear power reactors that research on irradiation damage really took off. Although the basic phenomena are now relatively well known, their complexity in industrial materials means that research is still active. Moreover, the increase in the life span of reactors requires a better understanding of the aging mechanisms of materials. Even though it is only slightly discussed in this book, the future 4th generation reactors, especially fusion reactors, will impose much harder irradiation and operating temperature constraints, and therefore specific research. All this research takes full advantage of the advanced techniques of material studies, which are increasingly efficient, making it possible to finely probe the material, giving access to a more accurate vision of the structure of materials and their defects.

Nuclear Materials under Irradiation,
coordinated by Serge BOUFFARD and Nathalie MONCOFFRE. © ISTE Ltd 2023.

The objective of this book is to provide the basis for research on nuclear materials subjected to irradiation, with a view to contextualizing them in the industrial environment. The reader will find chapters on nuclear reactor materials (vessel and internal steels, fuel cladding alloys, neutron absorbers and, of course, nuclear fuel). A chapter is devoted to graphite, which played an important role in the first generation of reactors. The management of nuclear waste and the safety of long-term storage and disposal are essential points, so it is important that there is a chapter on storage glasses. Two types of materials are found both on the reactor side and in the management of waste: concrete and organic materials. These material chapters are completed by four others on the basic notions of defects, radiolysis, and irradiation and characterization tools.

We thank the CNRS – IN2P3, and in particular Sylvain David, for offering to coordinate this book, which only exists thanks to the work of all the co-authors and their expertise recognized in the academic or industrial world. We would also like to thank Emmanuel Balanzat, Nicolas Bérerd, Pascal Bouniol, Christine Delafoy, Christophe Domain, Muriel Ferry, Frederico Garrido, Aurélie Gentils, Stéphanie Jublot-Leclerc, Sophie Le Caër, Philippe Pareige, Laurent Petit, Yves Pipon, Jean-Philippe Renault, David Siméone, Patrick Simon and Magaly Tribet for accepting this additional work.

1

Irradiation Defects

Serge BOUFFARD[1] and David SIMÉONE[2]

[1] *CIMAP, CEA – CNRS – ENSICAEN, Université Caen Normandie, France*

[2] *CEA/Paris-Saclay, Service de Recherches Métallurgiques Appliquées, Gif-sur-Yvette, France*

1.1. Introduction

The creation of defects is at the origin of the history of irradiation aging of nuclear materials. Subjected to bombardment by charged particles (electrons, ions) or neutral particles (photons, neutrons), the atoms of solids can be displaced from their equilibrium sites. The concentrations of displaced atoms and vacancies created depend, of course, on the nature of the particles, their energy, the flux and the fluence. These concentrations are, in all cases, much higher than those existing at the thermodynamic equilibrium. This is why the behavior of solids under irradiation escapes the thermodynamic description, which is nevertheless widely used by engineers to describe the microstructure (phase morphology, etc.) and properties (elastic constants, etc.) of materials. After this phase of defect production, a series of steps will follow during which the defects will migrate, and interact with each other and with impurities and native defects (dislocations, cavities, macles, grain boundaries, etc.). These different steps are, in essence, material-dependent and much more complicated to describe in a general way. Indeed, the recombination kinetics of defects created under irradiation depend on the nature of the interatomic bonds, the crystal structure, the cohesion energy, etc. The study of the microstructure induced by irradiation is therefore part of the thermodynamics of so-called forced systems, developed by the Prigogine school in the 1980s (Walgraef 1996). The continuous injection of defects acts as an external "force", preventing the material from relaxing towards its thermodynamic equilibrium. The microstructure induced

Nuclear Materials under Irradiation,

by the irradiation depends on the conditions of this irradiation (the particle, its energy, the rate of production of the defects, the temperature, etc.) and on the initial state of the material (purity, density of dislocation and clusters, grain size, etc.). The new microstructure that is thus produced under irradiation will modify the properties of use, typically by degrading them. Outside the nuclear sphere, there are many cases in which the impact of irradiation can be positive. Some examples are given at the end of section 1.2.3.

This chapter is not intended to be a detailed course on the mechanisms of defect creation, nor on the different kinetics occurring in materials under irradiation, but it will give readers the elements to understand the following chapters on different materials.

1.2. Some basic data

Neutrons or any other fast particle lose their kinetic energy in matter during elastic or inelastic collisions. Elastic collisions imply the conservation of kinetic energy and the momentum of the projectile–target set. Collisions are said to be inelastic (conservation of the momentum, but not of the kinetic energy) when there is a change in the internal energy of the projectile and/or the target. The excited particle de-excites through the emission of X photons if the electronic system has been excited or gamma photons if it is the nucleus, or by a profound rearrangement of the nucleus with the emission of ions of high kinetic energy (recoil nucleus, alpha) or radiation β. All these conditions of irradiation are found in the nuclear power industry. This chapter gives a brief overview of the different elementary mechanisms of defect production induced by elastic and inelastic collisions. For more details on these mechanisms, as well as on the transport of particles in materials, see, for example, Agullo-Lopez et al. (1988), Balanzat and Bouffard (1993), Was (2017) and Ortiz et al. (2020).

1.2.1. *Radiative environments and nuclear materials*

In the nuclear industry, materials undergo irradiation from four major sources of irradiation: 1) the reactor core, which imposes the most severe conditions, 2) fuel extracted from the core that has reached its maximum burnup rate, 3) recycled fuel containing plutonium and uranium from reprocessing (MOX) and 4) nuclear waste packages, mainly those of high activity. These sources emit practically all types of particles: fast neutrons (1–2 MeV), fast heavy ions (fission fragments), low-energy heavy ions (recoil of the nucleus when an alpha is emitted), light ions (alphas of a few MeV), and, of course, gammas and betas that are emitted by actinides and fission products.

The reactor core: in operation, the core of a 1,300 MWe reactor has a nominal thermal power of 3,817 MWth, of which 6.5% comes from the decay of fission products. Each fission of ^{235}U releases an average of 202.8 MeV. The core of a 1,300 MWe reactor is thus the site of 1.1 10^{20} fissions per second, generating the emission of 2.8 10^{20} neutrons/second. The energy spectrum of the neutrons depends on their production rate, their capture cross-sections and their thermalization rate. The shape of the spectrum is schematically represented in Figure 1.1. Fast neutrons whose kinetic energy is greater than 0.1 MeV are responsible for the creation of defects. The flux of these neutrons is in the range of 10^{18} $m^{-2}.s^{-1}$.

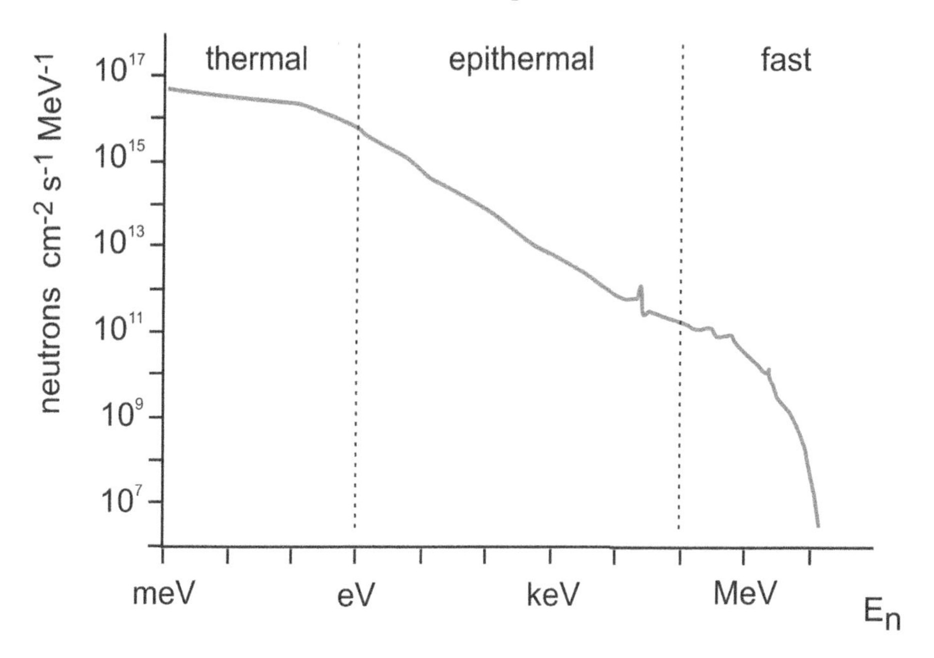

Figure 1.1. *Neutron spectrum of a PWR reactor (IAEA/CRP DPA data, 2019). The damage of materials is due to the fast part of this spectrum. For a color version of this figure, see www.iste.co.uk/bouffard/nuclear.zip*

Neutrons are the cause of damage to all the materials in the reactor vessel (internal structures, fuel, cladding, control rods, etc.) and to the vessel itself, but they are not the only particles. They are accompanied by a large flux of gammas, whose main action is to participate in the heating and, indirectly, to produce some defects by the emission of high energy electrons by Compton scattering. Within the fuel, the fission fragments are an important source of irradiation by fast heavy ions (a light fragment A = ~95, ~98 MeV and a heavy fragment A = ~138, ~68 MeV). They are also, with the alphas, the source of gas bubble creation at the origin of fuel swelling.

Spent fuel: when a 1,300 MWe PWR reactor is shut down, the residual power as a result of the activity of the fission fragments is equal to 265 MW, or about 7% of the thermal power of the reactor (23 MW after one day and 1.5 MW after one year). After one year of cooling, a fuel assembly[1] has a power of about 8 kW. After four years of cooling, an assembly of UOX enriched to 3.5% in ^{235}U and having reached a burn-up rate[2] of 33 $GWjt^{-1}$ has an activity of 24,000 TBq, of which 32% comes from actinides, 67% from fission fragments and less than 1% from the activation of metallic elements (Guillaumont 2004). This significant flux of particles requires radiation protection and transport precautions, the legal limit being fixed at 5.3 kW per container, but this radiative environment does not really pose an irradiation problem for the materials.

MOX fuel: the process of recycling spent fuel leads to the fabrication of mixed UO_2–PuO_2 fuel. Given that the majority of ^{239}Pu has a lifetime of 24,110 years, much shorter than that of ^{235}U ($7.04\ 10^8$ years), the plutonium oxide grains are relatively intense sources of alpha irradiation: $2.3\ 10^9\ \alpha/g_{Pu}/s$. However, with an isotropic emission, a non-negligible part of the alpha energy is deposited in the PuO_2 grains. Nevertheless, this activity must be taken into account in the MOX manufacturing process: aging of the transfer tubes of the oxide powder and of the various organic additives added, before the sintering operation.

Waste packages: the different categories of nuclear waste do not present the same problems of irradiation. Indeed, the vast majority of waste is low and intermediate-level short-lived waste (LILW-SL) and very low-level waste (VLLW), which do not present any problem of aging of materials under irradiation. In contrast, high-level waste (HLW) integrated into a glassy matrix is subjected to intense irradiation with a dose rate of 20 kGy/hour at one year and a cumulative dose of 10–20 GGy at 10,000 years (see Chapter 5 of this book). Intermediate-level long-lived waste (ILW-LL) may contain organic materials. For these materials, the focus is on the emission of dihydrogen and molecules that can corrode the container under irradiation (see Chapter 8 of this book).

1. A 1,300 MWe reactor core is composed of 193 assemblies of 264 rods.

2. Burn-up rates are given in units of energy produced per unit mass of fuel.

1.2.2. *Some notions about the transport of particles in matter*

1.2.2.1. *Ion transport*

Even though the vast majority of researchers use the SRIM software (Stopping and Range of Ions in Matter) to calculate the ion range, stopping powers and number of displacements (Ziegler and Biersack 1985), it is worthwhile to recall some data on ion transport in matter. First of all, let us give some reminders about SRIM. The values for the nuclear stopping power come from an analytical model based on the Firsov potential (see equation [1.1]) (Wilson et al. 1977)

$$-\left(\frac{dE}{dx}\right)_{nucl.} = \frac{8{,}462 Z_1 Z_2 m_1}{(m_1+m_2)\sqrt{Z_1^{2/3}+Z_2^{2/3}}} \frac{\ln(1+\varepsilon)}{2\varepsilon(1+0{,}1071\varepsilon^{0{,}37544})} \qquad [1.1]$$

with $\varepsilon = \dfrac{32{,}53 m_2}{Z_1 Z_2 (m_1+m_2)\sqrt{Z_1^{\frac{2}{3}}+Z_2^{\frac{2}{3}}}} E;$

Z_1, m_1 are the atomic number and mass of the projectile (Z_2, m_2 for the target).

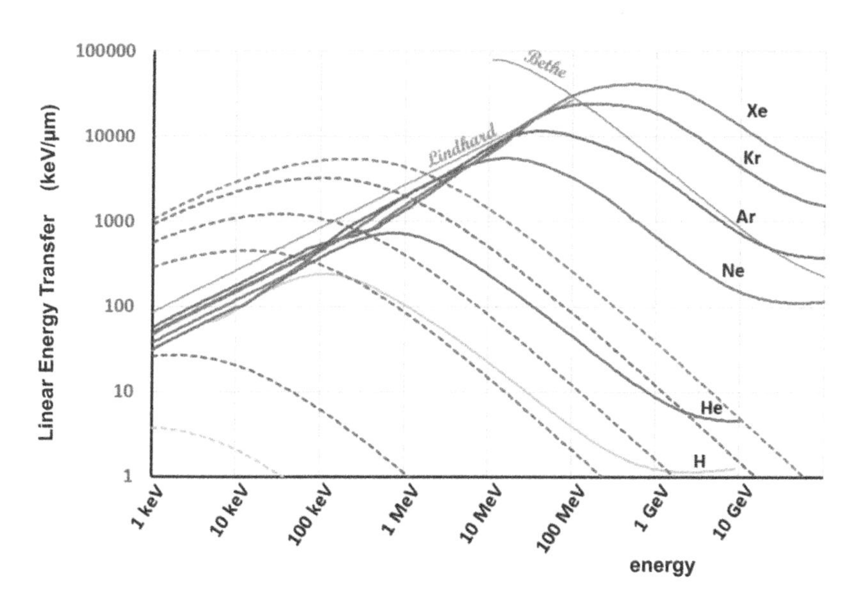

Figure 1.2. *Stopping power of iron for different ions. The nuclear stopping powers are shown using dashed lines, and the electronic stopping powers are shown using solid lines, according to the SRIM software. The electronic stopping powers of argon are shown using light green lines, according to the formulas of Lindhard and Bethe. For a color version of this figure, see www.iste.co.uk/bouffard/nuclear.zip*

As for the electronic stopping power, it comes from adjustments on the experimental data, energy domain and atomic number class[3]. Indeed, there is no theory or model to calculate the electronic stopping power in the whole energy range. We find a perturbative calculation by Bethe (1930), which only applies in the high energy range[4], as shown in Figure 1.2. It can be extended to low energies by considering that the charge of the ion decreases with its energy, Z_1 is replaced by a Z_1^*, an effective charge

$$-\left(\frac{dE}{dx}\right)_{electronic} = 4\pi N Z_2 \frac{Z_1^2}{m_e v_1^2}\left(\frac{e^2}{4\pi\varepsilon_0}\right)^2 ln\left(\frac{2m_e v_1^2}{I}\right) \quad [1.2]$$

with m_e being the mass of the electron, v_1 being the speed of the projectile, N being the number of target atoms per unit volume, and I being the mean ionization energy ($I \approx 10Z_2$).

The electronic stopping power is, at low energy, proportional to the speed of the ion. There are several models based on different interaction potentials (Firsov 1959; Lindhard and Scharff 1961). The loss of energy by electronic excitation dominates at high energy, but it will not necessarily be the dominant mechanism for the creation of defects. Indeed, the yield of this process can be zero.

For all these calculations, the ions are assumed to be at their equilibrium charge, but near the surface, this is not always verified. This is particularly the case for experiments with multi-charged slow ions. Indeed, when a highly charged slow ion (e.g. Ar^{18+}) approaches a surface, it locally captures a large number of electrons, thus changing the stability of the material (Haranger et al. 2006; Aumayr et al. 2011).

With knowledge of the stopping powers, it is possible to calculate the ranges:

$$R_{(E)} = \int_E^0 -\frac{dE}{\left(-\frac{dE}{dx}\right)_E} \quad [1.3]$$

This calculation gives the mean length of the trajectories of the ions (R). However, it is the depth of penetration of the ions, known as the projected range (Rp), that is the quantity of interest for the irradiations. Figure 1.3 shows the difference between R and Rp. The projected range and the range distribution can be calculated analytically in the case where the electronic excitations are negligible

3. Hence, the existence of a few slope disruptions that can be observed in Figure 1.2 at approximately 100 keV.

4. A relativistic version has been proposed since 1932.

(Winterbon et al. 1970); otherwise, it can be calculated more generally using Monte Carlo-type simulations.

The mechanisms of energy transfer are described in great detail and clarity in Peter Sigmund's books (Sigmund 2006, 2014).

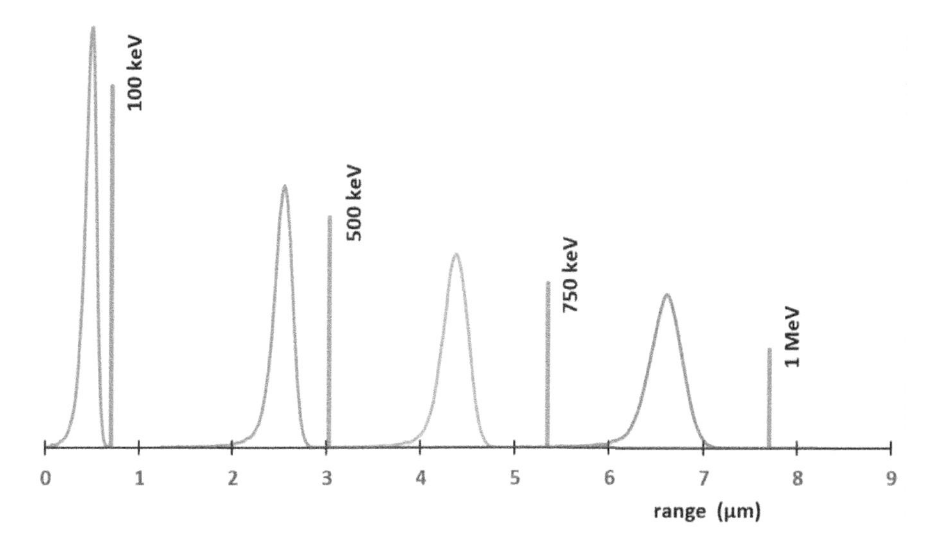

Figure 1.3. *Range in iron of a proton beam for four energies. The bell-shaped curves represent the statistics of the ranges projected on the direction of incidence (Rp), and the vertical green lines represent the ranges along the trajectories (R). For a color version of this figure, see www.iste.co.uk/bouffard/nuclear.zip*

1.2.2.2. *Electron transport*

The main differences for the transport of electrons come from their speed and mass. Indeed, from a few hundred keV, they must be considered as relativistic and be treated as such: relativistic formulas, introduction of the Bremsstrahlung effect and Cherenkov emission. As a result of their low mass, they will undergo much larger angular deflections during elastic collisions, with a high probability of being backscattered. The difference between R and Rp is maximal.

1.2.2.3. *Neutron transport*

During inelastic collisions, part of the kinetic energy of the neutrons is converted into excitation of the nucleus, which will de-excite by emitting a gamma or another particle. The neutrons can also be captured to give a heavier isotope, usually following the reaction ^{A}X (n,γ) ^{A+1}X. From the irradiation perspective, these capture processes are the source terms for the introduction of impurities and gases. However, the main interaction of neutrons with the target atoms remains the elastic collisions. To a first approximation, the interaction potential can be approximated by a hard sphere potential, thus with an equiprobability of transferring energy between 0 and T_{max}, undergoing an angular deflection between 0 and 180°. Figure 1.4 shows the evolution of the energy of a neutron beam with an initial energy of 1 MeV as a function of the number of collisions with carbon atoms, calculated with a hard sphere potential. In a reactor core, neutrons are found at all stages of thermalization. Additionally, a particularity of neutron irradiation is that materials are always irradiated by a spectrum of neutrons, as shown in Figure 1.1.

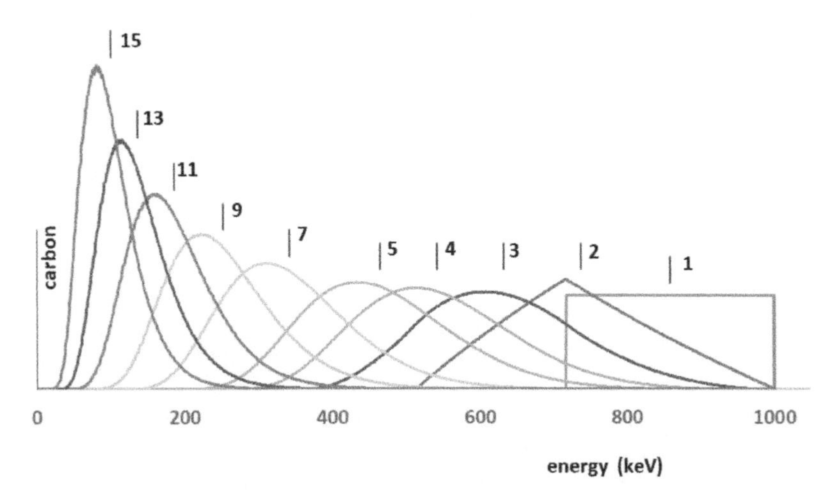

Figure 1.4. *Evolution of the energy distribution of a neutron beam with an initial energy of 1 MeV as a function of the number of collisions (from 1 to 15) in graphite. The average energy (vertical lines) decreases as $E((m-1)/(m+1))^n$. For a color version of this figure, see www.iste.co.uk/bouffard/nuclear.zip*

1.2.2.4. *Available software*

There are a large number of codes to describe the transport of charged particles in matter. These codes are all based on the solution of the linear Boltzmann equation. This equation is either solved numerically (DART code) or via Monte Carlo-type simulations. Although the most widely distributed code in the community is SRIM, there are generally a few codes developed to simulate particle physics experiments and for radiation protection purposes.

– **PENELOPE** (Penetration and ENErgy LOss of Positrons and Electrons) is a Monte Carlo code for the simulation of electron and photon transport in materials in the energy range of 50 eV to 1 GeV. The code is based on data from first principles calculations, semi-empirical models and databases. PENELOPE tracks particles in complex geometries limited by quadratic surfaces (Baro et al. 1995). PENELOPE is distributed by the OECD-Nuclear Energy Agency at http://www.oecd-nea.org/lists/penelope.html.

– **MCNP** (Monte Carlo N-Particle transport) can be used for the transport of neutrons (<20 MeV), ions, photons and electrons from 1 keV to 1 GeV in a three-dimensional geometry (Forster and Godfrey 1985). MCNP is distributed by the Los Alamos National Laboratory at https://mcnp.lanl.gov/.

– **DART** makes it possible to estimate the rate of production of vacancies, interstitials and anti-site defects produced by a beam of neutrons, ions and electrons in the material. The main interest of this code is to use different libraries of neutron-isotope cross-sections, in order to calculate the damage production rates induced by all nuclear reactions for a given material (Lunéville et al. 2006). DART is distributed by the OECD-Nuclear Energy Agency at http://www.oecd-nea.org/tools/abstract/detail/nea-1885.

– **GEANT4** is a collaborative software that simulates the transport of particles in matter (Geant4 collaboration 2003). Initially dedicated to high-energy physics, it has been extended to low-energy physics, in particular for applications in particle therapy. The software is a bit too heavy to use for simple calculations of the path and dose distribution. GEANT4 is distributed by CERN at https://geant4.web.cern.ch/.

1.2.3. *The zoology of irradiation defects*

Obtaining a defect-free material is practically impossible, regardless of the method of production. There are several reasons for deviations from a perfect crystallographic structure, such as the presence of impurities, inhomogeneities and temperature fluctuations in the growth furnaces, thermal cycles or mechanical stresses applied to the material, and irradiation by cosmic and telluric radiation. Figure 1.5 gives some examples of defects. Defects can be classified according to their dimensionality: point defects or dimension 0 such as the vacancy, the interstitial or a foreign atom, dimension 1 involving dislocations, dimension 2 such as macles, stacking faults, grain boundaries, etc., dimension 3 such as pores, bubbles and precipitates, etc. Most of these defects exist at thermodynamic equilibrium in solids. It is, for example, possible to estimate the concentration of vacancies as a function of temperature and at constant pressure in a metal using the following formula: $C_e^{eq} = exp(S^f/k)exp(-H^f/kT)$. The latter is obtained by minimizing the Gibbs energy (k being the Boltzmann constant, S^f and H^f being the entropy and entropy of formation, respectively). Thus, in a metal like gold at a temperature equal to nine-tenths of the melting temperature, the equilibrium concentration of vacancies is equal to 10^{-5}. The mechanical stress of a material can also cause defects such as dislocations. However, under irradiation, a wider range of topological defects is generated for those existing at thermodynamic equilibrium, in atomic fractions of higher orders of magnitude.

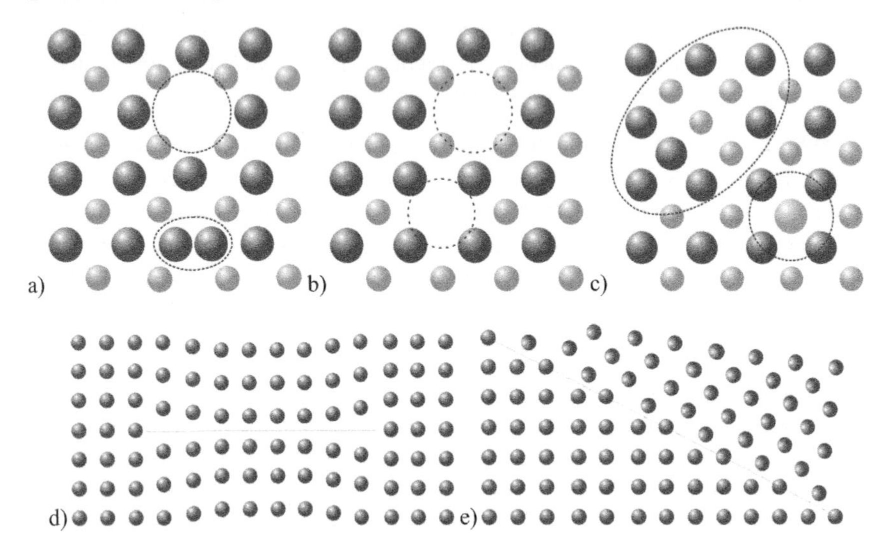

Figure 1.5. *Some examples of defects: a) Frenkel pair: vacancy and interstitial on the same sublattice, b) Schottky defect: an anionic vacancy associated with a cationic vacancy, c) an anti-site defect and an impurity in substitution, d) dislocation vacancy loop and e) macle. For a color version of this figure, see www.iste.co.uk/bouffard/nuclear.zip*

Moreover, in metallic materials, the defects formed are electrically neutral and the concentrations of vacancies and interstitials can therefore be different. On the other hand, in semi-conductors and insulators, electrical neutrality must be ensured locally, so that the formation of a single vacancy will be accompanied by a local redistribution of electrical charges. Depending on its environment, this vacancy may have a different charge, which is likely to modify its properties.

If the term "defect" with a negative connotation is used, it is in reference to a perfect crystalline order that is repeating itself in all directions to infinity (at least to the limits of the crystal). In reality, disorder governs many properties and improves a number of them:

– Electrical properties: the entire operation of microelectronics is based on the doping of silicon. The introduction of very small quantities of electron donor elements (P, As, Sb, etc.) or receivers (B, Ga, In, etc.) radically changes the electrical conductivity, as well as the transport mechanism, by electrons or by holes.

– Optical properties: alumina α-Al_2O_3, which is transparent when pure and monocrystalline, becomes a precious stone with the addition of impurities. It becomes a red ruby with about 1% of chromium, and a blue sapphire for titanium and iron. An equally important occurrence is the rare earth doping of crystals (YAG, $LiYF_4$, Y_2O_3, CaF_2, etc.) to obtain laser sources and optical amplifiers.

– Mechanical properties: the dislocation sliding is at the origin of the plasticity of metals, but their trapping by other defects induces hardening and embrittlement.

– Superconducting properties: the pinning of vortices by fast ion traces in type II superconductors makes it possible to increase the critical current.

– Magnetic properties: in ferromagnetic materials, the Bloch walls can be trapped by defects, thus increasing the coercive field.

Defects therefore play an important role in the properties of materials. Irradiations create a non-equilibrium situation by increasing the defect concentration, which accelerates the evolution of material properties.

1.3. Defect creation mechanisms

The creation of defects by irradiation is the result of the interaction of particles with the material. We can distinguish three types of particles that will interact with the constituents of the material: photons, neutrons and charged particles. Photons, UV, X-rays and γ are scattered or absorbed by the electronic system of the target, creating electron–hole pairs, which, in certain materials, can be the source of a radiation chemistry. Neutrons will only interact with the nucleus of the atoms of the material. Depending on their energy and the nature of the nucleus, the collision can be elastic (transfer of kinetic energy to the nucleus) or inelastic (with internal excitation of the nucleus, or even a nuclear reaction). In both cases, defects will be created. As for the charged particles, the ions and electrons, they interact with the ions and electrons of the target. The slowing down of the ions is thus composed of a succession of inelastic collisions of low energy transfer and elastic collisions with the nuclei, shielded by a certain number of electrons. In general, we can consider that the ions undergo a continuous slowing down by electronic excitation between two elastic collisions, which will induce an angular deviation. The two types of energy loss are therefore considered as independent. In reality, a collision with a nucleus can only disrupt the electron cloud. In the case of electrons, the situation is identical, except that their angular deviation will be much more significant as a result of their small mass.

The different processes we have just mentioned are characterized by their interaction probability and their total and differential sections in angles or energy.

	Neutrons	Protons	Electrons
Total elastic cross-section	~10^{-28} m²	~10^{-22} m²	~10^{-24} m²
Mean free path between two collisions	16 cm	160 nm	1.6 μm
Mean energy transferred	68 keV	220 eV	40 eV

Table 1.1. *Total elastic cross-section, mean free path between two collisions and mean energy transferred to a target atom for three 1 MeV particles in aluminum*

The total cross-sections in Table 1.1 give an indication of the mean distance between two elastic collisions. During one of these collisions, a target nucleus may receive enough energy to be displaced from its equilibrium position, creating a vacancy, before stabilizing in an interstitial position, in a normally unoccupied site. The vacancy-interstitial pair is called the Frenkel pair. If the atom is ejected with enough energy, it becomes a projectile that can create a succession of collisions leading to a localized energy deposit in a volume of a few tens of nanometers. These energy deposits, which occur randomly within the material, are called collision cascades. The electronic excitations produced during this localized energy deposition are transformed into heat in metallic materials, but can also lead to the formation of charged defects in insulators and semi-conductors.

Once the defects are created, they evolve according to complex processes that are similar to the thermally produced defects. Indeed, many of the defects formed are mechanically or chemically unstable and evolve spontaneously towards more stable configurations, which can lead to phase transitions and drastic changes in the properties of the materials. The strongly non-homogeneous nature of the energy deposit reflects the essence of difficulty in modeling and simulating (e.g. via a heat flux, a pressure wave or doping) the mechanisms of creation of defects induced by the irradiation and, a fortiori, the long-term evolution of materials and their properties under irradiation.

1.3.1. *Creation of defects by elastic collisions*

The creation of defects by elastic collision is a universal process that exists regardless of the structure and properties of the material. On the other hand, it is not always the dominant process of defect creation. For a defect to be created during an elastic collision, there must be a sufficient amount of energy transferred in order to extract the atom from its site and send it out of the spontaneous recombination volume, thus forming a Frenkel pair. This reflects the existence of a minimal energy that the target atom must receive for a defect to be created, the threshold energy of displacement. In reality, this threshold energy depends on the emission direction, so there is a displacement threshold distribution. The mapping of the displacement threshold energy has only been measured for a few metals, such as copper (King et al. 1981), but we will see that an average value is largely sufficient. Table 1.2 presents some values of displacement threshold energy.

Cu	Ni	Fe	Ag	Ti	Al	Mo	UO_2	SiC
22	24	24	28	29	32	37	40 (U) 20 (O)	25 (Si)

Table 1.2. *Some examples of displacement threshold energy in eV, for metals (Lucasson and Walker 1962), UO_2 (Soullard 1985) and SiC (Lefèvre et al. 2009)*

We consider that, for a transmitted energy T, the number of displacements is equal to $n(T) = 0\ si\ T < T_s$ and $n(T) = 1\ si\ T \geq T_s$, with T_s being the threshold displacement energy. If the transmitted energy is large in front of T_s, the ejected atom, known as PKA for primary knock-on atom, in turn becomes a projectile (see Figure 1.6). The number of displacements $n(T)$ thus increases with the transferred energy, $n(T) = T/2T_s$ if we consider that all energy is converted into displacement (Kinchin and Pease 1955) or, more realistically, if we consider that when $T \geq 2{,}5\ T_S$, a fraction of the PKA's energy is lost through electronic excitation: $n(T) = 0{,}8\hat{T}/2T_s$, where $\hat{T}$ is the damage energy, in other words, the energy of the primary minus the energy lost through electronic excitation by the primary (Norgett et al. 1975).

The number of displacements per atom produced by a particle flux[5] $\varphi(t)$ of energy E is given by:

$$n_{dpa} = \int_0^t \int_0^{T_{max}} n(T) \frac{d\sigma}{dT}(E,T)dT\varphi(t)dt \qquad [1.4]$$

5 Flux is the number of particles per unit area and time, usually per cm^2 per second, and fluence is the integral of the flux.

where $\frac{d\sigma}{dT}(E,T)$ represents the probability of creating an energy primary T. This probability depends on the interaction potential between the two particles (hard sphere, Coulombian, shielded sphere, etc.).

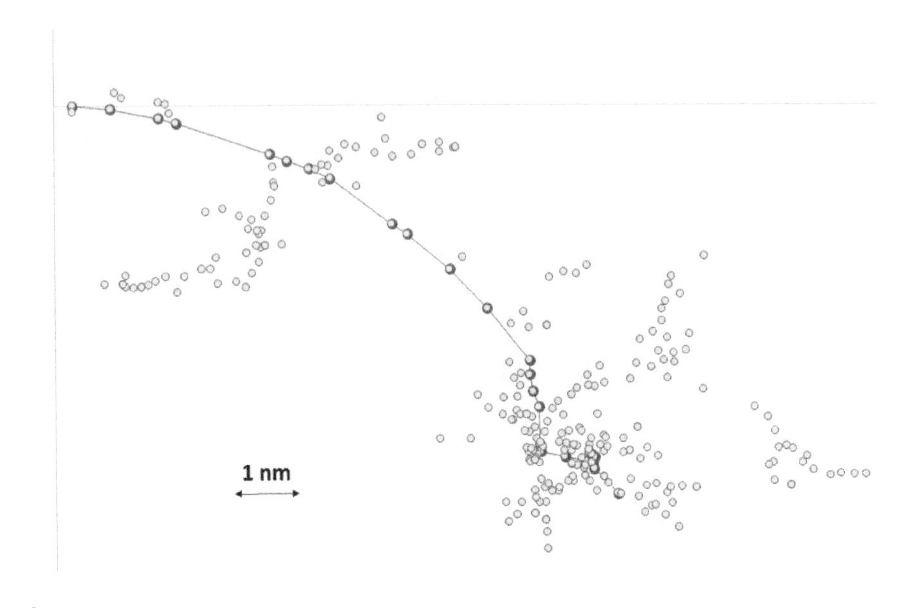

Figure 1.6. *A displacement cascade created by a 10 keV primary in iron from a SRIM simulation (Ziegler and Biersack 1985). The symbols indicate the location of a vacancy creation, and the solid symbols mark the path of the primary ion. For a color version of this figure, see www.iste.co.uk/bouffard/nuclear.zip*

It is important to keep in mind that **the number of dpa (displacements per atom) and the number of defects are two different quantities**. The dpa number represents the number of atoms that have been ejected from their site. As an atom can be displaced several times, the dpa number can be greater than 1. For example, the internal structures of reactor vessels nominally undergo damage in the range of 60 dpa in 60 years and, obviously, the number of defects cannot be greater than the number of sites, with the number even being much lower as a result of the recombinations between defects (with a recombination volume of 1.5 nm radius, the maximum concentration is about 4 10^{-3}). Furthermore, the number of defects is difficult to calculate because it depends on many parameters: the particle through the size of the collision cascades, the fluence reached, the temperature, the material through its structure, the migration energy of the defects and the presence of traps for the defects (surface, grain boundaries, dislocations, etc.).

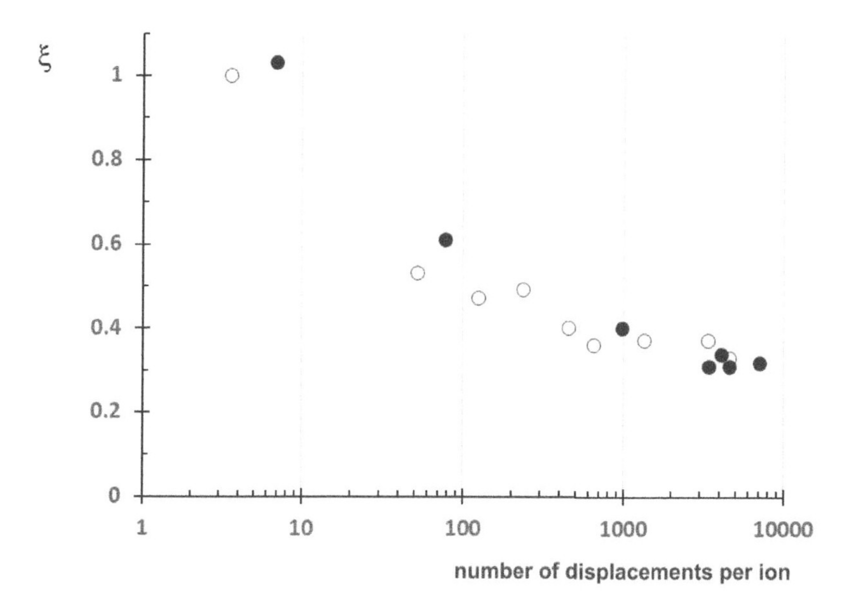

Figure 1.7. *Damage efficiency (ratio of measured to calculated number of Frenkel pairs) as a function of the number of displacements created by an ion in copper (solid dots) and silver (white dots), from Averback et al. (1978). For a color version of this figure, see www.iste.co.uk/bouffard/nuclear.zip*

On pure metals, some experiments at low temperature and low dose have shown that the fraction of non-recombined Frenkel pairs in the collision cascade decreases when the number of dpa increases[6] (see Figure 1.7). This possibility, in the case of metals and metal alloys, to relate the number of non-recombined Frenkel pairs to the number of displacements, regardless of the fact that this relationship is neither universal nor simple, is the reason why dpa has become the reference for quantifying elastic collision damage, despite its obvious limitations. The dpa is only an estimate of the quantity of atoms set in motion in a material during an irradiation. It does not make it possible to calculate the number of defects present over the long term (a few picoseconds) in the irradiated material. Through its simplicity of calculation and its

6. After determining the resistivity of the Frenkel pair (ρ_F =1,45±0,5 μΩ.cm/atom% for copper), the measurement at very low temperature of the increase of the electrical resistivity makes it possible to calculate the number of Frenkel pairs.

universal nature, it is an estimate characterizing the damage produced under different conditions (irradiations in reactor and with ion beams, for example). This is also the reason why the exact value of the energy displacement threshold does not matter; the fact that the community agrees on a standard value is sufficient. Even so, there are proposals to improve the representativeness of dpa (Nordlund et al. 2018).

The existence of a recombination volume imposes a limit on the concentration of point defects, and the probability of creating a vacancy or interstitial outside a recombination volume decreases as the defect concentration increases. In addition, thermal or athermal diffusion of defects increases their probability of encountering a trap in the form of a point or extended defect (Sizmann 1978).

1.3.2. *Amorphization*

Under irradiation, many crystalline materials of nuclear interest amorphize. This amorphization induces deep modifications of their intrinsic properties such as self-diffusion coefficients, their mechanical properties (elastic constants, etc.). The amorphization of a material has a well-defined meaning in solid state physics and is, strictly speaking, translated by the disappearance of long-range order, in other words, Bragg peaks in diffraction and the disappearance of scattering peaks in infra-red and Raman spectroscopy. It should be noted that geologists use the term *metamict* to designate amorphized minerals (Ewing 1994).

Amorphization under irradiation has been observed in practically all classes of materials, except pure metals and some ceramics with a mean fluorine-like structure such as UO_2 and MgO. In the case of electron irradiations that only produce point defects, amorphization is usually associated with the accumulation of local deformations induced by point defects. It is, for example, the case for quartz irradiated by electrons (Pascucci et al. 1983). On the other hand, during irradiations by ions, the damage has a heterogeneous aspect due to the presence of collision cascades. In a very qualitative way, it is possible to describe this heterogeneous amorphization via an empirical approach developed by Gibbons (1972). The amorphized volume fraction is related to the fluence through a formulation called k-impact. The material becomes amorphous in zones where k impacts overlap (see Figure 1.8):

$$f_a = 1 - \left(\sum_{k=0}^{n} \frac{(\sigma\varphi)^k}{k!} \exp(-\sigma\varphi t) \right) \qquad [1.5]$$

In order to explain the existence of a critical amorphization temperature that has been observed experimentally in pyrochlores, more elaborate empirical models involving annealing of amorphous zones during the development of collision cascades have been proposed (Weber 2000).

However, all these models are phenomenological and cannot predict the amorphization of a material under irradiation. At best, they are used to classify materials by characterizing them via an "effective amorphization cross-section" derived from experiments.

The basic mechanisms of amorphization remain largely uncomprehended, partly because of the difficulty in quantifying the amorphous nature of a material. The development of total scattering diffraction measurements should allow a better quantification of the amorphous state (Meldrum et al. 1998; Egami and Billinge 2012). On the theoretical side, a few attempts were made to relate the amorphization of materials to the connectedness of polyhedra in ceramics. This geometrical approach aims to explain the difficulty in amorphizing simple crystalline structures such as fluorines (Hobbs 1995). The main challenge in understanding irradiation-induced amorphization is essentially due to the difficulty of explaining the formation of an amorphous state, which is a major issue in materials physics.

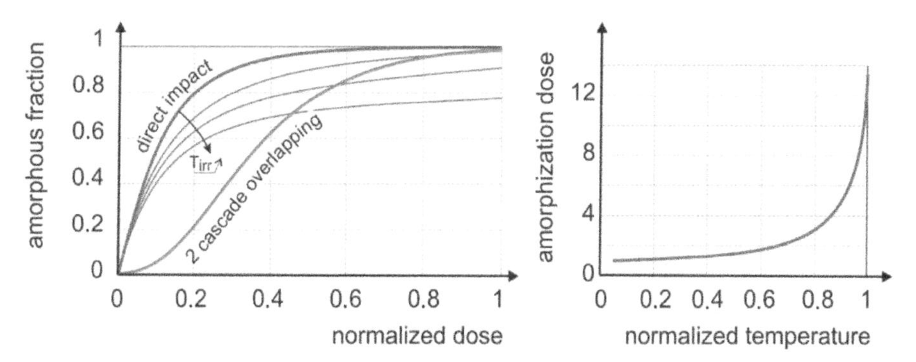

Figure 1.8. *On the left is the evolution of the amorphized volume fraction as a function of fluence for direct impact (k = 1 in the Gibbons formula) and double impact (k = 2) kinetics from the Gibbons formula, and on the right, the evolution of the amorphization dose as a function of temperature. For a color version of this figure, see www.iste.co.uk/bouffard/nuclear.zip*

1.3.3. *Creation of defects by electronic excitation*

The observation of the effects of electronic excitations on the color of certain materials was reported long before the origin was described. For example, in the 18th century, Schulze (1727) showed that the change in color of chalk that was soaked with a solution of silver in nitric acid, and exposed to sunlight, was due to the presence of silver. This is the basis of the photographic process. In the following century, Goldstein (1894) studied the origin of the coloring of sodium chloride subjected to cathode radiation. Establishing a link between phosphorescence and coloration, he introduced the hypothesis of atomic displacement. However, it was not until the 1930s that electronic excitations were associated with the creation of defects.

The structural modifications induced by electronic excitation result from more complex processes than those described for elastic collisions. In particular, their efficiency is very material-dependent, zero for metals, and maximal for organic matter. For an atomic displacement to occur, the energy absorbed by the electronic system must be transferred with high efficiency to the atomic lattice by a material-dependent mechanism. This mechanism is only described in detail in alkali halides. Therefore, it seemed interesting to devote a section to the creation of defects by electronic excitation in these materials, even if they are not found in nuclear materials.

Whatever the incident particle, the story begins with the creation of an electron–hole pair. In these materials, electrons and holes are strongly coupled to the lattice, resulting in strong atomic relaxation in their presence. In alkali halides, this relaxation of the atomic lattice, which is not very mobile, leads to a self-trapping of the hole on a pair of anions $[X_2^-]$. This configuration, known as the V_K center, can trap an electron to form a metastable species: a self-trapped exciton (STE). The de-excitation of the STE can be achieved in three ways: luminescences[7] π and σ or the creation of a F + H pair (see Figure 1.9), in other words, a Frenkel pair composed of an anionic vacancy that has trapped an electron (F center) and of a hole trapped on an anionic interstitial (H center). The yield of defect creation thus depends on the relative weight of these pathways. In some class I alkali halides, there is an anti-correlation in temperature between luminescence π and the creation of the F center. Luminescence dominates at low temperature and F-center creation at ordinary temperature. The F–H pair would result from a thermally activated conversion of the self-trapped exciton, following a potential surface of lower excitation. The yield will also depend on the distance at which the two elements of

7 Luminescence π comes from the recombination of the electron with the hole from the lower energy triplet state, whereas, for luminescence σ, the electron is in the singlet state.

the pair are created. The transport of the interstitial away from its vacancy is achieved by a sequence of replacement collisions along the <110> row of halogen ions. The efficiency of these collision sequences depends on geometric considerations: the diameter of the halogen atom D and the distance S between two halogens in the <110> direction. When S and D are very different, many close pairs are created; they will disappear during correlated recombinations in a few tens to hundreds of picoseconds. At 4 K, the energy to create an F center varies according to the ionic crystal from 10^3 to 10^7 eV, with a maximum efficiency for S/D between 0.5 and 1.5, in other words, approximately 1 (Townsend 1973). Noriaki Itoh has published many articles on this mechanism; the reference (Itoh and Tanimura 1990) is a good introduction.

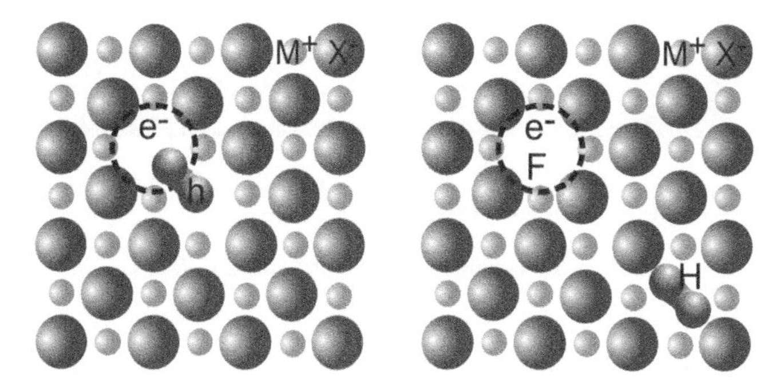

Figure 1.9. *Structure of a self-trapping exciton (left) and the F–H pair (right) in an alkali halide. The* $[X_2^-]$ *molecule can be shared on both halogen sites (on-center) or be localized on one site (off-center). In the latter case, the exciton prefigures the F–H pair. For a color version of this figure, see www.iste.co.uk/bouffard/nuclear.zip*

This example shows that both experimental and theoretical approaches have allowed us to understand the mechanisms of defect creation in these ionic crystals, but do not allow us to calculate a quantity comparable to the dpa of elastic collisions. The defect creation yield must therefore be determined experimentally. In order to consider defect creation by electronic excitation, it is necessary but not sufficient that four conditions are met:

1) the excitation must be localized on a limited number of sites;

2) its energy must be of the order of the threshold energy of displacement in the direction of displacement;

3) its life time must be of the order of the characteristic times of atomic vibrations;

4) there must be a non-radiative de-excitation pathway.

Metals do not meet the first three conditions, so they are considered insensitive to electronic excitations. For semi-conductors and insulators, electrons ejected to a high level of the conduction band relax to the bottom of the band. The energy available for the creation of a defect is then of the order of magnitude of the width of the band gap. Electronic excitations can therefore only create defects in large gap insulators.

The defect concentration depends on the amount of energy absorbed by the material per unit mass, in the form of electronic excitation and ionization. The coefficient that links the two quantities is called radiolytic yield. In the international system, it is expressed in moles per joule (mol/J). The legal unit of absorbed energy (dose) is the Gray [Gy], which is equal to 1 J/kg[8].

Materials	Silica 77K E' center	KBr 4K F center	Polystyrene cross-linking	Polyethylene creation H2
G 10^{-7} mol/J	~0.001	0.02	0.02	3.3

Table 1.3. *Radiolytic yield G for some defects*

Therefore, when this yield has been measured, the structural change can be expressed in defect concentration, unlike the calculation of the number of displacements produced by elastic collision, which does not allow us to calculate the number of stabilized defects. The defect concentration is related to the particle fluence by the following relation:

$$c_{(x)} = 1.6\ 10^{-15}\ G_{(mol/J)} \times LET_{(keV/\mu m)} \times \frac{1}{d_{(g/cm^3)}} \times \emptyset t_{(part./cm^2)} \qquad [1.6]$$

However, it is important to remember that only those species for which the signal was used to determine the efficiency are counted. Here, we speak of creation yield of F center, of cross-linking, etc. A defect which does not have an optical or magnetic signature could therefore not be taken into account. Given that only insulators are sensitive to electronic excitations, optical methods (infrared or

8. Within the volume of the material, the absorbed dose is equal to the dose given off by the incident particles. This is not necessarily true near a surface or an interface where secondary particles can escape from the material. For the energy absorbed by the material, we speak of the dose and KERMA (kinetic energy released per unit mass) for the energy lost by the particle. Both are expressed in J/kg.

UV–visible spectroscopies, etc.) are the preferred tools to follow the evolution of these materials under irradiation. In this case, it is necessary to know the molar absorption coefficients in order to access the absolute values of concentrations.

Moreover, as shown in Figure 1.10, the evolution of the defect concentration shows saturation at high doses. The yield therefore evolves continuously with the dose. The tabulated yields are generally the initial yields, at zero dose. There are also temporal variations of the yield. Indeed, time-resolved optical absorption experiments make it possible to measure the temporal evolution of the yield of transient species. In the case of radiolysis of water, 20 ps after excitation, the radiolytic yield of the solvated electron is equal to 4.2 10^{-7} mol/J., and after 200 ns, it is only 2.8 10^{-7} mol/J. In the long term, the solvated electrons all recombine and the yield tends to zero.

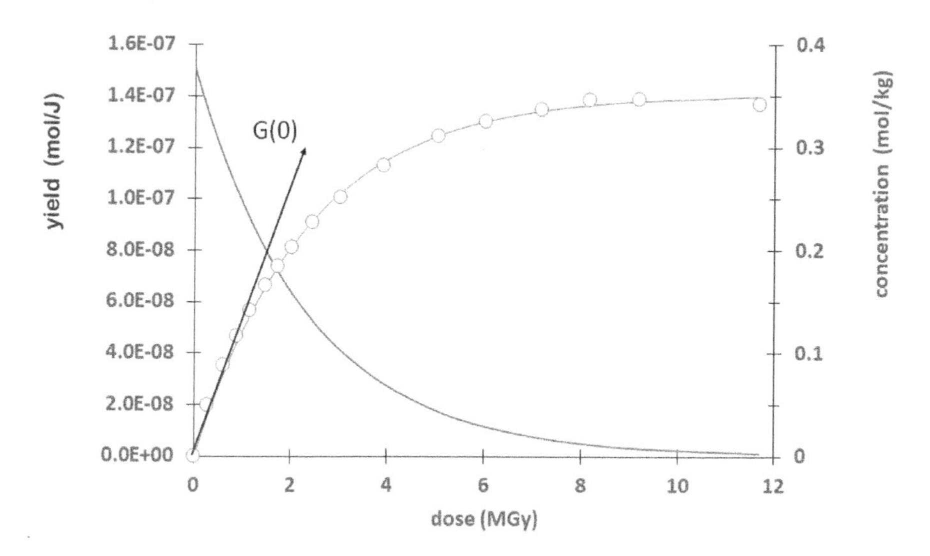

Figure 1.10. *Evolution of the trans-vinylene concentration as a function of the irradiation dose for polyethylene irradiated with 1 MeV electrons. The radiolytic yield is the derivative of the evolution of the defect concentration. The curves correspond to 1st-order kinetics. The experimental points are taken from Ventura et al. (2016). For a color version of this figure, see www.iste.co.uk/bouffard/nuclear.zip*

The density of electronic excitations is also a quantity that modifies the yields, as well as the mechanisms of the creation of defects, by electronic excitation.

1.3.3.1. *Specificities of high electron excitation densities*

These high electronic excitation densities are created in the wake of fast ions. In nuclear energy, fission fragments fall into this category, as well as alphas for the most sensitive materials. Although fission fragments were discovered in 1930, their effects on matter were not really studied until the late 1950s. Through different approaches, chemical attack (Young 1958) or electron microscopy in mica (Silk and Barnes 1959) and in UO_2 (Noggle and Stiegler 1960), these studies showed that the defects were rod-shaped objects, which they called tracks. Among all these works, those of Fleischer et al. (1965b) and Price and Walker (1962) played a significant role in the understanding and, in particular, the use of tracks. From their first publications, they proposed to use them in geological dating, or as particle detectors or filters for biological applications.

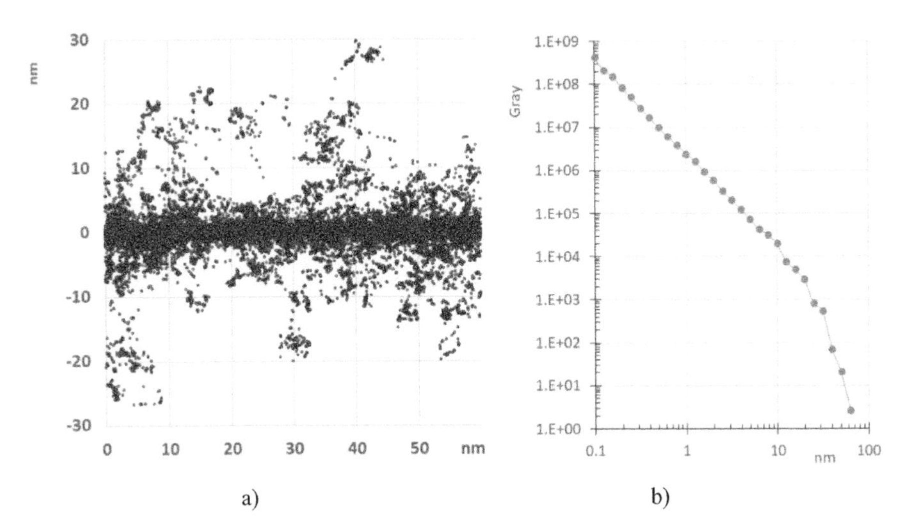

Figure 1.11. *(a) Distribution of ionizations around the trajectory of a 160 MeV Kr ion in UO_2 and (b) distribution of the dose as a function of the distance to the ion trajectory. For the Monte Carlo simulation, see Gervais and Bouffard (1994). For a color version of this figure, see www.iste.co.uk/bouffard/nuclear.zip*

What do we know now about these heavy ion tracks? First of all, they are the result of very high stopping powers by electronic excitation of swift heavy ions[9]. Monte Carlo simulations of the transport of these ions and of the electron cascades make it possible to realize the impressive ionization density around the ion trajectory

9. In the following text, the stopping power by electronic excitation will be called LET (linear energy transfer).

(see Figure 1.11). Within a radius of 1 nm around the trajectory of a 160 MeV krypton ion, 30% of the electrons are ejected. After the passage of the ion over a radius of 1 to 2 nm, all the atoms of the material are ionized, or even multi-ionized. As the electrons have been ejected with very little kinetic energy on average, many of them will be available for recombination after thermalization. Expressed in terms of energy deposited at the track core, the dose exceeds a hundred MGray.

Few materials remain insensitive to such an excess of energy. Tracks have been observed in minerals that are not sensitive to individual electronic excitations; this is the case, for example, for many oxides such as garnets ($Y_3Fe_5O_{12}$, etc.), as well as Al_2O_3 or UO_2 (Toulemonde 1999). More surprising is the appearance of damage for very strong electronic excitations in certain metals and metal alloys, such as iron (Dunlop et al. 1994). For polymers that are sensitive to individual excitations, new chemical bonds can appear, such as alkynes (Ngono-Ravache et al. 2007).

Structural modifications are only induced by strong electronic excitations above a threshold value, which depends on the material properties. When tracks are formed, their morphology evolves with the LET. Near the threshold, the tracks are discontinuous, probably due to fluctuations in the energy deposition. As the LET increases, the tracks become continuous and their diameter increases. The internal structure of the track depends on the material and the energy transfer; we can find disordered tracks, microcrystalline tracks, tracks from another phase, etc. Furthermore, the modification can take place by simple impact or during the overlapping of two impacts.

Considering the above for strong LETs, material damage does not come from the sum of individual events, but has a collective aspect. Collective effects were proposed very early on to describe the creation of extended defects. The first was a Coulomb repulsion model: the ion explosion spike model (Fleischer et al. 1965a; Fleischer 1981). In this approach, the projectile ion creates a small cylindrical zone containing a high density of positive ions. The Coulomb repulsion between all these ions ejects a certain number of them in the interstitial position. The authors consider that displacement occurs if the electrostatic stress is greater than one-tenth of Young's modulus. As the authors point out, this model is very simplified, particularly the electrostatic stress, which is calculated from a pair of ions even though it depends on all the forces between the charges created in the track.

In the case of metals, the life time of the space charge responsible for the pulse is less than a femtosecond, it is of the order of the inverse of the plasma frequency.

Despite this very short duration, the kinetic energy acquired by each ion is between 0.1 and a few eV, which is insufficient energy to move atoms. However, the spatial and temporal coherence of this response can be analyzed in terms of phonon excitations, which can be at the origin of a shock wave or the excitation of a plasmon soft mode (if it exists). Both processes can lead to atomic displacements in the form of phase change or defects. The few simple metals (Ti, Fe, Co, Zr) that are sensitive to strong electronic excitations all have several allotropic varieties, and their phonon spectrum has a soft mode associated with a displacive phase transition[10] (Dunlop et al. 1993; Lesueur and Dunlop 1993).

This Coulomb explosion model and its derivatives are not predictive or generic enough to be widely used. This model is nevertheless relevant in small isolated objects for which molecular physics uses a Coulomb explosion approach to interpret the results of ion–molecule or ion-aggregate collision experiments.

Dessauer (1923) explained the damage to irradiated biological cells by the creation of heat peaks. It was only after World War II that the "thermal peak" approach gained significance. In the 1960s, many groups developed thermal peak calculations to explain their experimental results, for example, Noggle and Stiegler (1960), Chadderton et al. (1963) and Izui (1965), a model that is included in the very comprehensive review article (Seitz and Koehler 1956). The basic assumptions of this model are:

– through its electronic stopping power, the ion transfers energy to the electrons of the target;

– these "hot" electrons will diffuse and thermalize initially by electron–electron interaction;

– this will ultimately lead them to transfer their energy to the atomic network by electron–phonon interaction;

– the transformed zone is assimilated to a molten zone, which, in the simulation, behaves like a liquid phase. The subsequent rapid quenching stabilizes the disorder of the liquid phase at lower temperatures.

By analogy with plasma physics, it is then possible to define a two-temperature model: a temperature T_e describing the behavior of an electron gas and a temperature T_a describing the temperature of the ionic network. The energy transfer by inelastic collisions towards the electrons is the heat source term of the electron

10 A displacive phase transition is characterized by a collective displacement of atoms inducing a symmetry break due to either a deformation of the crystal lattice or to atomic displacements within the lattice.

plasma. Finally, the electron–phonon coupling is described at the lowest order of the Frolich Hamiltonian by a coupling term g (Landau's approach), by analogy with what is done in plasma physics. This two-temperature model is described by the following set of equations:

$$C_e(T_e)\frac{\partial T_e}{\partial t}=\frac{1}{r}\frac{\partial}{\partial r}\left[rK_e(T_e)\frac{\partial T_e}{\partial r}\right]-g(T_e-T_a)+A_{(r,t)}$$
$$C_a(T_a)\frac{\partial T_a}{\partial t}=\frac{1}{r}\frac{\partial}{\partial r}\left[rK_a(T_a)\frac{\partial T_a}{\partial r}\right]+g(T_e-T_a) \quad [1.7]$$

where $A_{(r,t)}$ is the spatial distribution of the energy deposited on the electrons and $T_{e,a}, C_{e,a}, K_{e,a}$ is the temperature, the specific heat, the thermal conductivity of the electron and atomic sub-systems, respectively. Taking the macroscopic values of these last three quantities, the only adjustable variable is the coupling term g between the electronic and atomic sub-systems. Even if it is possible to calculate g (Kaganov et al. 1957), the value of this constant is generally obtained by adjusting the threshold of creation of the tracks and the evolution of their diameter with the LET. Toulemonde et al. (2006) give an overview of the experiments and their modeling by the thermal spike model. In his critical analysis of these simulations, Klaumünzer (2006) describes the limitations of such an empirical model:

– In insulators, it is the holes and electrons created by the projectile that transport heat in the electronic system. Their density therefore varies spatially; C_e and K_e should not be considered as constant.

– The radius of the trace is assimilated to that of the molten zone, but two phenomena can decrease this radius: overheating, which corresponds to a higher melting temperature, and a possible partial re-crystallization from the liquid–solid interface during cooling.

Are there any experiments or tests that allow us to confirm one hypothesis or another? Access to structural evolutions in the nanometric volume of a track with a femtosecond time resolution is currently out of reach. Moreover, the measurements must be recorded ion by ion, as current ion accelerators can only deliver ion bunches in the nanosecond range. Perhaps, we can hope for more with laser acceleration. On the other hand, the particles that are emitted by the irradiated surface can carry information on the potential of the track and its temperature. This is the case for electrons of the Auger transition (K^1VV) of carbon emitted about 10 fs after their creation, whose peak will be shifted in energy by the potential and widened by the temperature. This experiment, which is very delicate to analyze, gives very high electron temperatures for amorphous thin films of carbon or silicon: from 15,000 to 85,000 K (Schiwietz et al. 2004; Caron et al. 2006). The inelastic thermal spike model gives lower temperatures of about 10,000 K. The existence of a track potential is more questionable because it is only observed in insulators such as polypropylene and polyethylene terephthalate, both of which are very sensitive to irradiation. These experiments give a snapshot of the electronic temperature, but do not provide any information on the heating of the atomic lattice.

Information on the local heating can be obtained from the desorption of gold nanoparticles deposited on the irradiated surface. Transmission electron microscopy observation of a SiO_2 thin film, on which nanoparticles of 1 nm mean radius have been deposited, makes it possible to see the impact of an ion surrounded by a nanoparticle-free zone. The radius of this depopulated area is about 20% smaller than that calculated by a thermal point model (Nakajima et al. 2015).

The existence of the thermal peak cannot be doubted; the heating of the material is proof at the macroscopic level. However, the mechanism of creation of the tracks remains to be described in detail. Traditionally, the melting occurs from the surface by emission of vacancies towards the interior of the material. In the case of tracks, there is no source of vacancies, so the material should remain in the solid phase well above the melting temperature, unless the Coulomb explosion creates the vacancies.

1.3.4. *Synergy between elastic collisions and electronic excitations*

In the vast majority of studies, the effects of electronic excitations and elastic collisions are considered to be decoupled, and it is accepted that one of the pathways dominates the structural modification process. However, there are situations where the two processes act in parallel and their effects can be cumulative, counteracting or synergistic. Most of the results come from sequential irradiations, which is, in general, an irradiation in the field of elastic collisions followed by an irradiation in the field of electronic excitations. For example, fluorapatites that are heavily damaged by implantation rearrange epitaxially on undamaged areas when irradiated with high-energy helium ions (Ouchani et al. 1997). This antagonistic effect has also been observed in silicon carbide SiC (Benyagoub et al. 2006) or in silicon when irradiated with two simultaneous beams[11] (Thomé et al. 2020). Depending on the material and its irradiation conditions, this annealing of the damage can either come from a succession of local heatings that induce a short distance thermal diffusion, or from an accelerated diffusion of the defects induced by the electronic excitations. The question remains open. It should be noted that a mechanism of accelerated diffusion by oscillations between two states of charge of the defects was already

11. Double or triple beam irradiations are sequential irradiations. The probability of two ions interacting in the same area during a typical defect creation time (ns) is almost zero with commonly used fluxes.

proposed in the 1970s (Bourgoin and Corbett 1978). It was also as early as the 1960s that the annealing of irradiation defects under the electron beam of transmission electron microscopes was reported (Bonfiglioli et al. 1961).

However, this result cannot be generalized. Indeed, the opposite behavior can be observed. For example, Ni ions of 21 MeV, whose LET is lower than the threshold of track creation by electronic excitation in $SrTiO_3$, damage this material when it is pre-irradiated by Au ions of 900 keV. The authors explain this phenomenon in the context of a thermal spike model, where the scattering of electrons ejected by the high LET ion is reduced by pre-existing defects, and the energy density deposited near the ion's path would become large enough to create a track (Weber et al. 2015). This synergy between elastic and inelastic processes is particularly significant in AlN. For the strongest LETs, the amplification factor is of several thousands (Sall et al. 2013).

Iron is a good example in which both phenomena can be observed. At low LET, the damage efficiency of the ions decreases as the LET increases, the electronic excitations anneal the defects, while the creation of defects is exalted at higher LET (see Figure 1.12) (Dunlop et al. 1994).

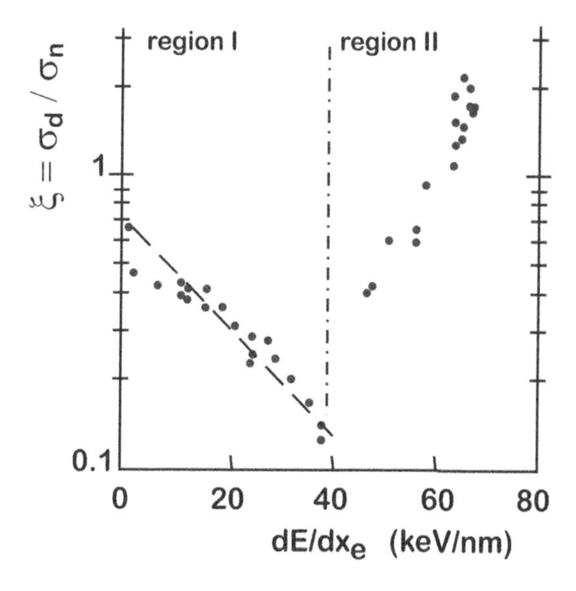

Figure 1.12. *Evolution of the damage efficiency in iron (ratio of the experimental (d) and theoretical (n) cross-sections, as a function of the LET of the ions, from Dunlop et al. (1994))*

The effects of the simultaneous presence of elastic collisions and electronic excitations are therefore very dependent on the properties of the materials. There is much to be achieved in this field.

1.4. Kinetics of defect evolution

Just as important as the defect creation phase, the dynamics of the defects, whether they are thermal and/or under irradiation, modifies the internal microstructure of the material thanks to the flux of vacancies and interstitials, possibly coupled with the flux of impurities, and the presence of extended defects that are pre-existing or created by the irradiation. Figure 1.13 gives an interesting illustration: the irradiation of aluminum by a high fluence of neutrons creates bubbles whose surface is covered with a thin layer of silicon, as well as silicon precipitates in volume. The silicon created in situ by the reaction $^{27}Al(n,\gamma)^{28}Al$, followed by a β decay to ^{28}Si, migrates to the surfaces or to the silicon clusters under the influence of irradiation (Farrell et al. 1977).

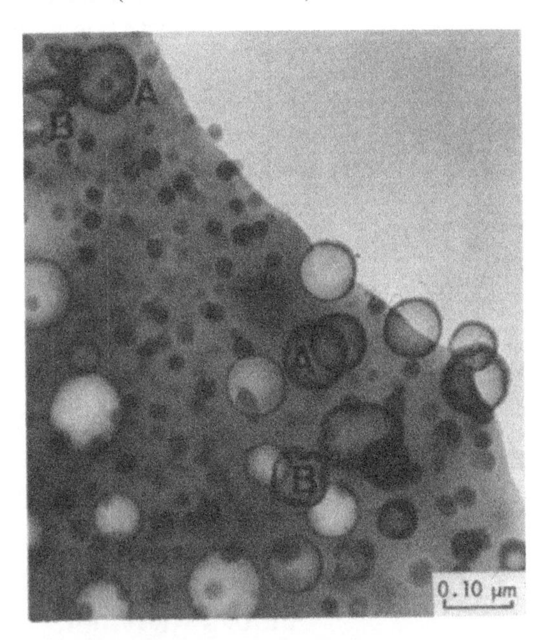

Figure 1.13. *TEM images of silicon cavities and precipitates in aluminum irradiated by 1.4 10^{27} neutrons/m² (~5.3% of Al is transmuted to Si). The cavities have a silicon-coated surface (Farrell et al. 1977)*

The knowledge of the defects produced under irradiation (vacancies, bi-vacancies, interstitials, dumbbell, etc.) and their spatial distribution can be obtained from molecular dynamics simulations. In these simulations, at each time step, the position and velocity of all atoms are calculated by applying Newton's second law $\sum \vec{f_i}(t) = m\vec{a}(t)$, with $\vec{f}$ being the forces applied to the considered atom and $\vec{a}$ being its acceleration. The time and space scales accessible by this numerical technique are in the range of a few picoseconds and a few hundred nanometers. This numerical technique can therefore describe the nature and distribution of defects produced during the development of a collision cascade. However, under the effect of thermal agitation, this distribution of defects will evolve on time and space scales in the range of a few microseconds and over distances in the micrometer range, the characteristic size of the grains of materials. An athermal diffusion is added to this thermal diffusion due to atomic collisions. This ionic mixing depends on the irradiation flux and can, at high flux, become predominant in the regime of elastic collisions. The contribution of electronic excitations to this athermal diffusion has not been clearly demonstrated.

Under the combined action of thermal and ballistic diffusion, the defects will recombine, agglomerate, trap themselves on pre-existing extended defects (grain boundaries, dislocations, etc.) or induce by irradiation (cavities, bubbles, etc.). This dynamic, which is slow and involves distances in the range of the micrometer, modifies the microstructure of the material and thus its properties of use. The first example of such a modification was the swelling of steels induced by the formation of clusters of vacancies under irradiation. By analogy with the work done in materials at thermodynamic equilibrium, a mean field approach was developed in the 1970s to describe the evolution of the microstructure induced by irradiation (Sizmann 1978). This approach, known as the reaction rate theory[12], allows us to simulate the evolution of the microstructure (cavity, dislocation loop, etc.) under irradiation. In the mean field approximation, which neglects the spatial correlations between defects produced under irradiation, the reaction rate theory allows us to define growth under irradiation of cavities or dislocations.

1.4.1. *Mean field approach: reaction rate theory*

This approach, which was developed in the 1960s and 1970s, is strongly inspired by the mechanisms of diffusion in crystalline solids. The diffusion of atoms in alloys occurs by a succession of random, more or less correlated jumps (Philibert 1985). Although there are mechanisms of jumps by exchange of atoms, the majority of the

12. In the literature, this theory is also called the "chemical rate theory".

diffusion processes occur during jumps with point defects. This theory of reaction rates is a hydrodynamic-type theory, in other words, point defects are only characterized by their average concentration. It therefore provides a way to calculate the distribution of point defects in the long term over a large scale of space. As the correlations between point defects are neglected, such an approach belongs to the class of mean field methods. The diffusion coefficient D_A under irradiation of an A atom can be written as:

$$D_A = f_v D_v c_v + f_i D_i c_i + f_{2v} D_{2v} c_{2v} + \cdots \qquad [1.8]$$

At thermodynamic equilibrium, the different terms associated with atomic diffusion can be calculated. The term f_v (f_i and f_{2v}, respectively) represents the correlation factor between the different jumps mediated by the vacancy (interstitial and bi-vacancy, respectively). It expresses the non-independence between successive jumps: after a jump in one direction, the vacancy has the possibility to make the jump in the opposite direction (f depends on the crystal structure and, more precisely, on the number z of the first neighbor, which is approximately 1-2/z). $D_v = exp(-H_v^M/kT)$ is the diffusion coefficient of the vacancy at a temperature T (H_v^M is the enthalpy of migration of the vacancy) and the concentration of the vacancy is equal to $c_v = \exp(S_v^F/kT)\exp(-H_v^F/kT)$, with S_v^F and H_v^F being the entropy and enthalpy of formation of the vacancy (respectively of the interstitial and the bi-vacancy). The idea developed by the English-speaking groups (Bullough et al. 1975) was to assume that these notions defined at thermodynamic equilibrium could be applied under irradiation. This implicitly assumes that the perturbations induced by irradiation do not prevent the definition of a thermodynamic potential $exp(-\Delta G/kT)$, where ΔG is the free enthalpy. Provided that this assumption is lawful, irradiation damage then opens up new diffusion paths via, for example, the interstitials produced by elastic collisions and whose concentration at thermodynamic equilibrium is extremely low. They also accelerate the thermal diffusion by increasing the number of vacancies well beyond its concentration at thermodynamic equilibrium.

The simplest approach to quantify the concentration of these point defects is to describe their temporal evolution through a set of kinetic equations, in a manner equivalent to the description of chemical reactions in the gas phase, hence the name reaction rate theory. In order to achieve such an approach, it is assumed that the relaxation time of defects created by irradiation, of the order of 10^{-6} seconds, is much longer than that of their formation in the collision cascades (of the order of 10^{-9} seconds). The defects produced in the collision cascades are then considered as a source term. In such an approach, all spatial correlations between collision cascades, or the separation of cascades into sub-cascades, are neglected. In other words, the morphology of the collision cascades is not considered in this approach. At the most elementary level, it is possible to write the evolution of point defect populations under irradiation by a set of "chemical" equations (Dienes 1966). The evolution of vacancies and interstitials can be described by the following set of chemical reactions:

$$\begin{aligned} & i + v \rightarrow 0 \\ & i + s \rightarrow s \\ & v + s \rightarrow s \end{aligned} \qquad [1.9]$$

This example of chemical equations reflects the fact that irradiation only produces vacancies and interstitials, and that these elementary point defects can either annihilate each other (see the first equation) or disappear on the existing wells of the microstructure (see the second and third equations). Obviously, the list of reactions must be extended to take into account all mechanisms of importance, such as $v + v \rightarrow v_2$. This elementary approach relies on the existence of two distinct time scales: one to describe atomic collision mechanisms and the other, three orders of magnitude slower, for species diffusion. However, there is no general method to determine all the diffusion paths activated by irradiation and it is therefore difficult to ensure the comprehensiveness of the list of kinetic equations.

By analogy with chemical reactions, the evolution of these equations is written as:

$$\begin{cases} \frac{dc_v}{dt} = K_0 - K_{iv}c_v c_i - K_{vs}c_s c_v \\ \frac{dc_i}{dt} = K_0 - K_{iv}c_v c_i - K_{is}c_s c_i \end{cases} \qquad [1.10]$$

In this set of equations, K_0 is the irradiation production rate of Frenkel pairs. $K_0 = \sigma^d \emptyset$, where σ^d is the effective cross-section of the Frenkel pair creation obtained from a binary collision or molecular dynamics simulations approach, and ϕ is the incident particle flux. The terms K_{iv}, K_{is}, K_{vs} are the "chemical reaction" constants that must be determined in an ad hoc manner. These equations derive from the application of a law of mass action, whose existence is not a priori self-evident in a material under irradiation. Indeed, the existence of a Gibbs energy under irradiation is not guaranteed. As expected for chemical reactions (Nicolis and Prigogine 1977), this set of differential equations is strongly nonlinear. Furthermore, K_{is} is typically different from K_{vs} and thus the concentrations of vacancies and interstitials are, in general, not identical.

Note that equation [1.10] assumes that the point defect concentrations are spatially homogeneous. For point defect recombination, this is simply explained by introducing a characteristic spontaneous vacancy-to-interstitial recombination distance r_{iv} of the order of the lattice parameter. The scattering distance $\sqrt{D_{i,v}(T) \times t}$ is always assumed to be larger than r_{iv}, thus leading to a spatially homogeneous distribution of these defects.

Within this approximation, it is possible to define different domains for point defect concentrations:

– at low temperature and low sink density ($4K_0 > K_{vs}K_{is}c_s^2/K_{iv}$), point defects are produced faster than they are destroyed on the sinks. We then obtain $c_i \approx c_v = \sqrt{K_0/K_{iv}}$;

– at high temperatures ($4K_0 < K_{vs}K_{is}c_s^2/K_{iv}$), the vacancies and interstitials disappear on the sinks. The concentrations $c_{i,v}$ reach a steady state $K_0/K_{is,vs}c_s$.

All of the cases are very well described in Sizmann (1978).

Figure 1.14 shows an example of induced precipitation in electron-irradiated lead glasses. The system of nonlinear differential equations (see equation [1.10]) is used to describe this precipitation. The solutions of many chemical equations describing the evolution of point defect populations have already been calculated by many authors (Damask, cited in Dienes (1967)).

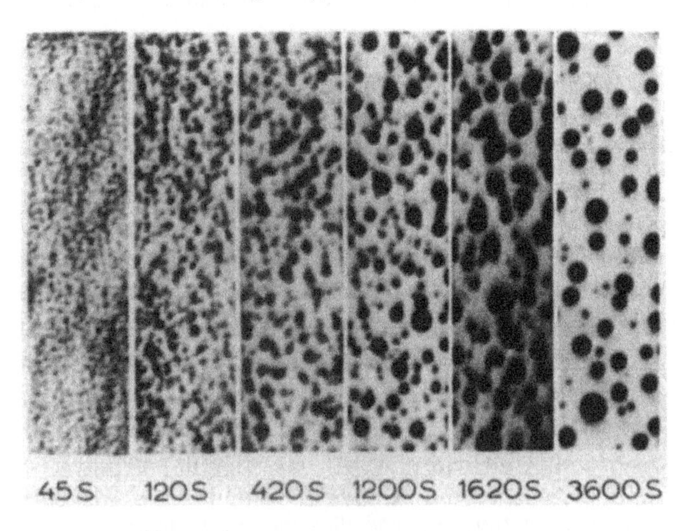

Figure 1.14. *Kinetics of lead precipitate formation induced by electron irradiation at 670 K in a lead-borate glass (flux of 4.8 10^{18} $cm^{-2}s^{-1}$) (Biron and Barbu 1988)*

In general, defect sinks such as dislocations or grain boundaries destroy the symmetry vacancies and interstitials $c_v \neq c_i$ because K_{sv} is different from K_{si}. Moreover, the presence of these sinks no longer ensures the assumption of spatial uniformity[13]. There is a point defect flux to the nearest sink. A term of the form $\nabla(D_{i,v}\nabla c_{i,v})$ must therefore be added to equation [1.4]. A set of partial differential equations must then be solved and boundary conditions imposed. The nonlinear nature of this set of partial differential equations makes the calculation of the evolution kinetics of the vacancies strongly dependent on the initial conditions and the boundary lines. Small changes in the operating conditions (irradiation flux, temperature, etc.) generate instabilities known as Turing instabilities, which are capable of modifying the solutions obtained in the long term.

Lastly, to conclude this brief point on reaction rate theory, it is obvious that the sinks for point defects, which are assumed not to vary in the reaction rate theory approach, can be modified by feedback. They will grow, disappear and their density will evolve. This evolution can be described by explicitly including new equations. Describing the evolution of a microstructure under irradiation by the reaction rate theory leads to the multiplication of the number of constants and coupled equations, quickly making the calculation inefficient.

1.4.2. *Evolution of extended defects: kinetics of clusters in the mean field*

The reaction rate method is fairly simple to implement, but it is unable to give the size distribution of extended defects. A more accurate description of the distribution of extended defects produced by irradiation can be obtained by using Becker and Döring's approach to nucleation, developed in the 1930s (Becker and Döring 1935; Debenedetti 1997). It is thus possible to describe the evolution of defect clusters under irradiation of different sizes, ranging from a few defects (bi-vacancies) to a few hundred vacancies (Kiritani and Takata 1978). In order to follow a very large number of defect clusters, it is common to replace the set of coupled differential equations by a Fokker–Planck-type equation (a linear partial differential equation that must be verified by the transition probability density) when the clusters exceed a certain size (a few dozen vacancies). The detailed description of each cluster is abandoned in favor of a probability of occurrence of clusters of defined size. The difficulty of such an approach lies in a good parameterization of the equations, as is the case for the reaction rate theory. What values should be given to the rates of absorption and emission of defects by a cluster, which are the building

13. Uniformity is maintained if the average distance between point defects is greater than that between sinks.

blocks of the modeling? At thermodynamic equilibrium, the detailed balance makes it possible to link the absorption rates to the emission rates, but under irradiation, nothing implies that such a simplification is possible.

1.4.3. *Kinetic Monte Carlo approach*

Different strategies were implemented to overcome this problem of parameterization. The kinetic Monte Carlo approach consists of solving a master equation that directly describes the evolution of the microstructure and not the set of trajectories, as is the case for molecular dynamics. This method exploits the fact that long-term dynamics consists of diffusive jumps from one state to another. Rather than following the trajectory through each vibrational period, as in molecular dynamics, the state-to-state transitions are treated directly. For this, the system is reduced to N distinct sites on which the vacancies are distributed. We are no longer interested in the trajectory of each atom but in the state of a macro-system, defined by a vector P of dimension N. This macro-system can, for example, be the vacancy population described by the list of i sites occupied by a vacancy among the N accessible sites. The knowledge of the vector P at a given time t represents the "configuration" of the vacancies in the system. The transition rates describing the jumps from vacancies to sites are thus a matrix N×N of components K_{ij}. The evolution of the configuration is then simply given by the conservation of probabilities:

$$dP(t)/dt = \sum_i \left(-K_{i \to j} P_i(t) + K_{j \to i} P_j(t)\right) \qquad [1.11]$$

Given that the transition mechanisms are Markov processes without memory, the K_{ij} are therefore independent of time. Formally, it is possible to compute the distribution of $p_{ij}(t, t_0)$ for each path between time t_0 and t:

$$p_{ij}(t, t_0) = K_{ij} exp\left(-K_{ij}(t - t_0)\right) \qquad [1.12]$$

The description of the system in the long term, that is, at the thermodynamic limit, is obtained when $dP(t)/dt = 0$. In general, it is impossible to perform the analytical calculation of the evolution of the system. Numerical methods such as the Monte Carlo method were introduced in the 1950s to numerically carry out this type of calculation (Metropolis et al. 1953; Lanore 1974). In particular, the Monte Carlo method makes it possible to numerically solve the master equation describing the evolution of the defects' organization. Contrary to molecular dynamics simulations, which calculate particle trajectories according to the "true" dynamics of the system, this method simulates fictitious dynamics, characterized by occasional jumps of atoms between long periods of "inactivity" of the system. Thus, the existence of atomic vibrations imposes the tracking of motion in molecular dynamics, with very short time intervals (10^{-15} s). Such a constraint totally disappears during Monte Carlo simulations. Indeed, the important point in this type of Monte Carlo simulation is to consider that the system has "forgotten" how it reached a configuration. When DM simulations are able to reach quasi-stationary distributions, which is not obvious under irradiation, they make it possible to correctly define all possible transition rates, making the kinetic Monte Carlo realistic. Although transition state theory (TST) provides the theoretical framework for calculating transition rates (Merrick and Fichthorn 2007), this theory does not give an exact solution, and does not allow us to know all possible transitions. These are the reasons why the results obtained by the kinetic Monte Carlo approach must be critically analyzed. In addition to this conceptual problem, a technical problem that is inherent to the method limits its use. When the energy barrier associated with the transition between two configurations is very low, the system can remain trapped in a subset of configurations and not explore the whole phase space. This is the case for the diffusion of vacancies in iron in the presence of a chromium atom, which has the effect of significantly lowering the energy barrier (~0.1 eV). A recent method seems to be able to circumvent this problem (Athènes et al. 2019).

Object Kinetic Monte Carlo (OKMC) is a more global approach that uses transition rates to describe the evolution of classes of objects (vacancy and interstitial clusters, dislocations, cavities, etc.). As for the atomic kinetic Monte Carlo, the motion of these classes can be described from a kinetic Monte Carlo approach of their center of gravity, by using a residence time algorithm and adding some additional rules. The problem of knowing all transition rates with this type of approach is even more crucial. Ad hoc dependencies introduced to describe cluster coalescence or annihilation, for example, are in general simplified (Voter 2007). Real dynamics can sometimes generate counterintuitive diffusion paths. If the catalog of all possible paths is not complete, the dynamics found may not represent reality.

A very good introduction to these kinetic Monte Carlo methods can be found in Voter (2007).

1.4.4. *Phase field approach*

The phase field approach is also based on the solution of the master equation (see equation [1.10]), but rather than trying to determine the vector P(t), this approach is interested in the evolution of the expectation of the variable of interest, such as the average number of vacancies. This expectation is calculated on a representative elementary volume by a process called coarse-graining. This coarse-graining process simplifies the problem by making a jump in scale by the arbitrary choice of a characteristic distance of coarse-graining ξ. Thus, by removing the atomic nature of the problem, the variables can evolve continuously between two nearby points in space. The physics of the process is then described by a continuous field $\eta(x,t)$, which can be scalar (liquid–gas transition), vector (magnetic transition) or tensor (nematic transition in liquid crystals). This field is commonly referred to as the order parameter in the usual terminology of phase transitions (Wadhawan and Puri 2009). Since irradiations induce a change in the order parameter, its temporal evolution $d\eta(\mathbf{r},t)/dt$ is therefore the result of the temperature effect $\delta F(\eta(\mathbf{r},t))/\delta\eta(\mathbf{r},t)$ (F is the free energy of the system) and irradiation $G(\eta(x,t))$. In addition, this approach also allows the inclusion of external force fields (gravity, thermal expansion, mechanical stresses, etc.) and the calculation of generalized susceptibilities as a function of temperature, irradiation flux and external forces.

The contribution of this phase field model is indisputable for describing the evolution of a microstructure near thermodynamic equilibrium. Although it uses few parameters, this model remains phenomenological and the interpretation at the microscopic scale of the parameters is sometimes difficult.

As an example, this approach was used to describe the de-mixing of an alloy under irradiation. In this example, the free energy $F(\eta(\mathbf{r},t))$ is replaced by a functional $L(\eta(\mathbf{r},t))$ including the ionic mixing, and the temporal evolution of the alloy composition depends on the irradiation temperature and flux, as well as the average concentration of the alloy (Lunéville et al. 2020). This approach allows the calculation of elastic constants of the material under irradiation, an example of generalized susceptibility (Simeone et al. 2020). Although this approach is promising, there is no evidence that it is always possible to describe the irradiation effects by a functional, as is the case for ionic mixing. Moreover, this hydrodynamic approach does not make it possible to describe the short-range order and thus the small precipitates produced under irradiation.

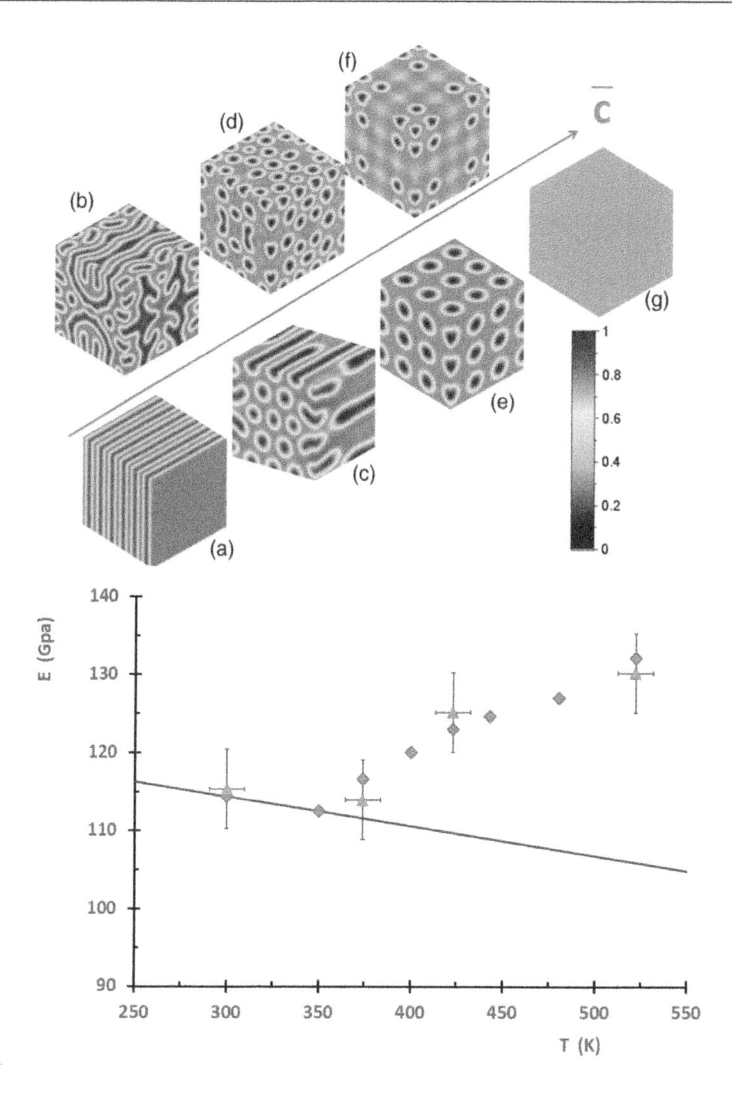

Figure 1.15. *The microstructures that can appear under irradiation in an immiscible AB alloy as a function of the average B concentration of the alloy, for a given flux and temperature (top figure). Comparison between the measured Young's modulus (triangles) and calculated by a phase field approach (diamonds) as a function of the irradiation temperature in an AgCu alloy, irradiated by Kr ions of 1.6 MeV with a flux of* 10^{12} $cm^{-2}s^{-1}$. *The red line represents Young's modulus of the alloy, assumed to be homogeneous. Figure extracted from Lunéville et al. (2020); Siméone et al. (2020). For a color version of this figure, see www.iste.co.uk/bouffard/nuclear.zip*

In conclusion, the dynamics of point defects and the evolution of the microstructures that it induces can only be described by using all the techniques that have been briefly presented. They are complementary and not competitive.

1.5. Open-ended problems

The results obtained during more than a century of research form a solid body of knowledge, allowing us to understand and predict the behavior of materials under irradiation. Nevertheless, the physico-chemical basis of all the phenomena is not known, and this is all the more evident as we move closer to the complexity of industrial materials. Thus, in a multi-scale approach to predict and simulate the behavior of materials under irradiation, each level is relatively well described, but many links between levels remain to be established. This section attempts to list some of the "hard points" that we feel are important to resolve:

– Under pressure from the nuclear power industry, the basic knowledge of irradiation defects in metallic materials has made remarkable progress since the 1950s. On the other hand, this knowledge is much more fragmented in the case of ionic–covalent materials. The displacement thresholds are often determined with great uncertainty, the configurations of elementary defects and small clusters are often poorly known, the migration mechanisms under and outside irradiation and the diffusion coefficients have rarely been studied, and the same is true for the defect-impurity interactions.

– The radiolysis of water is described in great detail; the reaction constants of more than 25 chemical reactions are known, in particular thanks to pulse radiolysis experiments and by modifying the chemical equilibrium by adding molecules (scavenger). The radiolysis of polymers is known in much less detail. The use of time-resolved spectroscopies could be an interesting contribution.

– The understanding of radiolysis of heterogeneous solid/water systems has made significant progress in recent years, especially with regard to radiolysis of confined water or with interfaces. The impact of the species created by radiolysis on the material is much less known. There is a vast field of research to explore.

– The main characteristic of defects produced under irradiation is a highly inhomogeneous spatial distribution, even within the same cascade. Despite some attempts to describe the spatial correlations between sub-cascades (Simeone et al. 2010), there is no formulation to refine the spatial distribution of defects, an essential input for all defect kinetics models.

– Molecular dynamics has emerged as the simulation technique of choice to better understand the nature and spatial distribution of defects produced in metals and metal alloys by ions belonging to the dominant elastic collision energy range. This method must be applied to semi-conductors and insulators with caution. Indeed, it is assumed that the charge of the ions set in motion during a collision remains constant. This approximation is probably correct for metallic materials in which free electrons very quickly spread the charges, but for ionic or covalent materials, the projectile ion, the primaries and the secondaries can change charge state during collisions and interact with other atoms with this new charge state. To take this phenomenon into account, it would be necessary to be able to adapt the interaction potential to each change of charge state.

– A fraction of the energy transferred to the primaries is dissipated in the form of ionizations and electronic excitations within the cascades. The fate of this energy is generally not taken into account. In metals, free electrons very quickly remove these charges and dissipate the excess energy. On the other hand, in insulators, this energy returning to the atomic lattice locally increases the temperature of the lattice. Can this brief and local rise in temperature promote the diffusion of species and modify the initial configuration of defects? The question is raised.

– The description of the track formation mechanism by intense electronic excitation remains phenomenological. The most used model is the thermal spike model, whose results are in congruence with the experiments. However, this model only correlates the calculation of a molten zone and the observation of a transformed zone, without giving any information on the mechanism at the origin of the creation of this track. Its coupling with a molecular dynamics simulation is a step further towards an understanding, even if the whole leaves substantial room for improvement. Indeed, in the thermal spike model, the electron (and hole) dynamics are only treated in the form of a free electron heat diffusion equation, whereas this model is often used for insulators. Moreover, the separation of charges must induce an electric field, creating a restoring force on the electrons. For molecular dynamics, the charge recombination is considered to be total, which is not obvious.

– All kinetic methods are based on the existence of a master equation, resulting from a coarse graining method. Implicitly, these approaches consider that there are two distinct types of variables: fast variables (vacancies, interstitials) that reach a stationary state, and slow variables (solutes, dislocations) that evolve slowly and define the microstructure. The evolution of the microstructure can, by feedback, modify the dynamics of the fast defects. This effect could be artificially taken into account by biases evolving with time. This coupling would make it possible to estimate realistic mobilities under irradiation for solutes.

– For UO_2-type oxides, it is always assumed that the rapid diffusion at high temperature of the oxygen anions is the cause of charge screening of the cations (U^{4+}, U^{5+}, for example). With this assumption, the calculation of the strength of the sinks is similar to that in metals. However, this approximation remains to be validated. Furthermore, its extension to other oxides such as storage matrices is far from being implemented.

– The phase field method seems promising for the calculation of changes in generalized susceptibilities induced by irradiation for materials subjected to external fields. Its limitations and its application to describe the behavior of non-metallic materials such as UO_2 or PuO_2, however, remain to be demonstrated.

1.6. Acknowledgements

We would like to thank Manuel Athènes and Jean-Pierre Bonal for their careful review of our document.

1.7. References

Agullo-Lopez, F., Catlow, C.R.A., Townsend, P.D. (1988). *Point Defects in Materials*. Academic Press, London.

Athènes, M., Kaur, S., Adjanor, G., Vanacker, T., Jourdan, T. (2019). Elastodiffusion and cluster mobilities using kinetic Monte Carlo simulations: Fast first-passage algorithms for reversible diffusion processes. *Physical Review Materials*, 3(10), 103802.

Aumayr, F., Facsko, S., El-Said, A.S., Trautmann, C., Schleberger, M. (2011). Single ion induced surface nanostructures: A comparison between slow highly charged and swift heavy ions. *Journal of Physics: Condensed Matter*, 23(39), 393001.

Averback, R.S., Benedek, R., Merkle, K.L. (1978). Efficiency of defect production in cascades. *Journal of Nuclear Materials*, 69–70, 786–789.

Balanzat, E. and Bouffard, S. (1993). Basic phenomena of the particle-matter interaction. *Solid State Phenomena*, 30–31, 7–74.

Baro, J., Sempau, J., Fernandez-Varea, J.M., Salvat, F. (1995). PENELOPE – An algorithm for Monte Carlo simulation of the penetration and energy-loss of electrons and positons in matter. *Nuclear Instruments & Methods in Physics Research Section B: Beam Interactions with Materials and Atoms*, 100(1), 31–46.

Becker, R. and Döring, W. (1935). Kinetische Behandlung der Keimbildung in übersättigten Dämpfen. *Annalen der Physik*, 24, 719.

Benyagoub, A., Audren, A., Thomé, L., Garrido, F. (2006). Athermal crystallization induced by electronic excitations in ion-irradiated silicon carbide. *Applied Physics Letters*, 89, 241914.

Bethe, H. (1930). Zur Theorie des Durchgangs schneller Korpuskularstrahlen durch Materie. *Annalen der Physik*, 397(3), 325–400.

Biron, I. and Barbu, A. (1988). Viscosity and stability of B_2O_3 PbO glasses under irradiation. *Nuclear Instruments and Methods in Physics Research Section B: Beam Interactions with Materials and Atoms*, 32(1), 279–282.

Bonfiglioli, G., Ferro, A., Mojoni, A. (1961). Electron microscope investigation on the nature of tracks of fission products in mica. *Journal of Applied Physics*, 32(12), 2499–2503.

Bourgoin, J.C. and Corbett, J.W. (1978). Enhanced diffusion mechanisms. *Radiation Effects*, 36, 157–188.

Bullough, R., Eyre, B.L., Krishan, K., Marshall, W.C. (1975). Cascade damage effects on the swelling of irradiated materials. *Proceedings of the Royal Society of London A: Mathematical and Physical Sciences*, 346(1644), 81–102.

Caron, M., Rothard, H., Toulemonde, M., Gervais, B., Beuve, M. (2006). Theoretical and experimental study of electronic temperatures in heavy ion tracks from Auger electron spectra and thermal spike calculations. *Nuclear Instruments and Methods in Physics Research Section B: Beam Interactions with Materials and Atoms*, 245(1), 36–40.

Chadderton, L.T., Montagu-Pollock, H.M., Bowden, F.P. (1963). Fission fragment damage to crystal lattices: Heat-sensitive crystals. *Proceedings of the Royal Society of London. Series A: Mathematical and Physical Sciences*, 274(1357), 239–252.

Debenedetti, P.G. (1997). *Metastable Liquids: Concepts and Principles.* Princeton University Press, New Jersey.

Dessauer, F. (1923). Über einige Wirkungen von Strahlen IV. *Zeitschrift für Physik*, 20(1), 288–298.

Dienes, G.J. (1966). *Studies in Radiation Effects: Series A: Physical and Chemical*, vol. 1. Gordon and Breach, New York.

Dienes, G.J. (1967). *Studies in Radiation Effects in Solids*, vol. 2. Gordon and Breach, New York.

Dunlop, A., Lesueur, D., Barbu, A. (1993). Evolution of microstructure resulting from high electronic excitation during swift heavy ion irradiations. *Journal of Nuclear Materials*, 205, 426–437.

Dunlop, A., Lesueur, D., Legrand, P., Dammak, H., Dural, J. (1994). Effects induced by high electronic excitations in pure metals: A detailed study in iron. *Nuclear Instruments and Methods in Physics Research Section B: Beam Interactions with Materials and Atoms*, 90, 330–338.

Egami, T. and Billinge, S. (2012). *Underneath the Bragg Peaks. Structural Analysis of Complex Materials*, vol. 16. Pergamon Press, Oxford.

Ewing, R.C. (1994). The metamict state: 1993 – The centennial. *Nuclear Instruments & Methods in Physics Research Section B: Beam Interactions with Materials and Atoms*, 91, 22–29.

Farrell, K., Bentley, J., Braski, D.N. (1977). Direct observation of radiation-induced coated cavities. *Scripta Metallurgica*, 11(3), 243–248.

Firsov, O.B. (1959). A qualitative interpretation of the mean electron excitation energy in atomic collisions. *Soviet Physics, JETP*, 36(9), 1076.

Fleischer, R.L. (1981). Nuclear track production in solids. *Progress in Materials Science*, Chalmers anniversary volume, 97–123.

Fleischer, R.L., Price, P.B., Walker, R.M. (1965a). Ion explosion spike mechanism for formation of charged-particle tracks in solids. *Journal of Applied Physics*, 36(11), 3645–3652.

Fleischer, R.L., Price, P.B., Walker, R.M. (1965b). Tracks of charged particles in solids. *Science*, 149(3682), 383–393.

Forster, R.A. and Godfrey, T.N.K. (1985). MCNP – A general Monte Carlo code for neutron and photon transport. *Lecture Notes in Physics*, 240, 33–55.

Geant4 collaboration (2003). Geant4 – A simulation toolkit. *Nuclear Instruments and Methods in Physics Research Section A: Accelerators, Spectrometers, Detectors and Associated Equipment*, 506(3), 250–303.

Gervais, B. and Bouffard, S. (1994). Simulation of the primary stage of the interaction of swift heavy ions with condensed matter. *Nuclear Instruments and Methods in Physics Research Section B: Beam Interactions with Materials and Atoms*, 88, 355–364.

Gibbons, J.F. (1972). Ion implantation in semiconductors. Part II: Damage production and annealing. *Proceedings of the IEEE*, 60(9), 1062–1096.

Goldstein, E. (1894). Über die Einwirkung von Kathodenstrahlen auf einigen Salze. *Sitzungsberichte der Königlich Preussischen Akademie der Wissenschaften zu Berlin*, 937–945.

Guillaumont, R. (2004). Éléments chimiques à considérer dans l'aval du cycle nucléaire. *Comptes rendus chimie*, 7(12), 1129–1134.

Haranger, F., Ban d'Etat, B., Boduch, P., Bouffard, S., Lebius, H., Maunoury, L., Rothard, H. (2006). Projectile charge and velocity effects on UO_2 sputtering in the nuclear stopping regime. *European Physical Journal D*, 38, 501–506.

Hobbs, L.W. (1995). The role of topology and geometry in the irradiation-induced amorphization of network structures. *Journal of Non-Crystalline Solids*, 182(1), 27–39.

Itoh, N. and Tanimura, K. (1990). Formation of interstitial-vacancy pairs by electronic excitation in pure ionic crystals. *Journal of Physics and Chemistry of Solids*, 51(7), 717–735.

Izui, K. (1965). Fission fragment damage in semiconductors and ionic crystals. *Journal of the Physical Society of Japan*, 20(6), 915–932.

Kaganov, M.I., Lifshitz, I.M., Tanatarov, L.V. (1957). Relaxation between electrons and the crystalline lattice. *Soviet Phyics JETP*, 4(2), 173–178.

Kinchin, G.H. and Pease, R.S. (1955). The displacement of atoms in solids by radiation. *Reports on Progress in Physics*, 18, 1–51.

King, W.E., Merkle, K.L., Meshii, M. (1981). Determination of the threshold-energy surface for copper using in-situ electrical-resistivity measurements in the high-voltage electron microscope. *Physical Review B*, 23(12), 6319–6334.

Kiritani, M. and Takata, H. (1978). Dynamic studies of defect mobility using high voltage electron microscopy. *Journal of Nuclear Materials*, 69–70, 277–309.

Klaumünzer, S. (2006). Thermal-spike models for ion track physics: A critical examination. *Matematisk-fysiske Meddelelser*, 52(1), 293–328.

Lanore, J.-M. (1974). Simulation de l'évolution des défauts dans un réseau par la méthode de Monte Carlo. *Radiation Effects*, 22(3), 153–162.

Lefèvre, J., Costantini, J.M., Esnouf, S., Petite, G. (2009). Silicon threshold displacement energy determined by photoluminescence in electron-irradiated cubic silicon carbide. *Journal of Applied Physics*, 105(2), 023520.

Lesueur, D. and Dunlop, A. (1993). Damage creation via electronic excitations in metallic targets. Part II: Theoretical model. *Radiation Effects and Defects in Solids*, 126, 163–172.

Lindhard, J. and Scharff, M. (1961). Energy dissipation by ions in the keV region. *Physical Review*, 124, 128.

Lucasson, P.G. and Walker, R.M. (1962). Production and recovery of electron-induced radiation damage in a number of metals. *Physical Review*, 127(2), 485–500.

Lunéville, L., Simeone, D., Jouanne, C. (2006). Calculation of radiation damage induced by neutrons in compound materials. *Journal of Nuclear Materials*, 353(1), 89–100.

Lunéville, L., Garcia, P., Siméone, D. (2020). Predicting nonequilibrium patterns beyond thermodynamic concepts: Application to radiation-induced microstructures. *Physical Review Letters*, 124(8), 085701.

Meldrum, A., Zinkle, S.J., Boatner, L.A., Ewing, R.C. (1998). A transient liquid-like phase in the displacement cascades of zircon, hafnon and thorite. *Nature*, 395(6697), 56–58.

Merrick, M. and Fichthorn, K.A. (2007). Synchronous relaxation algorithm for parallel kinetic Monte Carlo simulations of thin film growth. *Physical Review E*, 75(1), 011606.

Metropolis, N., Rosenbluth, A.W., Rosenbluth, M.N., Teller, A.H., Teller, E. (1953). Equation of state calculations by fast computing machines. *The Journal of Chemical Physics*, 21(6), 1087–1092.

Nakajima, K., Kitayama, T., Hayashi, H., Matsuda, M., Sataka, M., Tsujimoto, M., Toulemonde, M., Bouffard, S., Kimura, K. (2015). Tracing temperature in a nanometer size region in a picosecond time period. *Scientific Reports*, 5, 13363.

Ngono-Ravache, Y., Corbin, D., Gaté, C., Mélot, M., Balanzat, E. (2007). Alkyne creation in aliphatic polymers: Influence of side groups. *Journal of Physical Chemistry B*, 111(11), 2813–2819.

Nicolis, G. and Prigogine, I. (1977). *Self-Organisation in Nonequilibrium Systems*. John Wiley and Sons, New York.

Noggle, T.S. and Stiegler, J.O. (1960). Electron microspcope observations of fission fragment tracks in thin films of UO_2. *Journal of Applied Physics*, 31(12), 2199–2208.

Nordlund, K., Zinkle, S.J., Sand, A.E., Granberg, F., Averback, R.S., Stoller, R.E., Suzudo, T., Malerba, L., Banhart, F., Weber, W.J. et al. (2018). Primary radiation damage: A review of current understanding and models. *Journal of Nuclear Materials*, 512, 450–479.

Norgett, M.J., Robinson, M.T., Torrens, I.M. (1975). A proposed method of calculating displacement dose rates. *Nuclear Engineering and Design*, 33, 50–54.

Ortiz, C.J., Lunéville, L., Siméone, D. (2020). Binary collision approximation. In *Comprehensive Nuclear Materials*, Konings, R.J.M. and Stoller, R.E. (eds). Elsevier, Oxford.

Ouchani, S., Dran, J.C., Chaumont, J. (1997). Evidence of ionization annealing upon helium-ion irradiation of pre-damaged fluorapatite. *Nuclear Instruments & Methods in Physics Research Section B: Beam Interactions with Materials and Atoms*, 132(3), 447–451.

Pascucci, M.R., Hutchison, J.L., Hobbs, L.W. (1983). The metamict transformation in alpha-quartz. *Radiation Effects*, 74, 219–226.

Philibert, J. (1985). *Diffusion et transport de matière dans les solides*. Les Editions de Physique, Les Ulis.

Price, P.B. and Walker, R.M. (1962). Observation of fossil particle tracks in natural micas. *Nature*, 196, 732.

Sall, M., Monnet, I., Grygiel, C., Ban d'Etat, B., Lebius, H., Leclerc, S., Balanzat, E. (2013). Synergy between electronic and nuclear energy losses for color center creation in AlN. *Europhysics Letters*, 102(2), 26002.

Schiwietz, G., Czerski, K., Roth, M., Staufenbiel, F., Grande, P.L. (2004). Femtosecond dynamics – Snapshots of the early ion-track evolution. *Nuclear Instruments and Methods in Physics Research Section B: Beam Interactions with Materials and Atoms*, 225(1), 4–26.

Schulze, J.H. (1727). Scotophorus pro Phosphoro inventvs seu experimentum curiosum de effectu radiorum solarium. *Acta Physico-Medica Academiæ Cæsareæ Leopoldino-Carolina Natura Curiosorum exhibentia Ephemerides, Nürnberg*, 528–533.

Seitz, F. and Koehler, J.S. (1956). Displacement of atoms during irradiation. *Solid State Physics*, 2, 305–448.

Sigmund, P. (2006). *Particle Penetration and Radiation Effects, Volume 1: General Aspects and Stopping of Swift Point Charges*. Springer, Berlin.

Sigmund, P. (2014). *Particle Penetration and Radiation Effects, Volume 2: Penetration of Atomic and Molecular Ions*. Springer, Cham.

Silk, E.C.H. and Barnes, R.S. (1959). Examination of fission fragment tracks with an electron microscope. *Philosophical Magazine*, 4, 970–972.

Siméone, D., Lunéville, L., Serruys, Y. (2010). Cascade fragmentation under ion beam irradiation: A fractal approach. *Physical Review E*, 82(1), 011122.

Siméone, D., Garcia, P., Bacri, C.O., Lunéville, L. (2020). Symmetry breaking resulting from long-range interactions in out of equilibrium systems: Elastic properties of irradiated AgCu. *Physical Review Letters*, 125(24), 246103.

Sizmann, R. (1978). The effect of radiation upon diffusion in metals. *Journal of Nuclear Materials*, 69–70, 386–412.

Soullard, J. (1985). High voltage electron microscope observations of UO_2. *Journal of Nuclear Materials*, 135(2), 190–196.

Thomé, L., Gutierrez, G., Monnet, I., Garrido, F., Debelle, A. (2020). Ionization-induced annealing in silicon upon dual-beam irradiation. *Journal of Materials Science*, 55(14), 5938–5947.

Toulemonde, M. (1999). Nanometric phase transformation of oxide materials under GeV energy heavy ion irradiation. *Nuclear Instruments and Methods in Physics Research Section B: Beam Interactions with Materials and Atoms*, 156(1), 1–11.

Toulemonde, M., Assmann, W., Dufour, C., Meftah, A., Studer, F., Trautmann, C. (2006). Experimental phenomena and thermal spike model description of ion tracks in amorphisable inorganic insulators. *Matematisk-fysiske Meddelelser*, 52, 263.

Townsend, P.D. (1973). A new interpretation of the Rabin and Klick diagram. *Journal of Physics C: Solid State Physics*, 6(6), 961–966.

Ventura, A., Ngono-Ravache, Y., Marie, H., Levavasseur-Marie, D., Legay, R., Dauvois, V., Chenal, T., Visseaux, M., Balanzat, E. (2016). Hydrogen emission and macromolecular radiation-induced defects in polyethylene irradiated under an Iiert amosphere: The role of energy transfers toward trans-vinylene nsaturations. *Journal of Physical Chemistry B*, 120(39), 10367–10380.

Voter, A.F. (2007). Introduction to the kinetic Monte Carlo method. In *Radiation Effects in Solids*, Sickafus, K.E., Kotomin, E.A., Uberuaga, B.P. (eds). NATO Science Series, Springer, Dordrecht.

Wadhawan, V. and Puri, S. (2009). *Kinetics of Phase Transition*, 1st edition. CRC Press, Boca Raton.

Walgraef, D. (1996). *Spatio-Temporal Pattern Formation with Examples from Physics, Chemistry, and Materials Science (Partially Ordered Systems)*. Springer, New York.

Was, G.S. (2017). *Fundamentals of Radiation Materials Science. Metals and Alloys*. Springer, New York.

Weber, W.J. (2000). Models and mechanisms of irradiation-induced amorphisation in ceramics. *Nuclear Instruments & Methods in Physics Research Section B: Beam Interactions with Materials and Atoms*, B166–167, 98–106.

Weber, W.J., Zarkadoula, E., Pakarinen, O.H., Sachan, R., Chisholm, M.F., Liu, P., Xue, H., Jin, K., Zhang, Y. (2015). Synergy of elastic and inelastic energy loss on ion track formation in $SrTiO_3$. *Scientific Reports*, 5, 7726.

Wilson, W.D., Haggmark, L.G., Biersack, J.P. (1977). Calculations of nuclear stopping, ranges, and straggling in the low-energy region. *Physical Review B*, 15(5), 2458–2468.

Winterbon, K.B., Sigmund, P., Sanders, J. (1970). Spatial distribution of energy deposited by atomic particles in elastic collisions. *Matematisk-fysiske Meddelelser Det Kongelige Danske Videnskabernes Selskab*, 37(14), 1–73.

Young, D.A. (1958). Etching of radiation damage in lithium fluoride. *Nature*, 182(4632), 375–377.

Ziegler, J.F. and Biersack, J.P. (1985). The stopping and range of ions in matter. In *Treatise on Heavy-Ion Science*, Bromley, D.A. (ed.). Springer, Boston.

2

Metal Alloys

Philippe PAREIGE[1] and Christophe DOMAIN[2]

[1] *Groupe de Physique des Matériaux, Université Rouen Normandie, INSA Rouen Normandie, CNRS, France*

[2] *EDF R&D, Département Matériaux et Mécanique des Composants (MMC), Les Renardières, Moret-sur-Loing, France*

2.1. Introduction

The nuclear power industry is particularly concerned with controlling the material problems related to the specificities of its operation, for economic, safety and reliability reasons (Gras 2017). The context here is complex because it combines the impacts of irradiation, high temperature, contact with liquid media, and mechanical and thermal stresses, all over very long durations. Issues related to aging and corrosion of materials used in nuclear facilities are therefore of great significance. The understanding of the aging mechanisms is the trickiest to apprehend and occupies an important place in the studies and research on the behavior of metallic materials.

Studies on the improvement and optimization of the performance and durability of facilities (limitation of maintenance, extension of the service life, maintenance of properties over time, etc.) are, for the most part, closely related to research in materials science, and often call for new tools or generate instrumental research for the development of appropriate tools. Some of these studies are also based on original theoretical aspects. Indeed, numerical simulations are increasingly used to better understand the phenomena of irradiation and their effects on the properties of materials. The science of materials for nuclear energy began in the early 1950s with the development of civil nuclear power in France, the United States, the United

Nuclear Materials under Irradiation,
coordinated by Serge BOUFFARD and Nathalie MONCOFFRE.

Kingdom, Canada and the former Soviet Union. It is a multidisciplinary (chemistry, physics, mechanics, neutronics) and multi-scale science (from the angstrom to the meter). This approach is particularly complex, because its objective is to understand and control the behavior of materials in relation to their microscopic structure, which depends on all the previous manufacturing history and their conditions of use, and which only the operator has precise knowledge of. With regard to their functions, their locations within the reactor, as well as their possible replacements, the structural elements require a great diversity of materials, as illustrated in Figure 2.1.

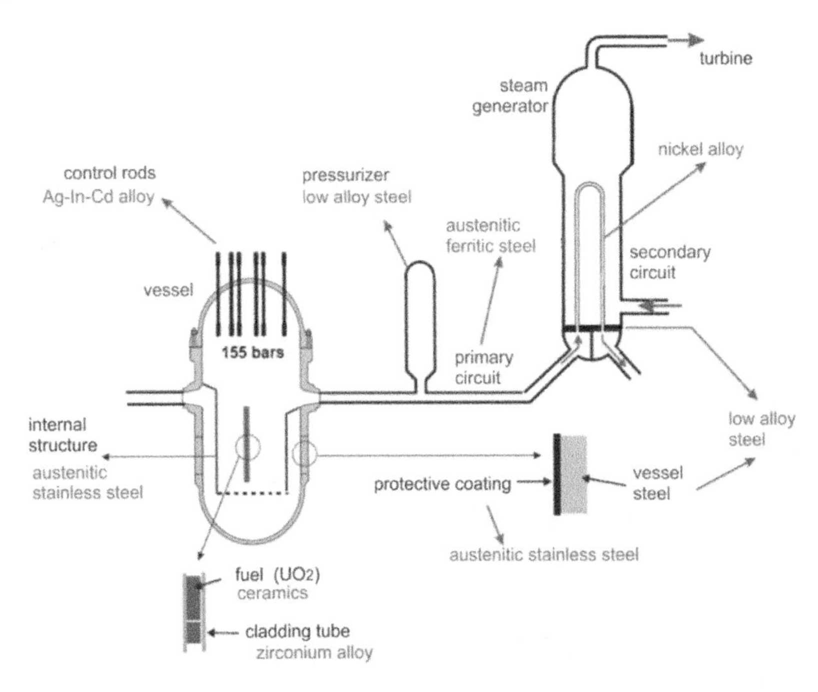

Figure 2.1. *Main metallic alloys used in pressurized water reactors (EDF figure). For a color version of this figure, see www.iste.co.uk/bouffard/nuclear.zip*

Most of these materials play a role in the safety of installations. Controlling and monitoring changes in the properties of the materials used throughout their life cycle is therefore a particularly demanding obligation. An important difference compared to the other industrial fields is that the nuclear installations are intended to function during long durations, ranging from 40 to 60 years, sometimes even more, as well as the fact that many materials are subjected to a specific constraint: irradiation. In this chapter, we are interested in the neutron irradiation of structural materials used in the reactor core: fuel cladding, internal structures and the vessel. Their characteristics and aging conditions are recalled in Table 2.1.

Component	Alloy	Chemical composition Metallurgical state	Conditions of service
Fuel cladding Function: safety barrier Type: consumable	Zircaloy 4	Zr-1.3%Sn-1,200 ppmO-0.2%Fe–0.1%Cr	320–400°C ±90 MPa 5 dpa (five years)
	Alloy M5	Zr-1%Nb–0.025%Fe-1,250 ppmO	
	Zirlo alloy	Zr-1%Nb–1%Sn-1,000 ppmO	
Internal structures Function: safety barrier Type: consumable	Austenitic steel	304L 18%Cr–10%Ni Hyper-tempered 316 17Cr–11Ni–2.5Mo Hardened	290–380°C 60 dpa (60 years) ~10 appm He/dpa
Vessel Function: safety barrier Type: non-replaceable	Low-alloy steel	Fe–Mn–Ni (Cu–P) Bainitic tempering	150 bars 290°C 0.2 dpa (60 years)

Table 2.1. *Main components of generation II and III nuclear reactors subjected to neutron irradiation, as well as their materials and operating conditions (Boutard et al. 2016)*

Under irradiation by energetic neutrons, as recalled in the reference Boutard et al. (2016), metallic alloys are permanently undergoing strong disruptions that can very quickly affect their crystalline structure in the long term, as well as their nominal chemical composition and the microstructure itself. These evolutions will inevitably have an impact on the properties of use of these materials, with the challenge being to understand and anticipate it. The origin of the structural modifications comes from the elastic and inelastic interactions of neutrons with matter.

Elastic interactions then drive the collision cascades and the creation of point, vacancy and self-interstitial defects, as well as disorder in atomic positions (Becquart et al. 2018). During elastic collisions, fast neutrons can eject atoms from their sites, creating vacancy – self-interstitial pairs, or Frenkel pairs. When an atom is ejected with sufficiently large kinetic energy, it can in turn displace other atoms. This sequence of collisions is called a collision cascade (see section 1.3 of Chapter 1). The number of displacements in relation to the number of atoms (dpa) allows us to characterize the damage. For example, each atom of a fuel rod cladding of a PWR is displaced an average of five times (5 dpa) during its stay in the reactor. On the other hand, only one out of 10 atoms in a PWR vessel is displaced during 40 years of service (0.1 dpa).

The inelastic interactions with the nuclei create new atomic species by nuclear reactions. The foreign atoms created by these nuclear reactions can stabilize in the solid or gas phase. Transmutation reactions, (n,α) or (n,p), for example, introduce helium and hydrogen atoms into structural materials[1]. Helium can play a role in the occurrence of swelling, as it promotes the germination of cavities.

At reactor operating temperatures, point defects are mobile and can either recombine (this is the mutual annihilation of a vacancy and a self-interstitial), or be eliminated on various defect sinks (grain boundaries, dislocations, phase interface), or agglomerate. Vacancies or self-interstitials can agglomerate in the form of dislocation loops, which are then called interstitial loops or vacancy loops. In the presence of gas, the vacancies agglomerate more frequently in the form of cavities. The formation of extended defects is all the more favorable when the concentrations of vacancies and self-interstitials are different, which is the case when a sink has a bias in favor of one of the point defects, such as dislocations in relation to self-interstitials. Indeed, the expansion of the crystal lattice around a self-interstitial is important. As a result, the self-interstitial interacts strongly with the elastic field of the dislocations and preferentially eliminates itself on the latter. In a work-hardened metal, the self-interstitials are eliminated on the dislocation lattice, creating an excess of vacancies. If, moreover, the conditions for the germination of cavities are realized (presence of helium), the surplus of vacancies through cavity elimination will allow them to grow.

The dynamics of microstructure evolution can therefore be driven by the permanent creation of defects (displacement cascades), the flows of defects and their possible interactions with solutes. The evolution of the system can no longer be governed by thermodynamics, but by kinetics. Under irradiation, a great variety of phase transformations has been observed: dissolution, amorphization of phases that were initially present (Laves $Zr(Fe,Cr)_2$ phase), precipitation of non-equilibrium phases, etc. This apparent variety where everything seems possible is, however, controlled by the fact that, under irradiation, the stability of a phase does not result from the minimum of the free energy, but from a stationary regime, resulting from the competition between the creation of defects generated by the irradiation and the elimination of these defects, thanks to diffusion. In all this, the irradiation

1. Since the neutrons produced by deuterium and tritium fusion have a higher energy (14 MeV) than those of fission (~2 MeV), the production of transmutation elements, notably helium and hydrogen, will be one to two orders of magnitude higher in fusion reactors.

temperature plays a significant role. At low temperatures, the ballistic effect is dominant (strong creation of defects and weak diffusion for recombination or elimination). This is a domain where thermally stable compounds can dissolve or become amorphous under irradiation. At higher temperatures, point defect elimination by diffusion is dominant, and the instabilities seen at lower temperatures disappear (Was 2017). However, the situation is not identical to that without irradiation. Indeed, supersaturation with point defects accelerates diffusive processes, and the coupling between solutes and point defect fluxes that are eliminated on the sinks produces non-equilibrium segregations.

Changes in properties of use under irradiation, including mechanical behavior, are thus controlled by the generation, migration and agglomeration of point defects and transmutation products created by irradiation, and, in some cases, by the interaction of the material with its surrounding environment (e.g. hydride formation in zirconium alloys as a consequence of water corrosion of the material).

It should be noted that these mechanisms extend over time scales ranging from a few tens of picoseconds, for collision cascades, to several years, for the evolution of the microstructure, and from a tenth of a nanometer for point defects to about 10 microns for the gliding of dislocations that controls plasticity. Thus, the physics that controls the aging of nuclear materials must be correctly described at the atomic scale, or even at the electronic structure level. It is at this scale that we can determine the formation and migration energies of point defects, the solution energies of transmutation elements or the thermal activation processes that control the mobility of dislocations. The physical modeling of macroscopic manifestations of aging will therefore be multi-scale (Adjanor et al. 2010; Becquart and Domain 2011; Nordlund 2019). In order to be reliable and predictive, it will also have to be conducted in close synergy with an experimental approach, so as to identify the physical mechanisms and validate the predictions. Experimental approaches on model materials with chemical compositions and microstructures of increasing complexity, from pure metals to industrial alloys, should make it possible to better understand the mechanisms of their aging, in order to extract what is a determining factor from the complexity of real materials, and lead to realistic and reliable physical models. This experimental section will have to be based on mechanical and physico-chemical characterizations (see Chapter 10), if possible up to the atomic scale and, in terms of the effects of irradiation, on the examination of materials irradiated by charged particles such as ions or electrons (see Chapter 9), and by neutrons in experimental or service reactors.

As indicated in this introduction, we subsequently present a synthesis of the problems related to the effects of neutron irradiation on the three structural materials, which mobilize a large part of the scientific community in the field, namely the cladding materials, the first barrier and in contact with the fuel, the materials of the internal structure of the reactor and, lastly, the reactor vessel itself, the master piece whose integrity must be absolutely guaranteed in service.

2.2. Fuel cladding

The fuel for water-cooled reactors is placed in the form of ceramic pellets stacked in rods, long tubes that are a few meters long. These rods, the claddings, are assembled in bundles held together by grids to form assemblies (see Figure 2.2). In these power reactors, the cladding is generally made of zirconium alloys. The choice of zirconium was determined by several of its properties: its very low thermal neutron absorption (in other words, transparency to neutrons), its good mechanical properties in the presence of some alloying elements, including oxygen to harden it, and its very good corrosion resistance at operating temperatures. The low temperature phase of zirconium is hexagonal compact (alpha phase), and a cubic centered phase is stable at high temperatures (beta phase) (transition temperature at 864°C).

Unlike the internals or the reactor vessel, the assemblies fall into the consumable material category. Changes in the properties of zirconium alloys under irradiation do not impact the life of the reactor. However, they affect performance throughout the life cycle, both in service and up to reprocessing or storage, with economic consequences. Thus, with the cladding being the first safety barrier, the usage properties of the cladding must meet the specifications throughout the life cycle of the assembly: in an accident situation (e.g. LOCA – Loss of Coolant Accident), in transport conditions (e.g. to the ORANO La Hague plant for EDF power plants), or in storage.

Various alloys have been developed and are still under development to optimize the performance of the claddings. For all these alloys, the control of the oxygen content is essential in order to increase the elastic limit. Among them, Zircaloy-4 has been widely used in EDF reactors and is still used worldwide with tin, iron and chromium as the main (substitutional) alloying elements. Tin significantly improves creep resistance. Iron and chromium have low solubility limits in zirconium (~150 ppm at 850°C); they precipitate in the form of intermetallic phases, $Zr(Fe,Cr)_2$ and Zr(Nb,Fe), which limit the grain size and improve corrosion resistance.

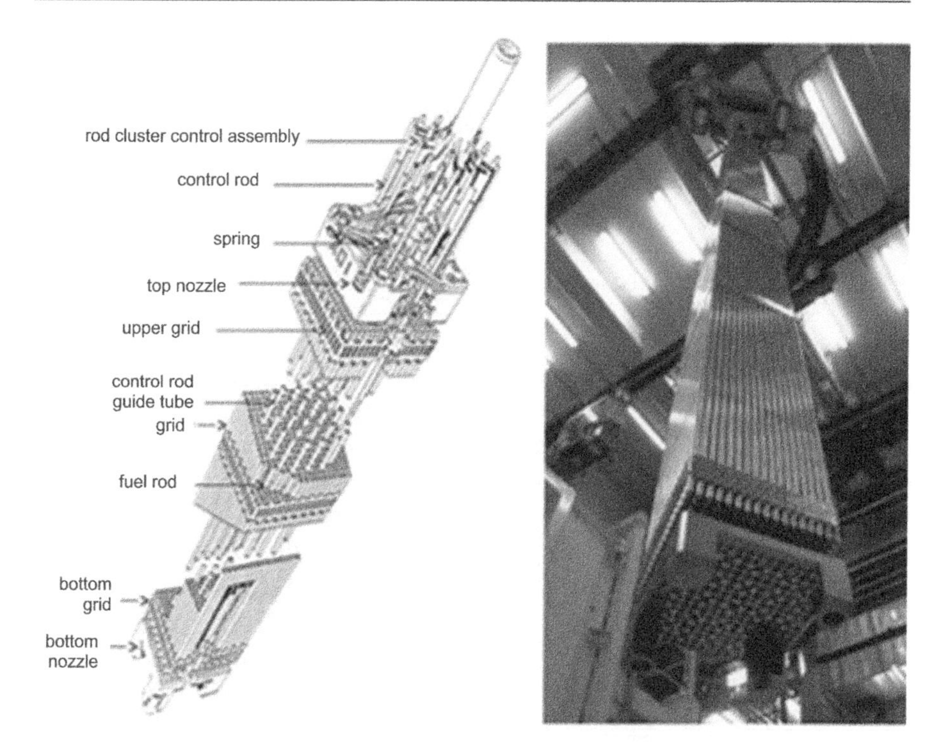

Figure 2.2. *Illustration of the fuel assembly (EDF figure). For a color version of this figure, see www.iste.co.uk/bouffard/nuclear.zip*

Niobium is the main element in a class of alloys developed to improve the corrosion and hydride resistance of fuel cladding. In zirconium, niobium forms precipitates in the alpha phase (hexagonal compact), while it is soluble in any concentration in the beta phase (cubic centered at high temperature). It allows control of the alpha–beta transition temperature through thermodynamic equilibrium. The main industrial alloys are Framatome's M5® and Westinghouse's ZIRLO®, for which the manufacturing range has been optimized: a recrystallized state for M5® and a partially recrystallized state for ZIRLO® (Tewari et al. 2020).

In addition, work is underway to improve the properties of the cladding and its performance in accident situations (ATF – Accident-Tolerant Fuel), particularly in terms of reducing hydrogen emissions. Indeed, in contact with water at 350°C, an oxidation phenomenon ($Zr + 2H_2O \rightarrow ZrO_2 + 2H_2$) takes place, with part of the hydrogen diffusing into the oxide and the metal. Given that the solubility limit of hydrogen in zirconium is low, hydrides can form and are likely to embrittle the cladding at low temperatures after it leaves the reactor, as well as affect the evolutions of the irradiation properties and damage by interacting with the irradiation defects in operation. To address this, different coating processes are being investigated: chromium or silicon carbide (SiC) (Terrani 2018; Chen et al. 2020).

2.2.1. *Growth and creep under irradiation*

In pressurized water reactors, when the fuel reaches its maximum burnup rate, the damage to the cladding is in the range of 15 dpa, with a dose rate of about 10^{-7} dpa/s. The irradiation defects cause alloy hardening and a ductility loss, as well as accelerated growth and creep phenomena. These last two mechanisms induce dimensional changes under neutron irradiation, which can have consequences for operation, such as the difficult unloading of assemblies.

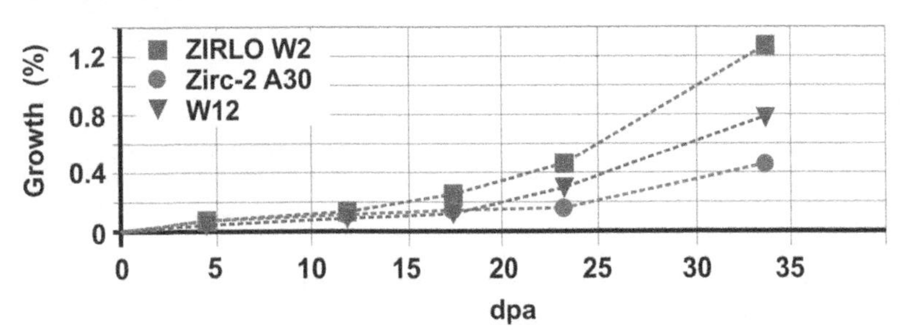

Figure 2.3. *Neutron irradiation growth at 320°C (BOR60 reactor) of different recrystallized Zr alloys, from Adamson et al. (2019), ZIRLO: $Zr_1Nb_1Sn_{0.1}Fe$ – Zirc-2: $Zr_{1.3}Sn_{0.07}Ni_{0.1}Cr$ – W12 : $Zr_1Nb_{0.05}Fe_{0.36}Cr$. For a color version of this figure, see www.iste.co.uk/bouffard/nuclear.zip*

The total deformation $\dot{\varepsilon}$ is usually broken down as the sum of the two contributions: growth and creep deformation.

Growth is an anisotropic dimensional change in the absence of applied stress. It is due to the anisotropy of the hexagonal structure and the texturing of the grains: an elongation of the claddings can be observed after an incubation time, which can lead to a macroscopic elongation of the claddings without a significant change in volume (see Figure 2.3). Irradiation creep leads to a greater deformation than without irradiation (thermal creep), a phenomenon observed in other materials such as austenitic steels.

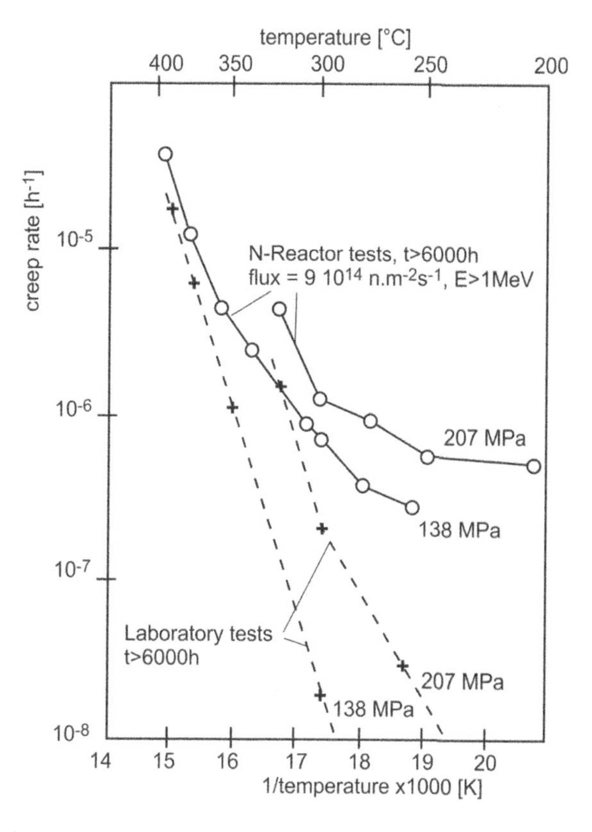

Figure 2.4. *Effect of temperature on the creep of hardened zircaloy-2, comparison of laboratory test and reactor irradiation (Fidleris 1988)*

All other conditions being equal, under irradiation, creep is therefore accelerated (see Figure 2.4). Irradiation creep is most often broken down as the sum of thermal creep (affected by irradiation) and irradiation-specific deformation. The effect of irradiation is different for each component. Creep under irradiation is the result of two competing phenomena: on the one hand, the presence of irradiation defects which, by hardening the material, reduce thermal creep, but the effect rapidly saturates with the dose, and, on the other hand, irradiation activates additional contributions to creep (mainly caused by the rise of dislocations). The thermal component can be easily determined experimentally by conducting a thermal creep test after a first irradiation creep test. Both thermal and irradiation creeps are anisotropic, due to the anisotropy of the compact hexagonal structure and the strong texturing of the material. Irradiation creep and associated deformation depend on many factors, such as the initial metallurgical state with a strong dislocation density and small grain size effect, as well as loading, flux, stress and temperature conditions (Adamson et al. 2019; Onimus et al. 2020).

2.2.2. *Stress corrosion cracking*

Inside the cladding, nuclear fission reactions create a large quantity of elements that are lighter than uranium. Volatile elements, such as helium, iodine or cesium, contribute to the increase in internal pressure of the cladding. Iodine, which is a volatile fission product that is not retained in the fuel pellet, could, under certain conditions, cause stress corrosion and lead to cladding cracking. In this case, the mechanical stress is due to the swelling and deformation (thermal expansion) of the diabolo-shaped fuel pellet, which comes to rest against the inner wall of the cladding once the initial gap is compensated. This phenomenon is called the pellet–cladding interaction (PCI) and the stress on the cladding increases with the increase in power. The more rapid the increase, the higher the burnup rate (Cox 1990). Once the desired power is reached, the stress reaches a plateau, and then the stresses relax by plastic deformation (Sidky 1998) (see Figure 2.5).

The PCI is not only mechanical. The increase in reactor power is accompanied by a rise in pellet temperature; the thermal expansion of the pellet causes it to crack, and it is essentially from these cracks that the fission products are released.

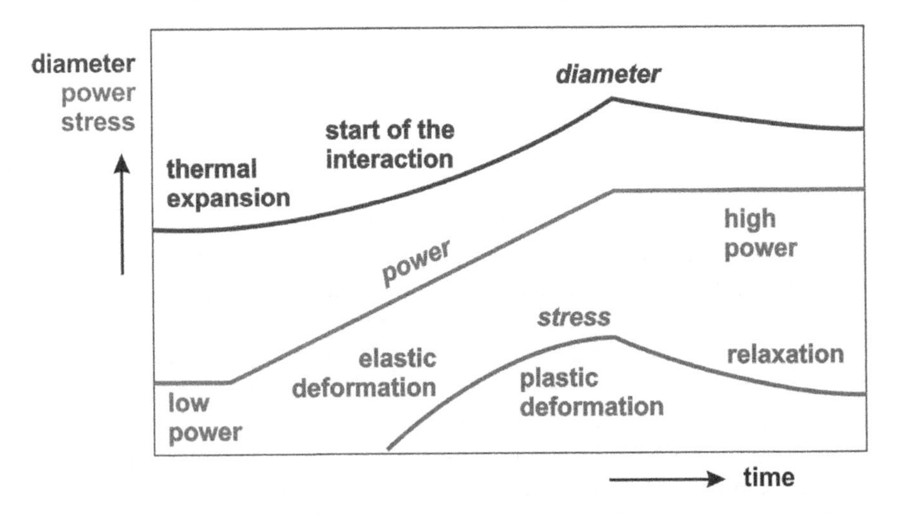

Figure 2.5. *Effect of a power ramp on the fuel pellet diameter and its stress on the cladding, from Sidky (1998). For a color version of this figure, see www.iste.co.uk/bouffard/nuclear.zip*

2.2.3. *The mechanisms of the evolution of the properties under irradiation*

The mechanisms governing the evolution of the properties under irradiation and the effect of the alloying elements are determined from experimental and modeling studies (Onimus et al. 2020). Claddings from reactor assemblies in service are being investigated, but the presence of fuel and fission products limits the feasibility of this work. Therefore, irradiations with charged particles (such as protons) are often performed to simulate these effects. Under irradiation, the measured growth (see Figure 2.6) can be broken down into three phases. Phase I is characterized by a rapid growth for fluences lower than $\sim 3\times10^{24}$ n.m^{-2}. In phase II (between 0.3×10^{25} n.m^{-2} and 2.5×10^{25} n.m^{-2}), a stationary regime is established.

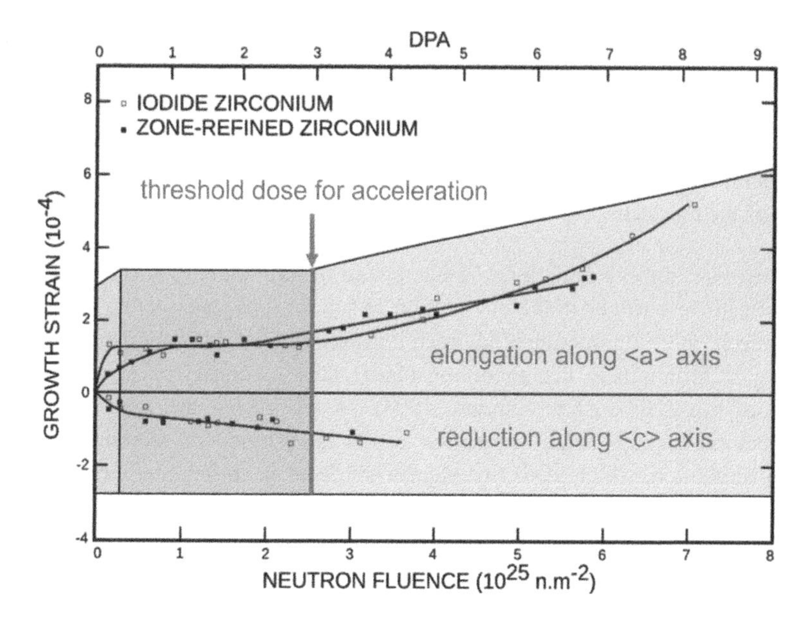

Figure 2.6. *Growth versus fluence of zirconium single crystals (figure adapted from Carpenter et al. (1988)). For a color version of this figure, see www.iste.co.uk/bouffard/nuclear.zip*

Phase III, known as breakaway, corresponds to a rapid acceleration for thresholds in the range of 3×10^{25} n.m^{-2}. These phases are correlated to the evolution of the microstructure. Irradiation defects have been studied by transmission electron microscopy since the 1970s, with a resolution-related size limit of the order of 2 nm (Griffiths 1988, 1993). The observed defects are mainly dislocation loops with

Burgers vector ⟨a⟩ (⟨1000⟩, ⟨0100⟩ or ⟨0010⟩ direction). Dislocation loops with a ⟨c⟩ component (⟨0001⟩ direction) are only observed at high fluence. Cavities are very rarely observed and only under very specific conditions that are not representative. The growth under irradiation is correlated to the appearance of ⟨c⟩ loops.

The ⟨a⟩ loops are present in the various commercial alloys between 250 and 400°C. These ⟨a⟩ loops are perfect, vacancy or interstitial in nature. The relative proportion depends on the irradiation temperature; the proportion of vacancy loops increases with temperature. They occupy a habit plane close to the prismatic plane and their diameter is of the order of 5–20 nm; their size increases with the irradiation temperature, while their density decreases.

The ⟨c⟩ loops appear for fluences above 5×10^{25} neutrons.m^{-2}; these loops are faulted and vacancy in nature (Northwood 1977). They are in the basal plane with Burgers vector 1/6 $\langle 20\bar{2}3 \rangle$ (Miller-Bravais 4 index writing). The ⟨c⟩ loops are much larger (diameter in the order of 100 nm) and less numerous than the ⟨a⟩ loops. The formation of these ⟨c⟩ loops depends on the chemistry of the alloy or the purity of the zirconium studied. The smallest ⟨c⟩ loops observed have a diameter of about 10 nm, and the mechanism of formation of these loops has not been directly observed.

Furthermore, under irradiation, the interaction of point defects with the alloying elements leads to changes in the distribution of solutes and precipitates. In Zircaloy 4, amorphization of the intermetallic precipitates $Zr(Fe, Cr)_2$ leads to the dissolution of iron and chromium in the zirconium matrix, and secondary precipitation may result from this (Gilbon and Simonot 1994). In Nb alloys, the most significant microstructural change under irradiation is the appearance of niobium-rich fine needle precipitation, which leads to a significant decrease in the niobium content in the matrix (Ribis et al. 2018).

2.2.4. *Simulations of growth and creep under irradiation*

Different mechanisms and different models of loop formation have been proposed in the literature to determine the growth under irradiation. The first model proposed to interpret the evolution of the microstructure is the diffusion anisotropy difference (DAD) model, based on the difference in the diffusion anisotropy of self-interstitials and vacancies (Woo 1988). The DAD model assumes that self-interstitials diffuse faster in the basal plane, while the diffusion of vacancies is isotropic. This assumption leads to a preferential capture of self-interstitials by prismatic loops (⟨a⟩ loops), while vacancies will preferentially be captured by ⟨c⟩ loops. This assumption on the diffusion of point defects has been invalidated by atomistic calculations (initial calculations based on the density functional theory).

Other models have been proposed, such as Bullough's 1975 model and, more recently, the production bias model (PBM), which assume that the numbers of moving vacancies and self-interstitials produced in the collision cascades are not equal (Barashev et al. 2015). This assumption is contradicted by observations under electron irradiations that lead to qualitatively similar microstructures, while these irradiations almost exclusively produce isolated Frenkel pairs, in other words, equality between the two types of defects. The new models, which are still being improved, are based on recent atomistic simulation results, such as the 1D and 2D mobility of small self-interstitial clusters and some ad hoc assumptions on loop germination (Barashev et al. 2015), or DFT calculations of single defects and small clusters of point defects, or even the anisotropy of vacancy scattering and a mechanism for ⟨c⟩ loop formation (Christiaen et al. 2019, 2020).

For creep under irradiation, several models have also been proposed: (i) models based on stress-induced preferential absorption (SIPA), leading to the climb of edge dislocations under stress, (ii) models based on an accelerated climb of dislocations, combining a climb due to the absorption of point defects and gliding due to creep-induced deformation. However, the elementary mechanisms responsible for irradiation creep are still not well known. The coupling between growth and irradiation creep in models does not seem to be necessary to reproduce experimental results (Onimus et al. 2020).

In summary, neutron irradiation leads to the formation of dislocation loops and to a modification of the distribution of solutes and the nature of precipitates. These modifications of the microstructure lead to cladding growth and to an accelerated creep. Some elements on the origins of these mechanisms have been given above. The consequences of irradiation also include the hardening of the material and the formation of slip bands. The effects of irradiation are potentially coupled with hydriding, which degrades the fracture properties of the material (hydrogen embrittlement), and corrosion with the environment throughout the life cycle of the component.

2.3. Internal structures in austenitic steel

The internal structures of the PWR vessel (see Figure 2.7) bear the weight of the core, hold and immobilize the fuel assemblies, and channel the flow of the coolant. These structures are subject to doses between 10 and 80 dpa over the life span of the reactor, depending on their location in the core.

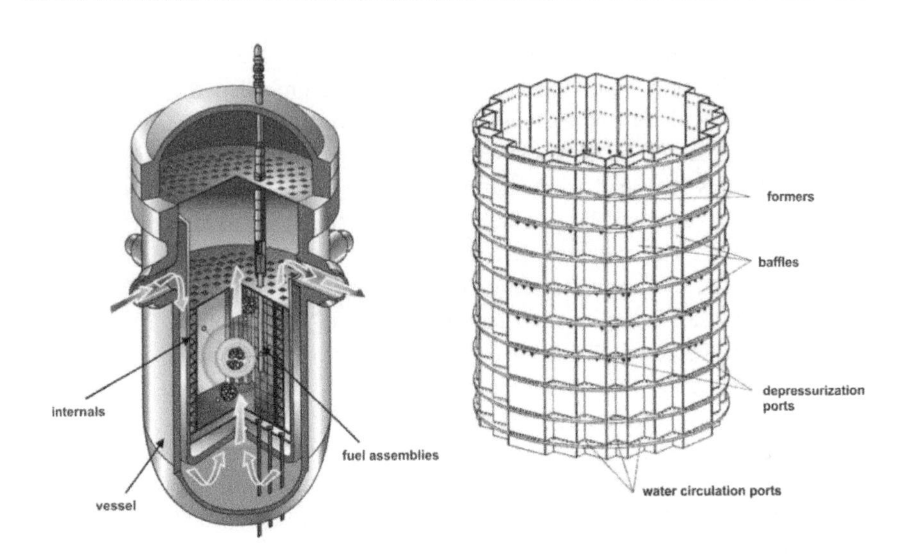

Figure 2.7. *On the left is a diagram of a pressurized water reactor and on the right is a diagram showing the baffles and formers of the internal structures (EDF figure). For a color version of this figure, see www.iste.co.uk/bouffard/nuclear.zip*

This internal equipment forms a construction approximately 10 m in height and 3 m in diameter, which must be kept intact and dimensionally stable. Alloyed steels containing more than 12% chromium, known as stainless steels, are resistant to aqueous corrosion. According to their structure and consequently their chemical composition, we can distinguish the following:

– ferritic stainless steels (e.g. ≈ 17%Cr, nickel-free steels);

– martensitic stainless steels, mainly used in bolting and for valves and fittings (e.g. 13-4, 17-4 PH or X5CrNiCuNb16-4 steels);

– austenitic stainless steels, mainly 18Cr–10Ni and 17Cr–12Ni–Mo (AISI 304L, 316L, 316NG, 308L, 321, 347, 309S);

– austenitic–ferritic stainless steels; these are the steels used to manufacture certain primary circuit castings for second-generation PWRs.

For the internal structures that we are interested in here, it is hyper-quenched 304 steel that is used for the partitioning and hardened 316 steel for the screws that ensure their assembly (see Figure 2.7).

Austenitic chromium–nickel stainless steels (AISI 304L, 316 and 316L) have long been known to be insensitive to stress corrosion in the primary environment (Gras 2017). However, on early design PWRs, fasteners that were particularly irradiated and stressed have cracked at the fillet between the head and shaft of the bolt. The expert reports of cracked bolts in 316 steel showed a strong hardening of the material, due to an important evolution of the microstructure induced by irradiation (disappearance of the initial network of dislocations, formation of dislocation loops, precipitation of second phases or intergranular segregation). Nano-cavities and helium nano-bubbles have been observed by transmission electron microscopy on samples taken from these bolts. These steels could therefore swell at high doses, with consequences being the dimensional variations of the structures and mechanical stresses induced by the swelling gradients, associated with thermal gradients. This could affect the dimensional stability and integrity of these internal components.

In order to understand these aging phenomena (Tanguy et al. 2015), studies are organized around the following three axes: (i) the conditions of appearance of degradation phenomena, (ii) the effects of aging on the performance of internal components and (iii) the mechanisms of aging. This work is based on the analysis of the experimental return of deposited reactor components and on irradiations in experimental reactors, whose neutron spectrum is representative of that of PWRs, such as the OSIRIS reactor of CEA/Saclay. Specific studies on model materials, for example, complete this work. In addition, to reach doses representative of 60 years of operation, fast spectrum reactors are also used, such as the Russian BOR-60 reactor, whose neutron flux is 10 times greater. In this way, 120 dpa can be reached in just a few years (Renault-Laborne et al. 2016).

Let us describe one of the effects mentioned above as an example, so as to illustrate the complexity of the phenomena induced by irradiation on the evolution of the microstructure: intergranular segregation.

2.3.1. *An effect of irradiation: intergranular segregation*

Radiation-induced segregation (RIS) by modifying the chemical composition of the grain boundaries, in particular by chromium depletion and silicon enrichment, has been identified as a phenomenon contributing to intergranular embrittlement. Indeed, couplings between the flux of point defects created by irradiation and the flux of solute atoms in the material can generate irradiation-induced segregation at the level of the defect sinks, and particularly at the grain boundaries. Irradiation-induced segregation depends on temperature and irradiation flux, as shown in Figure 2.8.

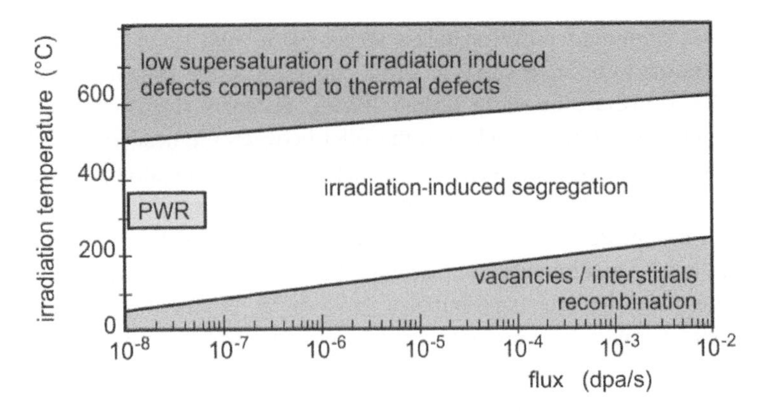

Figure 2.8. *Temperature and flux dependence (expressed here in dpa/s) of the irradiation-induced segregation occurrence zone in austenitic stainless steels. The PWR operating regime is represented by the green zone. According to Bruemmer et al. (1999). For a color version of this figure, see www.iste.co.uk/bouffard/nuclear.zip*

There are three regimes that depend on the irradiation conditions:

– At low temperatures (below 80°C for a PWR flux): the mobility of point defects is low and, consequently, there is little elimination on the sinks, and recombination between vacancies and self-interstitials is therefore predominant. Irradiation-induced segregation is low.

– At high temperatures (above 500°C for a PWR flux): the equilibrium concentrations of vacancy defects (created by thermal phenomena) are high. The supersaturation of vacancy defects generated by irradiation is therefore not very significant, which limits the flux of defects towards the sinks and thus the phenomenon of induced segregation. The concentration of self-interstitials induced by irradiation is always higher than the equilibrium content. However, these are very quickly recombined because of the very high content of vacancies.

– At intermediate temperatures (between 80 and 500°C for a PWR flux): the irradiation-induced segregation is maximal. The PWR irradiation conditions, represented by the green rectangle in Figure 2.8, are at the heart of the induced segregation domain.

A direct consequence is that the composition of the grain boundaries of PWR steels is modified by neutron irradiation. Figure 2.9 shows a concentration profile that has been obtained through a grain boundary in a sample taken from the Tihange reactor bolt (Edwards et al. 2003).

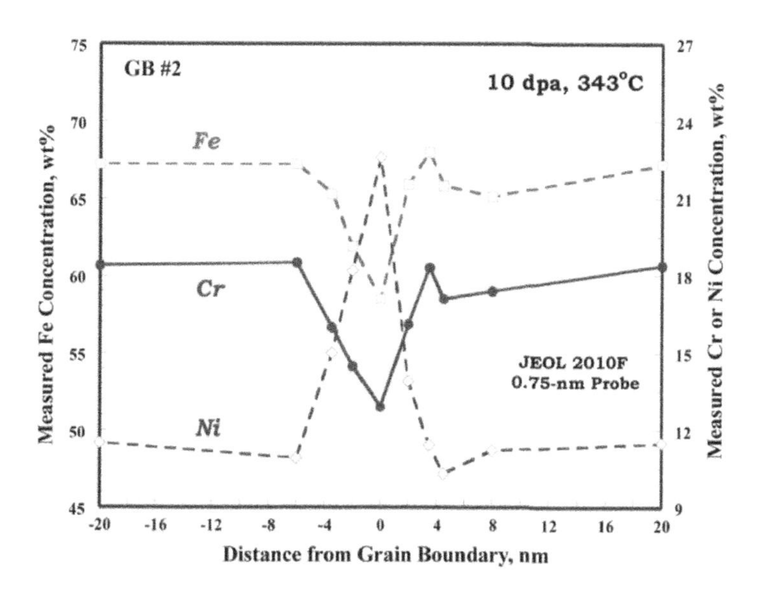

Figure 2.9. *Concentration profiles plotted across a grain boundary in a sample taken from a Tihange reactor bolt, irradiated at 10 dpa. Only the profiles of Cr, Fe and Ni are shown. These results were obtained by X-ray emission spectroscopy (Edwards et al. 2001). For a color version of this figure, see www.iste.co.uk/bouffard/nuclear.zip*

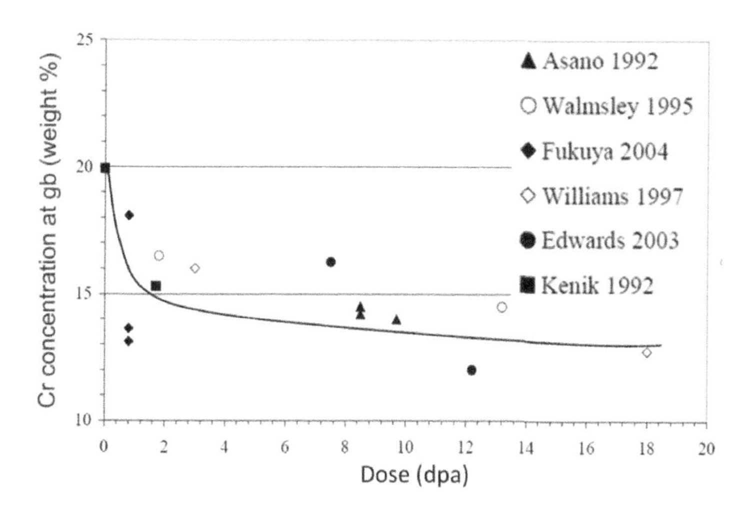

Figure 2.10. *Evolution of the Cr concentration at the grain boundaries as a function of the irradiation dose (expressed in dpa). The data come from 316 or 304 steels irradiated with neutrons (Etienne 2009)*

Chromium concentration decreases with increasing irradiation dose. Chromium depletion seems to saturate when the dose reaches 5–10 dpa (see Figure 2.10).

All these microstructural evolutions under irradiation that have been reported above cause modifications of the mechanical properties of the material. Here, we speak of hardening, loss of ductility, loss of tenacity and localization of plastic deformation. We will discuss these changes and their consequences in the following sections.

2.3.2. *Evolution of mechanical properties under irradiation*

Austenitic stainless steels used for vessel internals have a face-centered cubic structure. These alloys do not show a ductile–brittle transition. For doses below about 10 dpa, and in the temperature range of interest (290–350°C), the material shows a sharp increase in yield strength, accompanied by a marked loss in ductility (Tanguy et al. 2015). Above 10 dpa, these changes saturate. The sharp decrease in ductility due to irradiation is accompanied by a decrease in tenacity in these austenitic steels. The hardening results from the creation of clusters of defects under irradiation (dislocation loop), the formation of precipitates or second phases, or cavities that hinder the gliding of dislocations. As a result, the stress to be exerted to ensure the mobility of dislocations increases. For austenitic stainless steels, at the irradiation temperatures encountered in PWRs, the main defects are interstitial dislocation loops, cavities (and/or gas bubbles) and irradiation-induced precipitation, especially Ni_3Si (Renault-Laborne et al. 2015). Today, there is still a lot of work characterizing the nature, chemistry and characteristics of these induced nanostructures. The small size of the defects (2–3 nm) requires very high-resolution experimental techniques and the handling of radioactive samples, which makes the experiments complex.

Another consequence of these nanoscale transformations is the microscale deformation mode of these materials. As detailed in the thesis by Han (2012), the post-irradiation plastic deformation mechanisms are highly dependent on test temperature, stress, strain rate and irradiation dose. Here, pre-irradiated austenitic stainless steels subjected to a specific strain rate test exhibit localized deformation (specific zone). The main mechanisms of localized deformation are related to the formation of twins (stacking faults in the CFC structure related to the gliding of imperfect dislocations) and dislocation channeling (zones or cylinders within the grain in which perfect dislocations glide behind each other). For a low irradiation temperature (<100°C), the major mechanism of strain localization in austenitic stainless steels is the formation of bands of twins. At these temperatures, the stress required for these dislocations to glide (known as critical-resolved shear stress) is largely reached due to the increase in yield strength under irradiation. At higher temperatures, it is the increase in the number of dislocation channels that favors a localized deformation, because the critical-resolved shear stress becomes difficult to reach (indeed, the stacking fault energy increases with temperature). Here, we speak of a deformation mode transition. At the operating temperature of PWRs (280–380°C), the deformation mode depends on the strain rate. The deformation operates in channels at low strain rates (in the order of 10^{-6} s^{-1}), with very few twins observed. At medium strain rates (in the order of 10^{-4} s^{-1}), the deformation mode remains channel-based, but with a non-negligible presence of twins. Clear bands are channels free of irradiation defects (or with a very low defect density), "cleaned" by the gliding of perfect dislocations. In austenitic stainless steels, clear bands have a major role in localized deformation at PWR operating temperatures. The clear bands have a width between a few dozen to a hundred nanometers, with a thickness of a few dozen nanometers. The elimination of Frank loops (loops of sessile dislocations, i.e. which do not slide) in the clear bands generates a local softening, as well as a preferential zone for the gliding of perfect dislocations. The rest of the grain does not deform plastically, except by the appearance of new channels.

Thus, during plastic deformation of irradiated austenitic steels, clear bands free of Frank loops are observed. The formation of channels is considered as the main mechanism of localized deformation, which also leads to a softening at the macroscopic scale. All these concurrent transformations generate material behaviors, which are at the origin of its degradation. An example of this could be the joint effect of segregation at the grain boundaries and the clear bands, which contribute to the susceptibility to IASCC (see below) and cause a loss of tenacity. Frank loops play a significant role in the irradiation hardening and creep of materials. The formation of gas bubbles and void cavities can lead to macroscopic swelling. Without going into too much detail, which can be found in the literature, we will describe these characteristic effects of austenitic steels in a few words: creep, swelling, IASCC.

2.3.3. *Creep under irradiation*

Under neutron flux and subjected to a load below the elastic limit, the steels of the internal structures undergo a deformation that evolves over time, even though they are not or little sensitive to thermal creep at these temperatures (290–350°C). In practice, experience shows a decrease in the torque of the bolts with the dose, which can be explained by the appearance of the irradiation-induced creep phenomenon under neutron flux (Lemaire et al. 2011). The knowledge of the laws governing the kinetics of creep under irradiation (in the case of bolts, we speak of relaxation under irradiation, with the deformation being imposed at the time of the initial tightening) is particularly important in order to assess the level of constraint to which the bolts are subjected as a function of the dose, and if their properties remain within the framework of the initial specifications. The characteristics of this creep under irradiation are very different from those of thermal creep, with a quasi-linear dependence on stress and a low temperature dependence. There are different creep regimes under irradiation: after a short transient regime (approximately 1 dpa); the material enters a stationary regime where the creep deformation is proportional to the dose and the stress. Although experimental data are available and integrated in the parameterization of the material laws used for the calculations of the mechanical stresses in the internal structures, the mechanisms of creep under irradiation are not entirely established. The most classical mechanism is the creep related to the climb of dislocations by preferential absorption of interstitial defects, induced by the stress of the dislocation itself. However, modeling of this mechanism by the cluster dynamics technique shows that this mechanism alone cannot explain the experimental kinetics of strain as a function of dose (Garnier et al. 2011). Therefore, another mechanism comes into play. From this same modeling, it is shown that it is necessary to also consider, for example, the interactions of point defects with other structural defects (such as the initial dislocation network or the grain boundaries). These approaches make it possible to better reproduce the order of magnitude of the measured kinetics, as well as their sensitivity to the stress.

The main consequence of irradiation creep is a stress relaxation at the level of the bonding bolts (Lemaire et al. 2011), and is therefore likely to improve the resistance to irradiation-assisted stress corrosion (Tanguy 2011). In PWRs (900 MW CP0 level), this phenomenon of assisted corrosion has nevertheless resulted in the cracking of some baffle bolts. This is the result of a complex set of microstructural modifications and an oxidizing environment linked to the exposure to the primary fluid. These effects and their interactions are not yet fully understood.

2.3.4. *Swelling*

The current feedback from assessments of deposited PWR components (Edwards et al. 2003) shows that the microstructural evolutions observed on these components do not lead to significant macroscopic swelling. However, the swelling induced by irradiation is considered as a potential phenomenon that could limit the durability of these structures for high doses, mostly because of the dimensional changes that would be induced (swelling gradient between different components). Considering the high fluence levels that macroscopic swelling is associated with in fast neutron reactors (linked to a higher production of helium), this phenomenon has been little studied experimentally in the PWR environment. Most of the studies have been carried out in the RNR, where the conditions of temperature, neutron spectrum and doses received are more favorable for the development of swelling.

2.3.5. *Irradiation-assisted stress corrosion cracking*

Austenitic steels are, under irradiation, the seat of many processes that are more or less related: hardening, structural transformations, segregation under irradiation, cavities, swelling, creep, etc. One of the major problems encountered in PWRs is the intergranular cracking of stainless austenitic steel components of internal structures. The mechanism responsible for this cracking is called irradiation-assisted stress corrosion cracking (IASCC). This complex mechanism is still poorly understood and involves many variables such as stress, environment and irradiation effects (Wang et al. 2019).

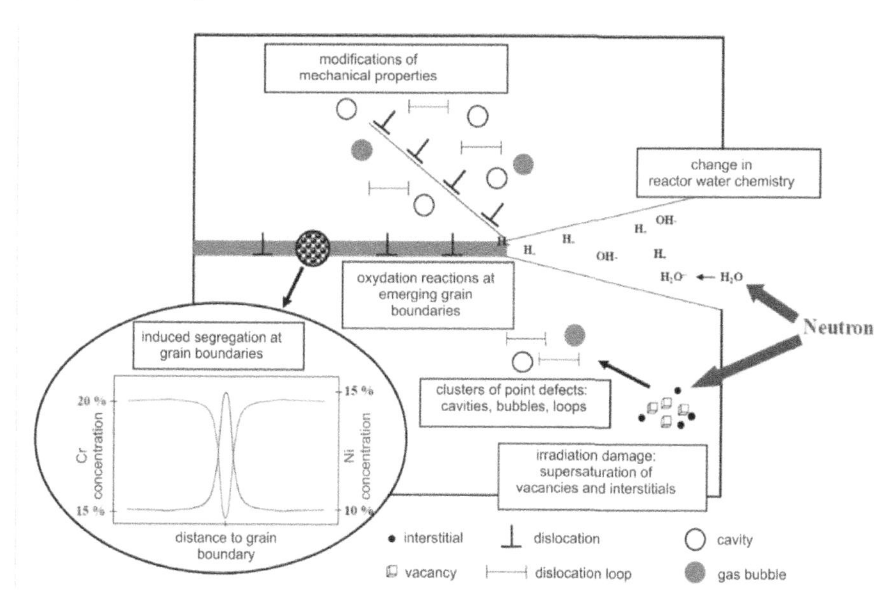

Figure 2.11. *The main processes that may be involved in irradiation-induced stress corrosion cracking (Bruemmer et al. 1999; Etienne 2009). For a color version of this figure, see www.iste.co.uk/bouffard/nuclear.zip*

Figure 2.11 schematizes the main processes that would be likely to cause IASCC:

– modification of the environment (radiolysis at the interface steel/water contained in the reactor) by neutron irradiation, making it more corrosive;

– segregation at the grain boundaries induced by irradiation, modifying their chemical composition, thus promoting intergranular corrosion;

– formation of cavities, gas bubbles, dislocation loops and possibly new phases blocking the dislocations movements, thus modifying the mechanical properties of the material.

According to several works, the combination of all phenomena is necessary to induce IASCC cracking mechanisms (Busby et al. 2002).

The phenomenon of IASCC was observed on a few bolts, with the development of intergranular cracking occurring in the fillets of these bolts for doses of about 5 dpa (see Figure 2.12).

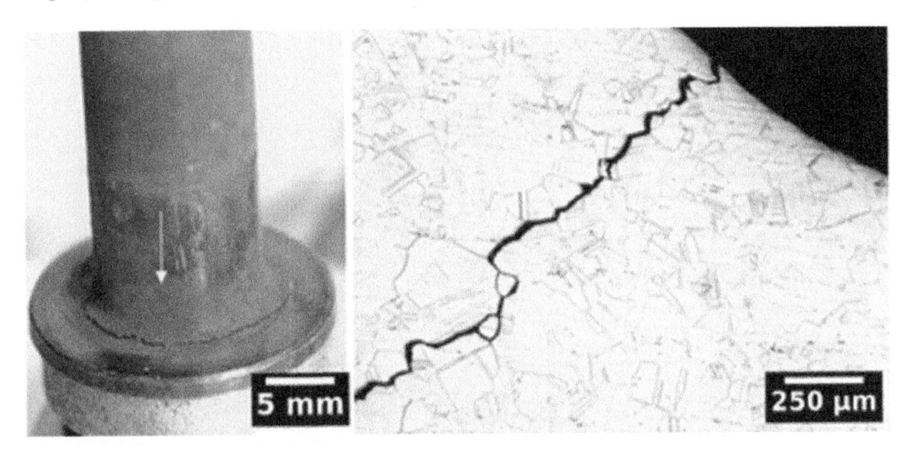

Figure 2.12. *Example of a crack at the head-to-shank connection of a bolt of an internal structure. Scanning electron microscopy image of a crack (EDF figure)*

The synergy of the mechanisms involved in IASCC damage has led to the definition of study strategies at several levels, both experimentally and for modeling. At the experimental level, the studies can be classified into three categories: (i) in situ stress corrosion tests (SC) in test reactors, (ii) post-irradiation SC tests on neutron-irradiated materials and (iii) SC tests on "representative" or "model" materials. The latter are materials likely to represent one of the microstructural modifications due to irradiation, such as hardening, intergranular chemical segregation or localized plastic deformation.

In summary, under PWR operating conditions, austenitic stainless steels undergo a change in their microstructure due to neutron irradiation. This change results in a restoration of the initial dislocation network, in the formation of irradiation defects (clusters of point defects, Frank loops, gas bubbles, intergranular and intragranular cavities and precipitates), in the presence of segregations on the defect sinks (dislocations, grain boundaries), as well as in the formation of clear bands during plastic deformation. Frank loops play an important role in the hardening and irradiation creep of materials. The Frank loop population only seems to be sensitive to the irradiation temperature, and its density increases with the dose and saturates for a dose of 5–10 dpa. On the contrary, the cavity population shows a high dependence on the neutron spectrum of the reactor (PWR, RNR, experimental), the temperature and the irradiation dose. Swelling probably does not appear in the internal structures of PWR vessels, according to the extrapolation of studies carried out under the RNR neutron spectrum. However, today, it is a challenge to carry out experiments under PWR conditions to explore this phenomenon, which could limit the operating time of the reactors. Thus, with a view to an operating life extended to 60 years, studies on the conditions of the appearance of swelling in PWR conditions have been undertaken. In parallel with the experimental studies, the development of a predictive model of the effects of irradiation on the evolution of the properties of these materials has been initiated on a European scale.

2.4. The vessel

Second-generation nuclear reactor vessels, commissioned in the 1960s in the United States, were fabricated from 175 mm thick carbon–manganese steel. With the increase in reactor power, C–Mn steels were no longer able to guarantee good tenacity for greater plate thicknesses. The vessels were then made of manganese–molybdenum steel, which made it possible to improve both the mechanical characteristics and the tenacity. A further evolution was necessary to suit the dimensions of the 900 MWe PWR vessels[2]. This led to the ASME SA533 grade B cl 1 (or SA508 cl 3), by adding nickel and molybdenum. This grade, designated 16MnNiMo05 in the ISO standard (formerly 16MND5), was used for all second-generation French PWR vessels, with some additional constraints on the

2. For the record, a 900 MWe reactor vessel has an internal diameter of 3,988 mm and a thickness of 200 mm (respectively 4,885 mm and 250 mm for an EPR).

carbon, sulfur, phosphorus and copper contents[3]. It was also the material used for the construction of the first EPR vessels, with even greater constraints on copper and phosphorus concentrations, as we will see later. The 16MnNiMo05 steel is used in the hardened and tempered state, with a bainitic structure. It represents an excellent compromise between good tenacity and relatively high mechanical properties (yield strength: ≈ 480 MPa and elongation at break: ≈ 25% at 25°C), allowing for limited component thickness. As manufactured today, the steel is very close to its optimum. If it were necessary to significantly improve the properties of the vessel steel, the 15CrMo9-10 steel (2.25%Cr–1%Mo), which is commonly used for the manufacture of petrochemical reactors, could be an interesting alternative: it has been shown to be very insensitive to neutron irradiation and has excellent tenacity. Variations of this steel have been used in the nuclear power industry in Eastern Europe. It should also be noted that in PWRs, the inner layer of the 16MnNiMo05 steel is coated with a deposit of AISI 308L stainless steel, so as to protect it from corrosion in water at high temperatures.

2.4.1. *Changes in tenacity and resilience*

Under the effect of neutron irradiation, it is experimentally observed that the toughness curve of the steel used for the vessel shifts towards higher temperatures (see Figure 2.13).

At a given temperature, the tenacity decreases with irradiation: the steel becomes brittle. As the reactor vessel is the second safety barrier, its integrity must be guaranteed throughout the life of the reactor. To achieve this, the steel tenacity must remain sufficiently high to guarantee the stability of any defects (cracks), whether they be real or postulated, subjected to the various loads considered in the design codes. The reactor vessel is, moreover, a non-replaceable component: its unserviceability means the end of the reactor life. The neutron flux undergone by the vessel is about 2×10^{18} n.cm^{-2}.year^{-1} (E > 1 MeV), that is, an irradiation damage of the order of 3×10^{-3} dpa.year^{-1} (in other words, still ≈ 0.1 dpa in 30 years). This causes an increase in the brittle–ductile transition temperature, which results in embrittlement of the ferritic steel (see Figure 2.14).

3. Composition of 16MnNiMo05 steel: C max 0.2 – Mn 1.15/1.55 – Ni 0.5/0.8 – Mo 0.45/0.55 – Cr max 0.25 – Si 0.1/0.34 – P max 80 ppm – S max 80 ppm – Cu max 800 ppm – Co max 300 ppm.

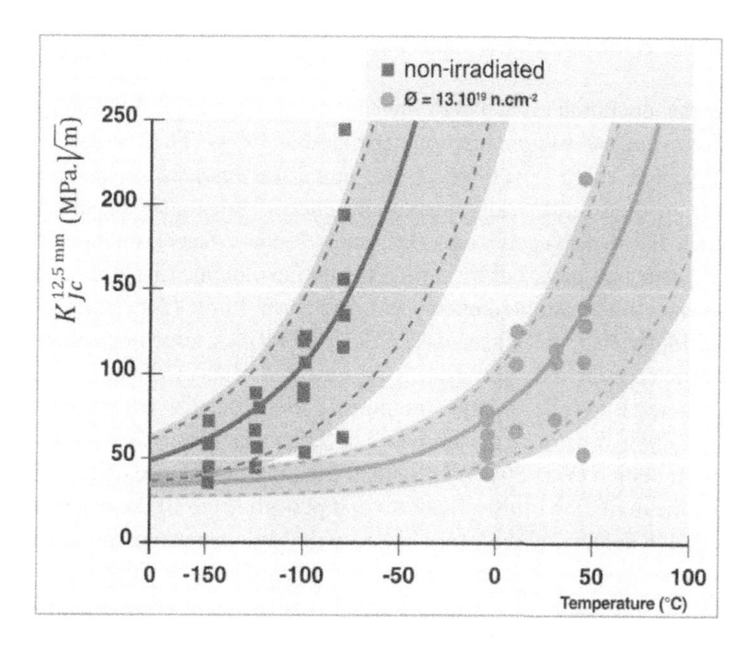

Figure 2.13. *Tenacity of pressurized water reactor vessel steel as a function of temperature, before (blue) and after (green) irradiation (Boutard et al. 2016). For a color version of this figure, see www.iste.co.uk/bouffard/nuclear.zip*

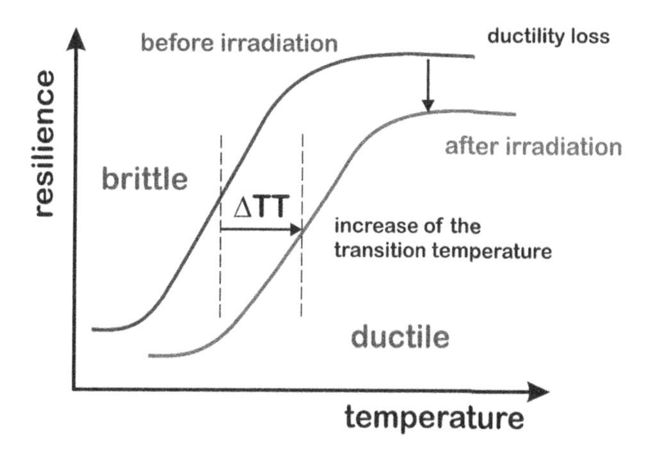

Figure 2.14. *The effect of irradiation on the resilience and the increase of the ductile–brittle temperature (symbolized by the delta in temperature ΔTT). For a color version of this figure, see www.iste.co.uk/bouffard/nuclear.zip*

2.4.2. *Cluster dynamics*

During the operation of a PWR, the atoms in the vessel are bombarded by neutrons from nuclear fission reactions (Brinkman 1954). The collisions between neutrons (average energy of 1 MeV) and atoms (at rest) cause the latter to be dislodged from their sites, with a maximum energy transmitted of about 70 keV (front shock). These displaced atoms (known as Primary Knock-on Atoms (PKA)), with a low mean free path, will in turn generate new atomic displacements around them as soon as the transmitted energy is higher than 40 eV (American Society for Testing and Materials (ASTM) standard (ASTM 2009)). A large population of point defects (self-interstitial and vacancies) is thus created. This is where issues start for the vessel material. While a large fraction of these defects will recombine and disappear, a small fraction will survive. The latter fraction is 4% in the case of 1 MeV ion irradiation (Was 2017). This small fraction, maintained over time, will be the cause of the evolution of the properties and performances of the material. On the one hand, point defects of the same nature will agglomerate to form vacancy or interstitial-type clusters of point defects, or agglomerate with solutes to form complex clusters of point defects and solutes. These different clusters are part of the "matrix damage". On the other hand, these surviving point defects can also accelerate or induce (depending on the mechanism) the mobility of solutes, as well as promote the formation of intragranular precipitates or solute clusters, or segregation towards defect sinks (dislocations, interphase interfaces, grain boundaries). From the simple increase of the diffusion coefficient to the more complex mechanisms of "defects/solutes" flux coupling, the addition elements such as Mn, Ni and Si, for example, or the impurities Cu and P, will be redistributed within the microstructure (Meslin-Chiffon 2007). The changes in the microstructure, which result from these effects of irradiations, impact the mechanical properties (see the previous section). It is therefore important to understand the dynamics of defects and their effects. Four main families of induced defects can evolve within the ferritic structure of these bainitic steels: (a) point defect clusters, solute/point defect clusters, nanocavities and dislocation loops; (b) copper-rich precipitates/clusters (always associated with the elements Mn, Ni, Si and P); (c) Mn- and Ni-rich precipitates or clusters (often associated with Si and P); (d) local segregations on dislocation lines, or at grain boundaries or interphase interfaces. We will describe them in the following sections.

2.4.2.1. *Point defect clusters, solute/point defect clusters, nanocavities and dislocation loops*

Positron annihilation spectroscopy (PAS) performed on model alloys (Fe–Cu, Fe–Mn–Ni, Fe–Cu–Mn–Ni) and irradiated 16MND5 steels reveals the presence of small clusters of a few vacancies. Copper atoms are often associated with these clusters (Toyama et al. 2007; Meslin et al. 2010). Interstitial clusters are much more complex to detect. Nanocavities are formed by long-range diffusion of defects. These defect clusters are very detectable by transmission electron microscopy in irradiated pure iron (0.19 dpa at 300°C) (Hernández-Mayoral and Gómez-Briceño 2010). Dislocation loops are easily observed in Ni^{2+}-irradiated A533B steels (Fujii and Fukuya 2005) and in neutron-irradiated model alloys. In pure iron that is irradiated with neutrons at 300°C (from 0.026 to 0.19 dpa), the loops are mostly interstitial in nature with a majority (>80%) having a Burgers vector of ⟨100⟩, the others being ⟨110⟩ (Hernández-Mayoral and Gómez-Briceño 2010).

2.4.2.2. *Copper-rich precipitates and clusters*

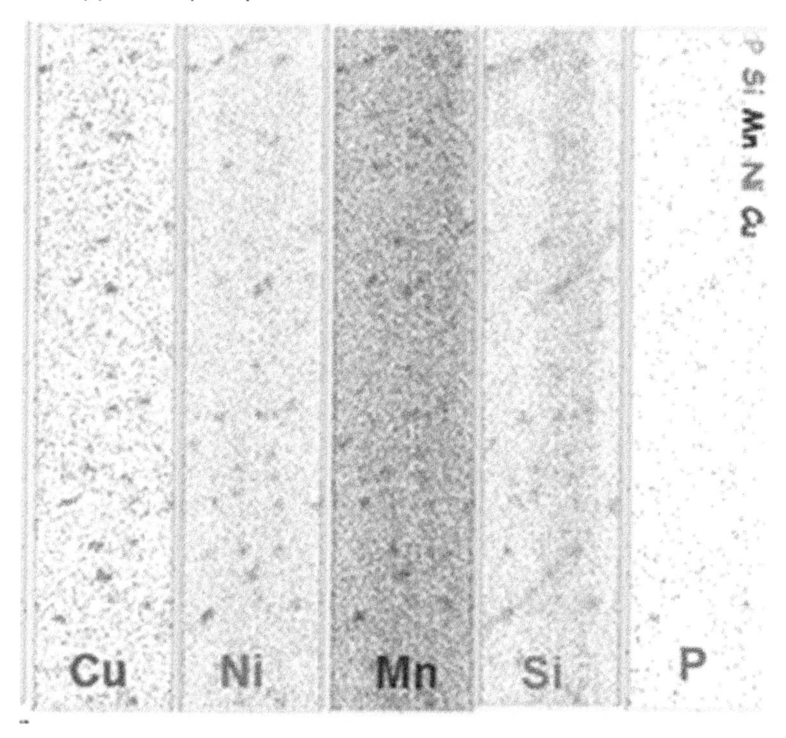

Figure 2.15. *Cu-, Ni-, Mn-, Si-, P-rich solute clusters in a French steel 16MND5 (290°C, 7.3×10^{23} n/m^2). The volume represented (obtained by atom probe tomography) is 30×30×140 nm^3 (GPM-EDF figure). For a color version of this figure, see www.iste.co.uk/bouffard/nuclear.zip*

Copper is a very poorly soluble impurity (0.007% weight at 300°C) and highly mobile in iron under irradiation (enhanced diffusion). It therefore tends to precipitate rapidly and consequently modify the mechanical properties of the material (Kuri et al. 2009). Copper contents range from 0.05 to 0.30% weight and are currently controlled to be as low as possible (below 0.05% weight). The presence of copper particles, or clusters of copper and solutes (Mn, Ni, Si and P) of small sizes (1–2 nm in diameter), is always observed when the copper content exceeds 0.1% weight (Miloudi 1997; Kuri et al. 2009). While the exact composition of these particles is still a matter of debate, iron is often observed in these clusters when the steel has a level of Cu < 0.1% weight, and average Ni (0.7% weight) and Mn (1–1.15% weight) levels (Pareige and Miller 1996). Studies on model alloys (FeCuMnNiP and FeMnNiP) irradiated with ions show that the clusters are only observed in the alloy containing copper (Radiguet et al. 2009). Copper therefore has a major role and its concentration must be controlled during the elaboration and assembly of the reactor vessel parts. Figure 2.15 shows these copper-rich solute clusters in a 16MND5 steel.

2.4.2.3. *Mn–Ni-rich precipitates/Ni–Mn clusters*

Since their first observation (Odette 1995), Ni- and Mn-rich clusters have been widely reported in the literature (Miller et al. 2009). Depending on the nominal composition of the steel, other elements may be associated (Si and P) (Pareige et al. 1993; Miloudi 1997; Miller et al. 2009). In steels with low copper content but irradiated at high fluence, the presence of these Mn, Ni and Si clusters becomes significant (depending on the nominal contents). Under these high fluence conditions, if the nominal copper content is higher, then the two families of particles rich in copper and Ni–Mn can coexist (Odette and Lucas 2001). This can be explained if we look at the portion of the ternary phase diagram presented in Figure 2.16 (Liu et al. 1997). In the presence of a high nominal copper concentration, copper precipitates first. Then, as it is depleted, the composition of the matrix shifts to the lower portion of the diagram, thus favoring the appearance of phases rich in Mn and Ni.

However, it is still a real debate today. These two particle populations can be generated and coexist, due to solute-defect flux couplings towards the point defect clusters that are initially generated by the lower fluence irradiation. This is clearly shown by recent experiments (Meslin et al. 2013) in undersaturated FeMn alloys, which, under irradiation, highlight the presence of Mn clusters. The literature is dense on this subject, but the congruence is weak. This is still an aspect of irradiation effects in these materials that needs to be clarified.

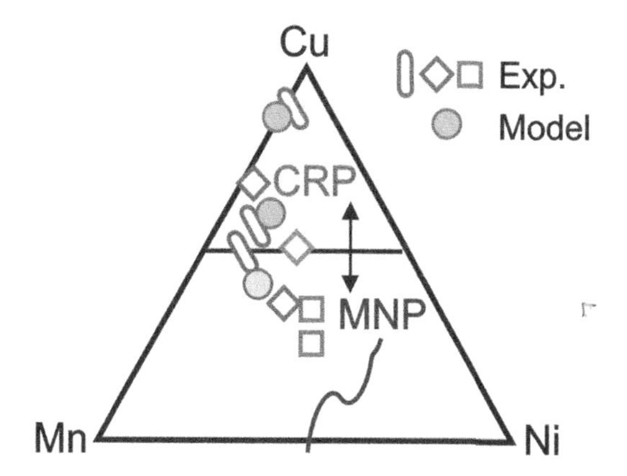

Figure 2.16. *Cross-section of a simplified Cu–Mn–Ni ternary phase diagram showing the transition from copper-rich precipitates (CRP) to Mn–Ni-rich precipitates (MNP) (Liu et al. 1997). For a color version of this figure, see www.iste.co.uk/bouffard/nuclear.zip*

Simulation works from the last few years have provided elements by considering two main points of view: (i) a precipitation related to a thermodynamic driving force and a surplus of vacancies related to irradiation, thus accelerating the diffusion to reach thermodynamic equilibrium (Mamivand et al. 2019); (ii) a preponderant role of the point defects coming from the irradiation and, in particular, the self-interstitials (Ngayam-Happy et al. 2013; Castin et al. 2020). The latter work is based on a large number of initial calculations of the different small point defect clusters with the different solutes, and they show the significant role of interstitials in the formation of solute-rich clusters.

2.4.2.4. *Segregation of solutes and impurities on 1D and 2D defects*

The segregation of solutes or impurities, particularly at grain boundaries, has been studied since the 1960s because of its possible impact on the mechanical properties of structural materials. As a result of its very harmful effects, phosphorus has been under the microscopic spotlight for many years (Naudin et al. 1999). First examined by Auger spectroscopy on fracture surfaces (Akhatova 2017), it is now quantified by coupled EBSD/FIB/TKD/APT approaches[4] (Mandal et al. 2014; Akhatova et al. 2019). The link between segregation rate, under the given aging conditions, and the crystallographic structure of the grain boundary itself is now accessible (see Figure 2.17).

4. EBSD: electron backscatter diffraction, FIB: focused ion beam, TKD: transmission Kikuchi diffraction, APT: atom probe tomography.

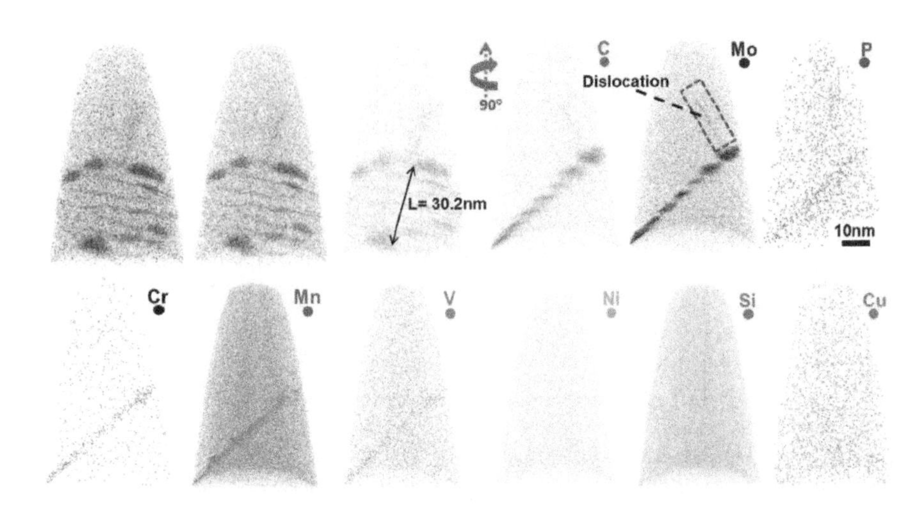

Figure 2.17. *Three-dimensional reconstruction of a 3° low angle grain boundary* $[5\bar{2}3](5\bar{4}\bar{3})/(3\bar{3}4)$. *This tilt grain boundary is composed of a stacking of edge dislocations that are 5.0 nm apart (Zhang et al. 2019). For a color version of this figure, see www.iste.co.uk/bouffard/nuclear.zip*

In order to assess the influence of non-hardening embrittlement with respect to the overall embrittlement of the material, specific studies have been conducted. For example, Figure 2.18(d) shows a quasi-linear relationship between the evolution of the ductile-to-brittle transition temperature (DBTT) and irradiation hardening (Nishiyama et al. 2008). However, the material with the highest phosphorus content shows a very large DBTT shift. These correlations between phosphorus content, phosphorus segregation at grain boundaries, irradiation hardening and DBTT evolution show a weak intergranular segregation effect with respect to hardening mechanisms, unless there is a high nominal phosphorus content and a relatively high fluence (Nishiyama et al. 2008).

The four effects produced by irradiation in the bainitic steels of reactor vessels, which we have just described, are often grouped into three families in the specialized literature: (i) clusters or precipitates mainly formed of solutes and resulting from homogeneous or heterogeneous processes, (ii) matrix damage such as clusters of point defects, clusters of defects/solutes, as well as nanocavities, and (iii) interstitial or vacancy loops. Some of these clusters (ii) are called stable matrix defects, which can evolve in time; others are called unstable matrix defects, generally of sub-nanometer size, with a vacant nature. They can dissolve by re-emission of defects. More extensive defects (>1 nm, defect/solute clusters, nanocavities, loops) are rather stable damage.

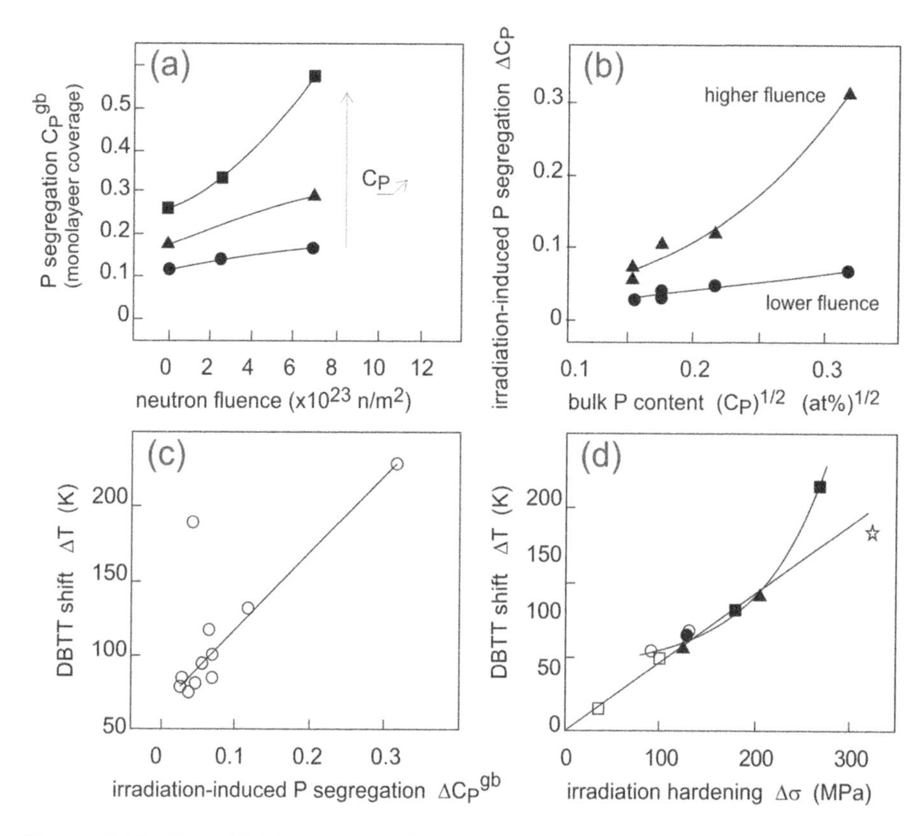

Figure 2.18. *From Nishiyama et al. (2008): (a) level of phosphorus segregation at grain boundaries as a function of neutron fluence; (b) rate of phosphorus segregated at grain boundaries as a function of nominal phosphorus content; (c) evolution of DBTT as a function of the phosphorus segregation rate; and (d) relationship between DBTT and hardening. The weight contents of phosphorus in the materials range from 40 to 570 ppm*

2.5. Perspectives

The reasons for the choice of certain specific materials for the construction and operation of a nuclear reactor vessel and its internal structures were presented, as well as the effects of the severe environment in which they evolve on a daily basis for several decades. This environment of high temperature, stress, corrosion and irradiation is one of the most complex of all large industrial architectures. The research carried out on these different materials in the context of constant improvement, through the recreation of the conditions of use parameter by parameter, or globally, is at the origin of the dynamism of research in physical metallurgy that has been taking place since the 1950s. As Brechet (2013) pointed out, the number of materials available to the engineer is of the order of 100,000. New materials are continually being developed, but only a very limited number of them make it out of the laboratory: it is rare for a material to only be used for one of its properties. However, the case is different for a material that has a combination of properties, and if there is a possibility of implementing it on an industrial scale. Thus, in each application field, particularly for the nuclear reactor vessel and all the metallic components it contains, a very limited number of families of materials are used: bainitic and austenitic steels, as well as zircaloy in the case of a nuclear reactor. The understanding of the evolution of metals and metal alloys in these complex environments, especially under irradiation, has reached a level well above that achieved for other classes of materials (ionic–covalent, organic, etc.). However, many points remain to be clarified if we want to further optimize the materials and be able to extrapolate their behavior in the long term.

This remarkable progress is the result of continuous improvements of experimental methods and of the possibility of coupling some of them. Thus, for example, scanning electron microscopy (SEM) associated with electron backscatter diffraction gives access to the crystalline orientation of the constituent grains of the alloy, and is essential in the preparation of samples for higher-resolution techniques. High-resolution transmission electron microscopy (HRTEM or STEM with HAADF detector[5]) complemented by analytical methods, based on energy-dispersive X-ray spectroscopy or electron energy loss spectroscopy (EELS), and atom probe tomography (APT), which gives a 3D image of the chemical composition of clusters, segregation and precipitates of sub-nanometer size at atomic scale, are all techniques that are continuously being improved (in terms of resolution, for example), and provide relevant information on the scales of the effects produced by irradiation. These very high-resolution techniques, which probe very small volumes of matter, have been supplemented by methods that allow sampling of larger volumes, such as small angle neutron scattering (SANS) and positron annihilation. The irradiation techniques that are necessary to simulate materials under radiation (electrons, ions, protons), such as those proposed by the EMIR&A[6] accelerator network, are also necessary. They are not perfectly representative of a real neutron irradiation, but allow the understanding of fundamental mechanisms (see Chapter 9 of this book). The use of irradiations in experimental reactors, as well as the EDF surveillance program, are therefore important, because neutron irradiations such as those in a PWR cannot be simulated. The flux effect between the production reactor

5. HAADF: high-angle annular dark-field detector.

6. EMIR&A: a French national network of accelerators for the irradiation and analysis of molecules and materials (https://emira.in2p3.fr/).

and the experimental reactor is also a widely discussed issue; nothing can be left to chance. In return, all these experiments require expertise on irradiated materials, which are generally radioactive and can only be handled in controlled laboratories such as the GENESIS platform of the Groupe de Physique des Matériaux laboratory at the University of Rouen Normandie[7] or the LECI[8] at CEA Saclay.

These experimental methods must be combined with simulation and modeling techniques, and their multi-scale coupling, as initiated by the REVE (Réacteur Virtuel d'Études – Virtual Reactor Training) project in the 2000s, as well as numerous collaborative projects, including European projects (Malerba 2020). At the atomic scale, it is the calculations of electronic structures (ab initio), which are currently state of the art, that give access to fundamental quantities such as the cohesion energy of phases, the energy of formation and of migration of point defects and their clusters, the solution energy of the transmutation atoms or the activation energy of the dislocation gliding, etc. When we are interested in the collision cascades induced by nuclear collisions and the resulting free point defects, or in the modeling of the gliding of a dislocation and its interaction with various obstacles, classical molecular dynamics is used with interatomic potentials. The kinetics, controlled by the diffusion–elimination of point defects, are approached using the so-called kinetic Monte-Carlo methods (Becquart et al. 2020) or cluster dynamics. The transition from the irradiation microstructure (dislocation loops, cavities, helium bubbles, precipitation, segregations, etc.) to the mechanical behavior is, today, mostly addressed by dislocation dynamics (DD) codes, and crystal plasticity models must take over to describe the behavior of a metal grain; homogenization methods are then applied to describe the plastic behavior of real solids that are polycrystalline.

Experimental techniques and numerical tools for physical metallurgy, particularly for nuclear materials here, feed off one another. An experimental observation will validate a numerical prediction or contribute to the understanding of a mechanism, which must then be reproduced by simulation. For more than 20 years, this "nuclear material" context has been a real driving force in understanding the structural evolution of structural materials. Of course, this progress feeds the whole international community of physical metallurgy. Despite the remarkable progress over the last few decades, a certain number of unknown or poorly known effects persist. For example, the mobility of point defect clusters, their interaction with solutes, their lifespan, etc., are still to be clarified if we want to be

7. GENESIS: Groupe d'Études et Nanoanalyses des Effets d'Irradiations (Group of Studies and Nanoanalyses of the Effects of Radiation).

8. LECI: Laboratoire d'Études du Combustible Irradié, CEA Saclay (Laboratory for the Study of Irradiated Fuel).

able to extrapolate the behavior of these materials in the long term. In particular, the simulations highlight small clusters of defects, which are precursors of the more easily observed multi-nanometer defects. This shows, for example, the interest of revisiting irradiated materials by X-ray diffraction techniques (Ehrhart 1994) for this specific study of nanometric defects below the detection limit of TEM. These approaches are of particular interest for austenitic steels (determination of the nature of "black dots" – clusters of poorly identified defects) and zirconium alloys (mechanism of ⟨c⟩ loop formation and growth). These analyses allow us to see the primary damage as closely as possible, without having to wait for the formation of large defects that are more easily observed. IASCC is also still only partially understood. This metallurgical problem is one of the most complex ones there is. Similarly, the role of the chemical composition on swelling or the exact mechanism of irradiation creep for both austenitic steels and zirconium alloys is still to be confirmed. In the case of vessel steels, the effects of heterogeneities (carbides, inclusions) are at the heart of current research. These macroscopic objects, which are the seat of atomistic reactions, still have a lot to tell us about the links between the microstructure and mechanical properties, and more particularly the fracture of materials.

These aspects related to the prediction of material aging and the knowledge of aging mechanisms are of major interest, in order to gain operating margins and in the context of extending the operating life of facilities (LTO – Long-Term Operation). They are also necessary to accelerate the validation of new materials such as high entropy alloys (for smaller components, at least initially), or new manufacturing processes such as powder metallurgy or additive manufacturing.

2.6. References

Adamson, R.B., Coleman, C.E., Griffiths, M. (2019). Irradiation creep and growth of zirconium alloys: A critical review. *Journal of Nuclear Materials*, 521, 167–244.

Adjanor, G., Bugat, S., Domain, C., Barbu, A. (2010). Overview of the RPV-2 and INTERN-1 packages: From primary damage to microplasticity. *Journal of Nuclear Materials*, 406(1), 175–186.

Akhatova, A. (2017). Méthodologie instrumentale à l'échelle atomique pour une meilleure compréhension des mécanismes de ségrégation intergranulaire dans les aciers : application au phosphore. PhD Thesis, Université de Rouen Normandie.

Akhatova, A., Christien, F., Barnier, V., Radiguet, B., Cadel, E., Cuvilly, F., Pareige, P. (2019). Investigation of the dependence of phosphorus segregation on grain boundary structure in Fe-P-C alloy: Cross comparison between atom probe tomography and auger electron spectroscopy. *Applied Surface Science*, 463, 203–210.

ASTM (2009). Standard practice for neutron radiation damage simulation by charged-particle irradiation (ASTM E521-96). American Society for Testing Materials Report, West Conshohocken, PA.

Barashev, A.V., Golubov, S.I., Stoller, R.E. (2015). Theoretical investigation of microstructure evolution and deformation of zirconium under neutron irradiation. *Journal of Nuclear Materials*, 461, 85–94.

Becquart, C.S. and Domain, C. (2011). Modeling microstructure and irradiation effects. *Metallurgical and Materials Transactions A*, 42(4), 852–870.

Becquart, C.S., De Backer, A., Domain, C. (2018). Atomistic modeling of radiation damage in metallic alloys. In *Handbook of Mechanics of Materials*, Schmauder, S., Chen, C.-S., Chawla, K.K., Chawla, N., Chen, W., Kagawa, Y., Hsueh, C.-H. (eds). Springer, Singapore.

Becquart, C.S., Mousseau, N., Domain, C. (2020). Kinetic Monte Carlo simulations of irradiation effects. In *Comprehensive Nuclear Materials*, 2nd edition, Konings, R.J.M. and Stoller, R.E. (eds). Elsevier, Oxford.

Boutard, J.L., Bonin, B., Cappelaere, C. (2016). Introduction. In *Les matériaux du nucléaire – Modélisation et simulation des matériaux de structure*, Direction de l'énergie nucléaire (ed.). Le Moniteur Editions, Gif-sur-Yvette.

Brechet, Y. (2013). La science des matériaux : du matériau de rencontre au matériau sur mesure (Extraits de la leçon inaugurale prononcée le 13 janvier 2013). *La lettre du Collège de France*, 36, 7.

Brinkman, J.A. (1954). On the nature of radiation damage in metals. *Journal of Applied Physics*, 25(8), 961–970.

Bruemmer, S.M., Simonen, E.P., Scott, P.M., Andresen, P.L., Was, G.S., Nelson, J.L. (1999). Radiation-induced material changes and susceptibility to intergranular failure of light-water-reactor core internals. *Journal of Nuclear Materials*, 274(3), 299–314.

Busby, J.T., Was, G.S., Kenik, E.A. (2002). Isolating the effect of radiation-induced segregation in irradiation-assisted stress corrosion cracking of austenitic stainless steels. *Journal of Nuclear Materials*, 302(1), 20–40.

Carpenter, G.J.C., Zee, R.H., Rogerson, A. (1988). Irradiation growth of zirconium single crystals: A review. *Journal of Nuclear Materials*, 159, 86–100.

Castin, N., Bonny, G., Bakaev, A., Bergner, F., Domain, C., Hyde, J.M., Messina, L., Radiguet, B., Malerba, L. (2020). The dominant mechanisms for the formation of solute-rich clusters in low-Cu steels under irradiation. *Materials Today Energy*, 17, 100472. doi: 10.1016/j.mtener.2020.100472.

Chen, H., Wang, X., Zhang, R. (2020). Application and development progress of cr-based surface coatings in nuclear fuel element I: Selection, preparation, and characteristics of coating materials. *Coatings*, 10(9), 808.

Christiaen, B., Domain, C., Thuinet, L., Ambard, A., Legris, A. (2019). A new scenario for ‹c› vacancy loop formation in zirconium based on atomic-scale modeling. *Acta Materialia*, 179, 93–106.

Christiaen, B., Domain, C., Thuinet, L., Ambard, A., Legris, A. (2020). Influence of vacancy diffusional anisotropy: Understanding the growth of zirconium alloys under irradiation and their microstructure evolution. *Acta Materialia*, 195, 631–644.

Cox, B. (1990). Pellet-clad interaction (PCI) failures of zirconium alloy fuel cladding – A review. *Journal of Nuclear Materials*, 172(3), 249–292.

Edwards, D.J., Garner, F.A., Simonen, E.P., Bruemmer, S.M. (2001). Characterization of neutron-irradiated 300-series stainless steels to assess mechanisms of irradiation-assisted stress corrosion cracking. *Core Components*, EPRI, Palo Alto, CA, 1001497.

Edwards, D.J., Simonen, E.P., Garner, F.A., Greenwood, L.R., Oliver, B.M., Bruemmer, S.M. (2003). Influence of irradiation temperature and dose gradients on the microstructural evolution in neutron-irradiated 316SS. *Journal of Nuclear Materials*, 317(1), 32–45.

Ehrhart, P. (1994). Investigation of radiation damage by X-ray diffraction. *Journal of Nuclear Materials*, 216, 170–198.

Etienne, A. (2009). Etude des effets d'irradiations et de la nanostructuration dans des aciers austénitiques inoxydables. PhD Thesis, Université de Rouen Normandie.

Fidleris, V. (1988). The irradiation creep and growth phenomena. *Journal of Nuclear Materials*, 159, 22–42.

Fujii, K. and Fukuya, K. (2005). Characterization of defect clusters in ion-irradiated A533B steel. *Journal of Nuclear Materials*, 336(2), 323–330.

Garnier, J., Bréchet, Y., Delnondedieu, M., Renault, A., Pokor, C., Dubuisson, P., Massoud, J.-P. (2011). Irradiation creep of SA 304L and CW 316 stainless steels: Mechanical behaviour and microstructural aspects. Part II: Numerical simulation and test of SIPA model. *Journal of Nuclear Materials*, 413, 70–75. doi: 10.1016/j.jnucmat.2011.02.058.

Gilbon, D. and Simonot, C. (1994). Effect of irradiation on the microstructure of zircaloy-4. In *Zirconium in the Nuclear Industry, 10th International Symposium, ASTM STP 1245*, Garde, A.M. and Bradley, E.R. (eds). ASTM International, West Conshohocken, PA.

Gras, J.-M. (2017). Matériaux du nucléaire revue et introduction, v2. *Techniques de l'ingénieur*, BN3679.

Griffiths, M. (1988). A review of microstructure evolution in zirconium alloys during irradiation. *Journal of Nuclear Materials*, 159, 190–218.

Griffiths, M. (1993). Evolution of microstructure in hcp metals during irradiation. *Journal of Nuclear Materials*, 205, 225–241.

Han, X. (2012). Modélisation de la fragilisation due au gonflement dans les aciers inoxydables austénitiques irradiés. PhD Thesis, l'École Nationale Supérieure des Mines de Paris. HAL Id: pastel-00818326 [Online]. Available at: https://pastel.archives-ouvertes.fr/pastel-00818326.

Hernández-Mayoral, M. and Gómez-Briceño, D. (2010). Transmission electron microscopy study on neutron irradiated pure iron and RPV model alloys. *Journal of Nuclear Materials*, 399(2), 146–153.

Kuri, G., Cammelli, S., Degueldre, C., Bertsch, J., Gavillet, D. (2009). Neutron induced damage in reactor pressure vessel steel: An X-ray absorption fine structure study. *Journal of Nuclear Materials*, 385(2), 312–318.

Lemaire, E., Monteil, N., Massoud, J.P., Pokor, C., Ligneau, N., Courtemanche, G. (2011). Lessons learned from baffle bolt replacements regarding irradiation-induced creep behaviour. *Revue générale nucléaire*, 1, 43–49.

Liu, C.L., Odette, G.R., Wirth, B.D., Lucas, G.E. (1997). A lattice Monte Carlo simulation of nanophase compositions and structures in irradiated pressure vessel Fe-Cu-Ni-Mn-Si steels. *Materials Science and Engineering: A*, 238(1), 202–209.

Malerba, L. (2020). Large scale integrated materials modeling programs. In *Comprehensive Nuclear Materials*, 2nd edition, Konings, R.J.M. and Stoller, R.E. (eds). Elsevier, Oxford.

Mamivand, M., Wells, P., Ke, H., Shu, S., Odette, G.R., Morgan, D. (2019). CuMnNiSi precipitate evolution in irradiated reactor pressure vessel steels: Integrated cluster dynamics and experiments. *Acta Materialia*, 180, 199–217. doi: 10.1016/j.actamat.2019.09.016.

Mandal, S., Pradeep, K.G., Zaefferer, S., Raabe, D. (2014). A novel approach to measure grain boundary segregation in bulk polycrystalline materials in dependence of the boundaries' five rotational degrees of freedom. *Scripta Materialia*, 81, 16–19.

Meslin, E., Lambrecht, M., Hernández-Mayoral, M., Bergner, F., Malerba, L., Pareige, P., Radiguet, B., Barbu, A., Gómez-Briceño, D., Ulbricht, A. et al. (2010). Characterization of neutron-irradiated ferritic model alloys and a RPV steel from combined APT, SANS, TEM and PAS analyses. *Journal of Nuclear Materials*, 406(1), 73–83.

Meslin, E., Radiguet, B., Loyer-Prost, M. (2013). Radiation-induced precipitation in a ferritic model alloy: An experimental and theoretical study. *Acta Materialia*, 61(16), 6246–6254.

Meslin-Chiffon, E. (2007). Mécanismes de fragilisation sous irradiation aux neutrons d'alliages modèles ferritiques et d'un acier de cuve. PhD Thesis, Université de Rouen Normandie.

Miller, M.K., Chernobaeva, A.A., Shtrombakh, Y.I., Russell, K.F., Nanstad, R.K., Erak, D.Y., Zabusov, O.O. (2009). Evolution of the nanostructure of VVER-1000 RPV materials under neutron irradiation and post irradiation annealing. *Journal of Nuclear Materials*, 385(3), 615–622.

Miloudi, S. (1997). Étude du dommage d'irradiation dans les aciers de cuve des réacteurs à eau pressurisée. PhD Thesis, Université de Paris-Sud.

Naudin, C., Frund, J.M., Pineau, A. (1999). Intergranular fracture stress and phosphorus grain boundary segregation of a Mn-Ni-Mo steel. *Scripta Materialia*, 40(9), 1013–1019.

Ngayam-Happy, R., Becquart, C.S., Domain, C. (2013). First principle-based AKMC modelling of the formation and medium-term evolution of point defect and solute-rich clusters in a neutron irradiated complex Fe–CuMnNiSiP alloy representative of reactor pressure vessel steels. *Journal of Nuclear Materials*, 440(1), 143–152. doi: 10.1016/j.jnucmat.2013.04.081.

Nishiyama, Y., Onizawa, K., Suzuki, M., Anderegg, J.W., Nagai, Y., Toyama, T., Hasegawa, M., Kameda, J. (2008). Effects of neutron-irradiation-induced intergranular phosphorus segregation and hardening on embrittlement in reactor pressure vessel steels. *Acta Materialia*, 56(16), 4510–4521.

Nordlund, K. (2019). Historical review of computer simulation of radiation effects in materials. *Journal of Nuclear Materials*, 520, 273–295.

Northwood, D.O. (1977). Irradiation damage in zirconium and its alloys. *Atomic Energy Review*, 15(4), 547–610.

Odette, G.R. (1995). Radiation induced microstructural evolution in reactor pressure vessel steels. *Materials Research Society*, 373, 137–147.

Odette, G.R. and Lucas, G.E. (2001). Embrittlement of nuclear reactor pressure vessels. *The Journal of the Minerals, Metals & Materials Society (TMS)*, 53(7), 18–22.

Onimus, F., Doriot, S., Béchade, J.L. (2020). Radiation effects in zirconium alloys. In *Comprehensive Nuclear Materials*, 2nd edition, Konings, R.J.M. and Stoller, R.E. (eds). Elsevier, Oxford.

Pareige, P. and Miller, M.K. (1996). Characterization of neutron-induced copper-enriched clusters in pressure vessel steel weld: An APFIM study. *Applied Surface Science*, 94–95, 370–377.

Pareige, P., Van Duysen, J.C., Auger, P. (1993). An APFIM study of the microstructure of a ferrite alloy after high fluence neutron irradiation. *Applied Surface Science*, 67(1), 342–347.

Radiguet, B., Pareige, P., Barbu, A. (2009). Irradiation induced clustering in low copper or copper free ferritic model alloys. *Nuclear Instruments and Methods in Physics Research Section B: Beam Interactions with Materials and Atoms*, 267(8), 1496–1499.

Renault-Laborne, A., Gavoille, P., Malaplate, J., Pokor, C., Tanguy, B. (2015). Correlation of radiation-induced changes in microstructure/microchemistry, density and thermo-electric power of type 304L and 316 stainless steels irradiated in the Phénix reactor. *Journal of Nuclear Materials*, 460, 72–81.

Renault-Laborne, A., Garnier, J., Malaplate, J., Gavoille, P., Sefta, F., Tanguy, B. (2016). Evolution of microstructure after irradiation creep in several austenitic steels irradiated up to 120 dpa at 320°C. *Journal of Nuclear Materials*, 475, 209–226.

Ribis, J., Doriot, S., Onimus, F. (2018). Shape, orientation relationships and interface structure of beta-Nb nano-particles in neutron irradiated zirconium alloy. *Journal of Nuclear Materials*, 511, 18–29.

Sidky, P.S. (1998). Iodine stress corrosion cracking of Zircaloy reactor cladding: Iodine chemistry (a review). *Journal of Nuclear Materials*, 256(1), 1–17.

Tanguy, B. (2011). Corrosion sous contrainte assistée par l'irradiation des aciers inoxydables austénitiques (IASCC). *Revue de métallurgie*, 108(1), 39–46. doi: 10.1051/metal/2011022.

Tanguy, B., Sefta, F., Joly, P. (2015). Le vieillissement des internes de cuve. Programme de recherche en support à la durée de fonctionnement des réacteurs REP. *Revue générale nucléaire*, May–June, 3, 56–63.

Terrani, K.A. (2018). Accident tolerant fuel cladding development: Promise, status, and challenges. *Journal of Nuclear Materials*, 501, 13–30.

Tewari, R., Mani Krishna, K.V., Neogy, S., Lemaignan, C. (2020). Zirconium and its alloys: Properties and characteristics. In *Comprehensive Nuclear Materials*, 2nd edition, Konings, R.J.M. and Stoller, R.E. (eds). Elsevier, Oxford.

Toyama, T., Nagai, Y., Tang, Z., Hasegawa, M., Almazouzi, A., van Walle, E., Gerard, R. (2007). Nanostructural evolution in surveillance test specimens of a commercial nuclear reactor pressure vessel studied by three-dimensional atom probe and positron annihilation. *Acta Materialia*, 55(20), 6852–6860.

Wang, M., Song, M., Lear, C., Was, G. (2019). Irradiation assisted stress corrosion cracking of commercial and advanced alloys for light water reactor core internals. *Journal of Nuclear Materials*, 515, 52–70. doi: 10.1016/j.jnucmat.2018.12.015.

Was, G.S. (2017). *Fundamentals of Radiation Materials Science, Metals and Alloys*. Springer, New York.

Woo, C.H. (1988). Theory of irradiation deformation in non-cubic metals: Effects of anisotropic diffusion. *Journal of Nuclear Materials*, 159, 237–256.

Zhang, L., Radiguet, B., Todeschini, P., Domain, C., Shen, Y., Pareige, P. (2019). Investigation of solute segregation behavior using a correlative EBSD/TKD/APT methodology in a 16MND5 weld. *Journal of Nuclear Materials*, 523, 434–443.

3

Ceramics within PWRs

Christine DELAFOY[1], Frederico GARRIDO[2] and Yves PIPON[3]

[1] *Fuel Design, Framatome, Lyon, France*

[2] *Laboratoire de Physique des 2 Infinis Irène Joliot-Curie (IJCLab), CNRS, Université Paris-Saclay, Orsay, France*

[3] *Institut de Physique des 2 Infinis de Lyon, Université Claude Bernard Lyon 1, CNRS/N2P3, Villeurbanne, France*

3.1. Introduction

In this chapter, we will discuss two ceramic materials used in pressurized water reactors (PWRs), which represent 61% of the world's reactors and were the only technology deployed in France in 2022. The materials in question are the uranium dioxide fuel[1] and the boron carbide present in the reactor's reactivity control rods.

In the core of a nuclear reactor, the fuel is the active part and the seat of fission of heavy uranium atoms, and therefore, the source of heat that ultimately allows for the production of electricity. The design of the fuel and its properties have a direct impact on the performance and safety of the reactor. Although the fuel contributes relatively little to the price of the kilowatt-hour produced, its proper use represents an important economic issue. In particular, fuel failure, if it were to lead to a reactor shutdown, would represent a very high additional cost. The aim of the design and the studies on the sizing of fuel elements is therefore to guarantee their performance, by anticipating the various operating situations.

1. There is also mixed oxide fuel UO_2-PuO_2 (MOX), whose specificities will not be discussed here.

Nuclear Materials under Irradiation,
coordinated by Serge BOUFFARD and Nathalie MONCOFFRE.

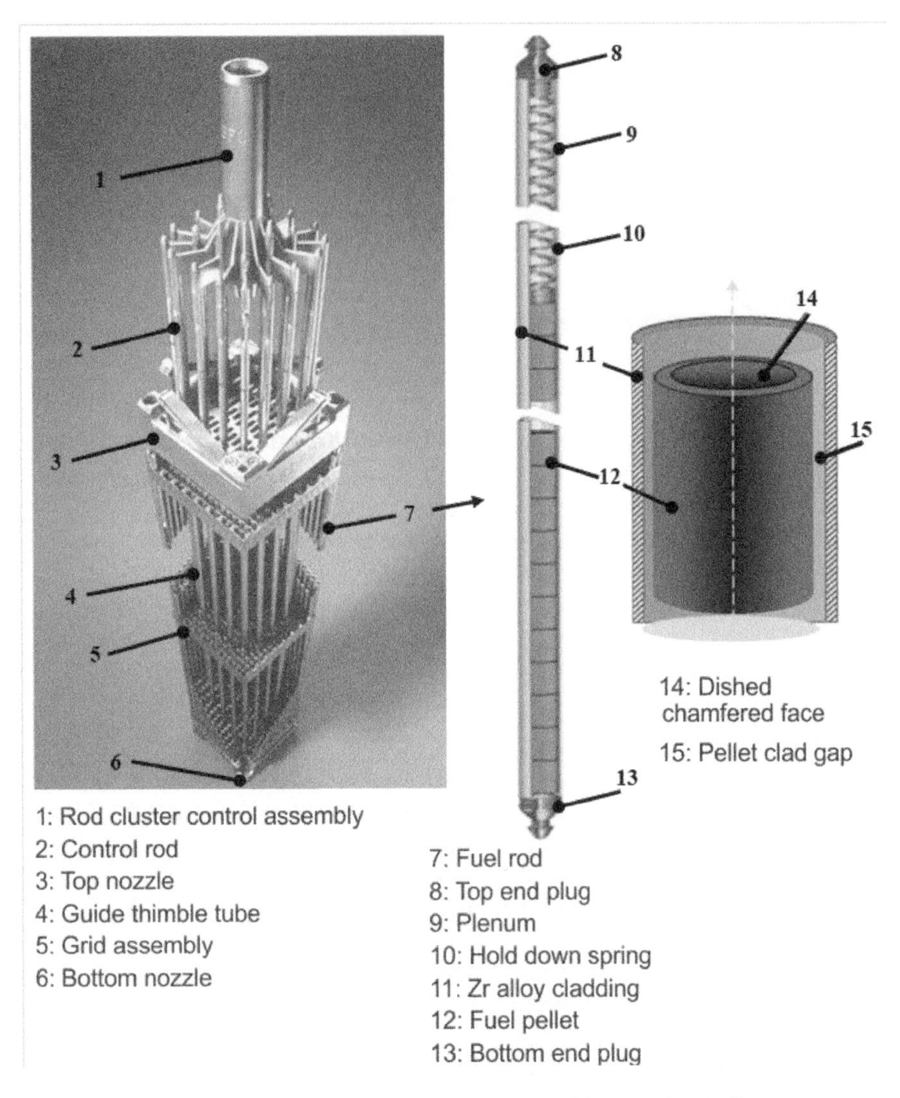

Figure 3.1. *From left to right: fuel assembly with control rod; diagram of a fuel rod; fuel pellet in its cladding. Figure partly inspired by the CEA/DEN monograph "Les combustibles nucléaires". For a color version of this figure, see www.iste.co.uk/bouffard/nuclear.zip*

Depending on its power, the core of a nuclear reactor includes anything between 157 and 241 fuel assemblies for 900 MWe PWRs and for EPRs, or a mass of uranium of about 71–125 tons. Each assembly remains in the reactor for about four years before reaching its operating limits; the reactors themselves have a lifespan of 40–60 years. At each shutdown (every 12–18 months), one-third or one-quarter of the core is renewed, which makes the fuel the consumable part of the reactors.

There are various assembly structures, the most common of which consists of 264 rods supported by an orthogonal 17x17 square lattice structure, with sides of 21.4 cm and a height of nearly 4 m for the active part (see Figure 3.1). The fuel rod is a closed system, similar to a pressurized vessel, which must remain leak-tight in order to contain the fissile material and the fission products. The rod thus constitutes the first containment barrier. The fissile material is packaged in the form of cylindrical pellets, which are hollowed at the ends with chamfers. The pellets are 13.6 mm high and ~8.2 mm in diameter, for a mass of about 8 g. Oxide has become the preferred fuel material for water reactors, notably because of its good resistance to the consequences of irradiation, its high melting point and its low chemical reactivity. The 300 or so pellets per rod are stacked in a zirconium alloy cladding[2] with a thickness of 0.6 mm and an external diameter of 9.50 mm. The rod is completed with an upper expansion chamber to accommodate the release of fission gases from the fuel. A stainless-steel helical spring is positioned there to maintain the fissile column during handling and transport prior to loading into the reactor. Finally, after pressurization with helium – a gas that promotes heat exchange between the pellets and the cladding – the tube is closed at both ends using plugs, which are pressed in and welded to guarantee the tightness of the fuel rod.

The operating conditions of fuel materials are extreme in many ways (Bailly et al. 1998). Most of the phenomena that develop during irradiation depend on the temperature and the local fission density. They are therefore directly correlated to the evolution of the linear power during the different irradiation cycles. In nominal regime, PWR fuel rods operate at average linear powers between 150 and 250 $W.cm^{-1}$, which represents a power release of about 400 $W.cm^{-3}$ of fuel. The production of heat comes from the energy released by the fission of uranium atoms. It is typically 200 MeV and results from 80%[3] of the kinetic energy of the two fission products (FP) emitted during each fission. FPs dissipate their energy by slowing down in the matrix, which is a source of heat, and displaces atoms, creating numerous defects (CEA 2008). Each gram of fully fissioned ^{235}U produces

2. Zirconium was chosen for its very low neutron capture cross-section. Alloying elements (such as Nb, Fe, Cr, O) are added to ensure satisfactory physico-chemical properties in service (see Chapter 2 of this book).

3. The rest of the energy comes from the β and γ emissions, and from the kinetic energy of neutrons.

~24 10^3 kW.h or ~1 MW.d[4]. The energy dissipated within the material, as well as the exchange conditions with the coolant at the cladding surface, lead to high temperatures and a large radial thermal gradient in the fuel, an oxide with poor thermal conductivity. Moreover, due to the fission reactions and the accompanying production of fission products, the composition of the fuel changes very significantly. At the end of its use in the reactor, the UO_2 fuel (burnup rate of 60 $GW.d.t_U^{-1}$) contains about 12% of new atoms. Consequently, the physico-chemical properties of the fuel and its geometry will have significantly changed compared to its initial state.

In all nuclear reactors, control and shutdown systems occur at the core level, whether it be for power control and planned shutdowns, or in the event of an incident or accident. In the case of PWRs, these systems are located in mobile units known as control rods. The reactivity of the core is also controlled by boron, which is dissolved in the cooling water in the form of boric acid (neutron absorption mainly takes place at the level of the boron isotope ^{10}B, whose natural content is 19.8%).

The absorbing materials of the control rods are chosen according to different criteria, but first and foremost for their effective neutron absorption cross-section, which must be greater than 100 barns for thermal neutrons. It is also preferable that the neutron absorption spectrum be as uniform as possible over the entire neutron spectrum, so as to have good control of the core reactivity. The materials selected are as follows:

– SIC metal alloy: 80% silver–15% indium–5% cadmium; a material common to all PWRs in France;

– boron carbide (B_4C). Given its change in service (see section 3.3.2), this material is positioned in the upper part of the control rods in areas with low neutron flux; the lower part is made of SIC. This combination of absorbers is only used in 1,300 MWe, 1,450 MWe and EPR reactors.

The control rods are inserted into about every third fuel assembly (see Figure 3.1). For this purpose, 24 positions of the assembly receive open tubes, called “guide tubes”, in which the absorbing rods will slide. The differentiated movements of the rods in the core make it possible to optimize the spatial power distribution, in

4. The ^{235}U content of the fuel of a 1,300-MWe PWR is reduced from 4.2% to 0.7% in 3×18 months. The reduction of 1% in ^{235}U corresponds to the production of 15.3 GW.j per ton of total uranium (^{238}U+^{235}U). This characterizes the rate of utilization of the fuel, also known as the “burnup”, a technological term expressed in $GW.j.t_U^{-1}$. This represents the thermal energy (1 GW.j = 8.6410^{13} J) extracted from one ton of uranium.

order to control the power variations of the reactor and to regulate the average temperature of the primary circuit water. The control rods are therefore essential to the control and safety of the reactor.

From what is mentioned above, it appears that despite their evolution under irradiation, the properties of UO_2 and B_4C must continue to meet their specifications. In order to predict the in-service behavior of these two materials, it is necessary to have a thorough understanding of the physico-chemical phenomena that govern their evolution. Modeling them in fuel rod performance codes makes it possible to understand the physical quantities of interest (e.g. temperatures, deformations, release of fission gases, etc.), so as to carry out performance calculations and safety assessments under known or specific conditions (CEA 2008). This modeling is also essential for orientating the avenues of development to be investigated, in order to improve the performance of these ceramic materials, which are UO_2 and B_4C (see section 3.4).

3.2. Development and typical properties of UO_2 and B_4C ceramics

Ceramics are compounds formed from metallic and non-metallic elements. Most often, the non-metallic elements constituting ceramics are oxygen, carbon and nitrogen, forming what are called oxides (such as uranium oxide UO_2), carbides (such as boron carbide, B_4C) or nitrides (such as titanium nitride, TiN). Their mechanical and thermal properties differ from those of metals, mostly because of the iono-covalent nature of their atomic bonds.

3.2.1. *Development and structure of uranium dioxide*

The nuclear fuel for pressurized water power plants consists of uranium dioxide UO_2 enriched in ^{235}U, which has been processed through numerous stages from the extraction of uranium ore.

3.2.1.1. *Fuel, from the mine to the reactor*

Uranium from a mine is extracted by selectively dissolving the rock into a fine powder in an acid solution that is in the presence of an oxidant, whose role is to transform insoluble tetravalent uranium into soluble hexavalent uranium. After purification, precipitation in a basic solution (NaOH, MgO, NH_4OH), drying and then calcination, we obtain a diuranate that is known as a uranium ore concentrate or a yellow cake (with a composition of $Na_2U_2O_7$, MgU_2O_7 or $(NH_4)_2U_2O_7$), which is yellow in color (Patarin 2002).

An enrichment stage is then undertaken, in order to increase the ^{235}U isotopic fraction from 0.72% to a value between 3% and 5%, which is the required percentage for the operation of PWR reactors. This stage requires a uranium compound that is gaseous at a sufficiently low temperature (easy to implement on an industrial scale) and monoisotopic, so that the difference in mass comes from the uranium isotope alone. The choice fell on uranium hexafluoride (vaporization temperature of 340 K; ^{19}F is the only fluorine isotope). In order to obtain this UF_6 compound, the uranium concentrate is dissolved in nitric acid. The uranyl nitrate obtained is then purified by liquid–liquid extraction using an organic solvent (tributyl phosphate). After a denitration that transforms it into UO_3, the compound is reduced to UO_2 by dihydrogen. Two stages of fluorination of the purified uranium dioxide are then carried out to transform the powder into uranium tetrafluoride UF_4 (green powder), by reaction with hydrofluoric acid at 620 K, then from UF_4 into hexafluoride UF_6 by oxidation with difluor F_2 at 1,700 K. Pure crystals of UF_6 composition are then obtained by condensation at 250 K. Isotopic enrichment can be performed using different methods, based either on the mass difference between the isotopes (mass spectrometry, gas diffusion, centrifugation) or on the small energy differences of the different atomic orbitals of the different isotopes and selective ionization (laser methods). On an industrial scale, the process used is ultracentrifugation: in a centrifuge with a very high rotation speed, the centrifugal force separates the most massive ^{238}U-based molecules (statistically more at the periphery) from the lightest ^{235}U-based molecules (statistically more in the center). The slightly enriched gas is transferred to the next centrifuge, and the slightly depleted gas to the previous one. The operation must be repeated thousands of times to reach the desired isotopic fraction. The enriched UF_6 gas is then converted again into uranium dioxide powder UO_2 in a rotary kiln, in the presence of H_2 and H_2O gases (dry process):

$$UF_6 + 2\,H_2O + H_2 \rightarrow UO_2 + 6\,HF$$

With the help of additives, the UO_2 powder is shaped and compacted in a uniaxial press to reach the required dimensions, taking into account the shrinkage during sintering. This stage, carried out at 2,000 K under a reducing atmosphere of an Ar/H_2 mixture, makes it possible to weld the grains of the powder and to ensure their cohesion. The solid pellets, whose density is equal to 95% of the theoretical density, are then machined (ground) to reach the required diameter of 8.192±0.012 mm. Their quality is inspected to verify the absence of defects (cracks, chips). The examination of their microstructure reveals grains of a typical size of about 10 μm.

3.2.1.2. *Structure of uranium dioxide*

Uranium dioxide is a crystalline solid with a typical fluorite structure (CaF_2): the lattice is cubic, Bravais lattice face-centered cubic cell *F*, with a pattern consisting of U atoms located at atomic positions (0, 0, 0) and O atoms located at positions (¼, ¼, ¼) and (¾, ¼, ¼). The cubic cell parameter is 547.1 pm, which corresponds to a density of 10,951 kg.m^{-3}. From a crystallochemical point of view, the structure can be described as an arrangement of O^{2-} oxide ion cubes in which uranium U^{4+} ions fit once every other time, in a staggered pattern. The unoccupied oxygen cubes in the dioxide structure are potentially available for the incorporation of fission products: each ^{235}U fission creates a vacancy in the uranium sublattice, but two new atoms are generated that must fit into the crystal lattice. This oxide does not show any structural transformation except for the transition at 30.8 K to an antiferromagnetic state, which is out of the range of application for reactor fuels (Leask et al. 1963).

3.2.2. *Development and structure of boron carbide*

Boron carbide is used in PWRs in the form of dense cylindrical pellets, with dimensions similar to those of the fuel (diameter varying from 7.5 to 8.5 mm, depending on the reactor technology).

3.2.2.1. *Boron carbide, from powder to sintered pellets inserted in the reactor*

Two major industrial processes exist to produce boron carbide powder. The first process, known as magnesiothermic reduction, consists of reacting boron sesquioxide (B_2O_3) with magnesium in the presence of carbon according to the following reaction:

$$2\,B_2O_{3(s)} + 6\,Mg_{(s)} + 2\,C_{(s)} \xrightarrow{\sim 1270\text{ K}} B_4C_{(s)} + 6\,MgO_{(s)}$$

This is how the boron carbide, which was used as a neutron absorber in the French fast neutron reactors Phénix (1973–2010) and Superphénix (1984–1998), was manufactured. The creation of by-products such as magnesium borate explains why the overall yield is relatively low.

Nowadays, the most common process consists of the carbo-reduction of a starting reagent which can be boron sesquioxide, boric acid (H_3BO_3) or even borax ($Na_2B_4O_7$). The equation below presents the balance reaction from boric acid:

$$4\,H_3BO_{3(s)} + 7\,C_{(s)} \xrightarrow{\sim 2270\text{ K}} B_4C_{(s)} + 6\,CO_{(g)} + 6\,H_2O_{(g)}$$

The reaction product is purified, crushed and then ground to obtain a powder with the desired particle size. The majority isotope of boron, ^{11}B (~80%), has a smaller effective neutron absorption cross-section than ^{10}B for both thermal neutrons (3,850 barns for ^{10}B vs. 0.05 barn for ^{11}B) and fast neutrons. For use as a neutron absorber in the reactor, it is thus advisable to enrich the material in ^{10}B, particularly for use in a fast neutron reactor. Isotopic separation by distillation in a suitable column is one of the most common methods, based on the differences in vapor pressure between, for example, $^{11}BF_3$ and $^{10}BF_3$. The powder, which is purified and enriched in ^{10}B, must then be put in the form of compact pellets with the least possible porosity. Thus, if a mechanical cohesion is obtained during a thermal treatment from 2,070 K, the density is relatively low (60% of the theoretical density). However, the density of these pellets must be higher than 95% for fast neutron reactors and higher than 70% for PWRs. A better density is only possible at temperatures at least as high as that and under pressure. Two processes exist to reach such conditions: HP (Hot Pressing) and SPS (Spark Plasma Sintering). In the first case, the boron carbide is heated for a few hours at a temperature between 2,070 K and 2,470 K, at a pressure between 50 and 100 MPa. In the second case, a pulsed current is sent through the powder and heats it by the Joule effect. This process allows very high heating and cooling rates (up to 1,000 $K.min^{-1}$), which reduces the total process time to just a few minutes for a pressure of 50 MPa at a temperature of 2,070 K. As for the UO_2 pellets, the sintered B_4C pellets are ground to the required diameter and then subjected to physico-chemical monitoring.

3.2.2.2. *Structure of boron carbide*

The unit cell of boron carbide is composed of 15 atoms divided into an icosahedron (12 atoms) and a linear chain (3 atoms). Each icosahedron is placed at the top of a rhombohedral lattice with a trigonal symmetry (space group $R\bar{3}m$), whose cell parameter is 516 pm for an angle of 65.7°. It is also possible to locate the structure in a hexagonal lattice (a = b = 560 pm; c = 1,207 pm).

Each icosahedron has two types of sites, known as polar and equatorial. A linear chain, which is composed of three atoms, connects the icosahedra between their polar sites. Determining the position of each atom in the lattice has long been a challenge. Indeed, boron-11 and carbon-12 have a very close electronic and nuclear diffusion cross-section, making their differentiation difficult by most analytical techniques. Thanks to initial atomistic modeling calculations (Lazzari et al. 1999), it is now accepted that the most probable structure (at least from an energetic point of view) consists of an icosahedron made of 11 boron atoms and 1 carbon atom on the polar site (denoted as $B_{11}C^p$), as well as a linear C-B-C chain, as shown in Figure 3.2.

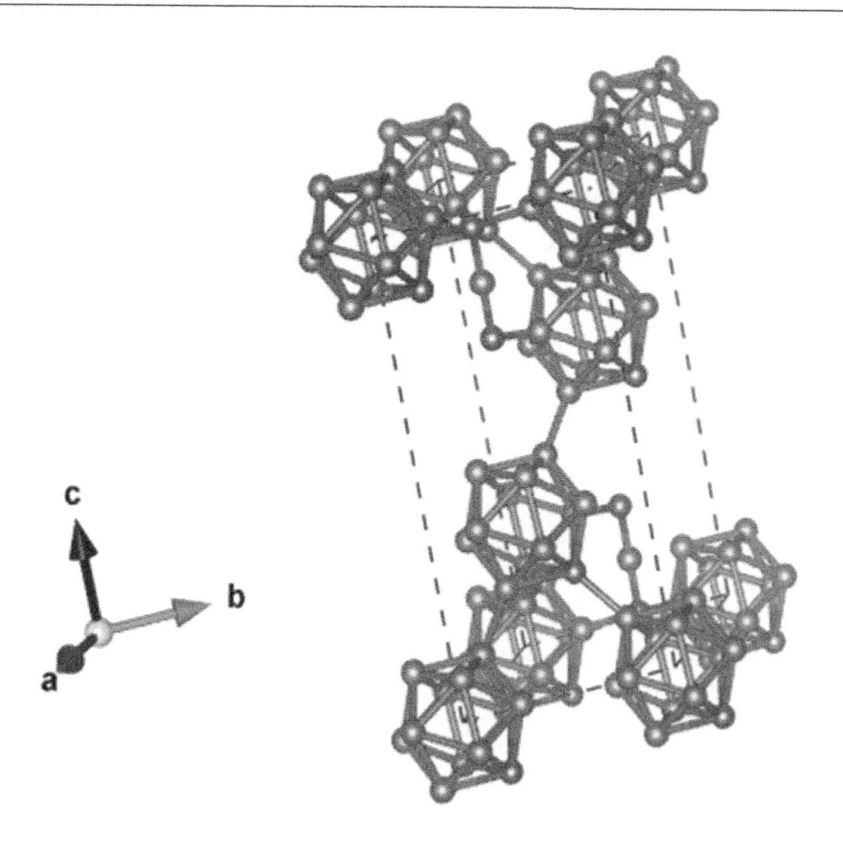

Figure 3.2. *Crystallographic structure of boron carbide in its form ($B_{11}C^p$)CBC in the hexagonal lattice (a=b≠c). The carbon atoms are in dark brown and the boron atoms are in green. For a color version of this figure, see www.iste.co.uk/bouffard/nuclear.zip*

In reality, experimentally developed boron carbide is a disordered material that can accommodate several polytypes. Thus, various studies report icosahedra formed of 12 boron atoms (B_{12}), or 10 boron atoms and 2 carbon atoms ($B_{10}C_2$), associated with linear chains CBB, CVC (V for vacancy) and BVB. X-ray diffraction (XRD) characterization studies have even shown that the central chain can be replaced by a B_B^BB quadrilateral (Yakel 1975). All these combinations allow a variety of boron and carbon concentrations, and just as many solid phases. What is known as "boron carbide" is therefore a variety of $B_{4+x}C$ ($0 \leq x \leq 6$) compounds, whose carbon concentration can vary from ~9%at. ($B_{10}C$), through 13.3%at. ($B_{6,5}C$), up to the limit of ~20%at. (B_4C).

Several properties are affected by this change of stoichiometry, in particular the density ρ (see equation [3.1]) (Bouchacourt and Thevenot 1981), which leads to consider a theoretical density of 2,520 kg.m^{-3} for B_4C. The cell parameters are also affected as well as the volume, which decreases almost linearly with the increasing carbon concentration (Gosset and Colin 1991).

$$\rho_{boron\ carbide}\,(\mathrm{kg.m^{-3}}) = 2422 + 4.8 \times [\mathrm{C}](\%at.) \qquad [3.1]$$

3.2.3. *Thermomechanical characteristics of B_4C and UO_2*

In contrast to metals, most ceramic materials are poor electrical conductors, with uranium dioxide and boron carbide both having an energy band gap of about 2.1 eV. It should be noted that UO_2 is a Mott insulator, characterized by two 5f electrons localized on each U^{4+} cation of the lattice and not delocalized over the whole solid, as is the case for conduction electrons (Baer and Schoenes 1980).

3.2.3.1. *Comparison of thermal properties*

The thermal properties of ceramic materials in PWRs are particularly important in order to estimate the stresses resulting from temperature gradients. Thus, for UO_2 fuel in normal operation, the temperature is about 1,270 K at the center of the pellet and about 720 K at its periphery. During a power transient, the central temperature can increase to 1,770–2,270 K, the periphery staying at about 720 K. For the neutron absorber B_4C in normal operation, the temperature is about 720 K and can increase to 1,400 K in accidental conditions.

These values can be obtained by a simplified calculation based on the analytical resolution of Fourier equations. This calculation, which follows a parabolic law (see equation [3.2]), makes it possible to estimate the temperature increase $\Delta T(r)$ (r = 0 and r = R correspond respectively to the center and the periphery of the pellet).

$$\Delta T(r) = \frac{P_l}{4.\pi.\lambda}\left(1 - \frac{r^2}{R^2}\right) = \Delta T_M\left(1 - \frac{r^2}{R^2}\right) \qquad [3.2]$$

In this equation, P_l is the linear power, λ is the thermal conductivity, and ΔT_M is the maximum temperature difference.

In a solid, the propagation of heat, measured by the thermal conductivity λ, whose central role appears in the previous equation, is either carried out by the propagation of atomic vibrations in the form of waves called phonons or by the propagation of particles that can move in the solid (e.g. electrons). The direct measurement of this parameter at a high temperature is difficult, and the experimental values are often deduced from the measurements of thermal diffusivity *D* and mass heat capacity C_P via the equation relation [3.3].

$$\lambda(T) = D(T).\rho(T).C_p(T) \qquad [3.3]$$

with ρ(T) being the density at a given temperature T.

Figure 3.3 shows the evolution of the thermal conductivity of UO_2 and B_4C as a function of temperature. For comparison, the thermal conductivity of iron is 80 $W.K^{-1}.m^{-1}$ at 300 K, and decreases to 30 $W.K^{-1}.m^{-1}$ at 1,250 K (melting temperature of iron).

In ceramics, at a low temperature (<1,000 K), thermal conduction mainly results from the propagation of phonons. For a perfect crystal of UO_2 or B_4C, its value decreases from room temperature to about 1,800 K. For UO_2, the thermal conductivity then increases until the melting temperature (3,120 K for UO_2). Indeed, the semiconducting behavior of UO_2 takes part in the increase of the thermal conductivity through the creation of U^{3+} and U^{5+} cations, which respectively play the roles of electrons and holes in traditional semiconductors.

Thermal conductivity depends on many parameters; globally speaking, it deteriorates under the effect of the presence of obstacles to the propagation of phonons with increasing concentration: crystalline defects, fission products, etc. The stoichiometry is also an essential parameter, since the value of the thermal conductivity decreases with it. The effect is more pronounced at a low temperature. Indeed, at room temperature, the conductivity is reduced from 7.4 $W.m^{-1}.K^{-1}$ in $UO_{2,0}$ to 2.5 $W.m^{-1}.K^{-1}$ for $UO_{2,1}$; whereas at 1,000 K, the values are respectively 3.3 and 2.5 $W.m^{-1}.K^{-1}$.

The specific heat of uranium dioxide is slightly lower (300–400 $J.K^{-1}.kg^{-1}$ between 300 K and 2,000 K (IAEA 2006)) than that of iron (450 $J.K^{-1}.kg^{-1}$ at room temperature). Ceramics therefore rapidly accumulate energy which, together with low thermal conductivity, creates strong thermal gradients and isotropic expansion. In the axial direction, the dished chamfered face compensates for this expansion, but in the radial direction, the expansion contributes to the modification of the gap between the pellet and the fuel cladding. Thus, at 1,000 K, the linear expansion of the fuel pellet is about 0.7%, while that of the cladding, at a lower temperature, is about 0.2%. The gap between the fuel pellet and the cladding is therefore initially reduced during operation, before eventually disappearing due to: (i) the swelling of the fuel under the effect of the presence of gaseous and solid fission products, as well as (ii) the action of the creep of the cladding, subject to the stress of the pressurized coolant, partially compensated by the helium pressure inside the cladding (see Chapter 2 of this book for the cladding behavior under irradiation). The linear expansion has a monotonically increasing variation up to the melting temperature: it is only 0.5% at 800 K (at the periphery), but exceeds 4% at the melting temperature (Fink 2000).

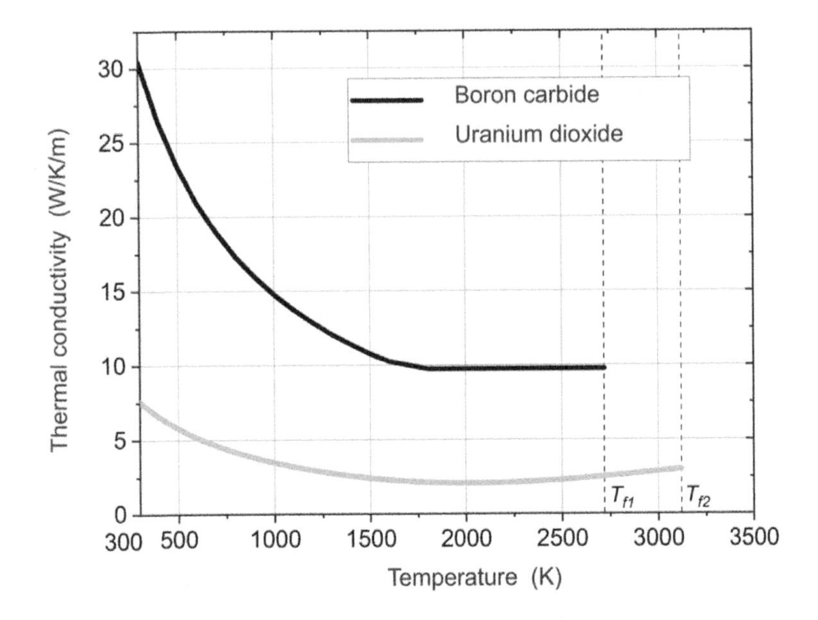

Figure 3.3. *Evolution of the thermal conductivity of boron carbide (T_{f1}=2,720 K) (IAEA 1997) and of uranium dioxide (T_{f2}=3,120 K) (IAEA 2006), from room temperature (300 K) to the melting temperature T_f*

On the contrary, since boron carbide has high specific heat values (750 $J.K^{-1}.kg^{-1}$ at 300 K and 1,880 $J.K^{-1}.kg^{-1}$ at 800 K) and a higher thermal conductivity than UO_2, the expansion of the material remains relatively low (0.4% at 800 K) and, under normal conditions, the thermal gradients within the material placed in PWR do not constitute a major problem.

3.2.3.2. *Comparison of mechanical properties*

From a mechanical point of view, most ceramics have a brittle nature, unlike metals which are ductile. This translates into a low tenacity (or resistance to crack propagation) with values ranging[5] from 2 to 4 MPa$\sqrt{m}$ for B_4C (Gosset and Provot 2001) and from 0.5 to 13 MPa$\sqrt{m}$ for UO_2 (Lankford 1982). For comparison, iron has a tenacity between 120 and 150 MPa$\sqrt{m}$.

Ceramics also have a high level of hardness. Thus, if the hardness of uranium dioxide is about 6.5 GPa (Vickers scale), then that of boron carbide (between 29 and 38 GPa) makes it one of the hardest materials on Earth after diamond (between 70 and 150 GPa, depending on its purity) and cubic boron nitride (~48 GPa). Boron carbide also has excellent impact resistance, with the highest Hugoniot[6] yield strength value of all ceramics (17–20 GPa). Anecdotally speaking, it should be noted that these excellent mechanical and elastic properties, combined with the low density of the material, make boron carbide an ideal candidate for bulletproof vest design.

On the contrary, in the presence of a thermal gradient, these mechanical properties are at the origin of the fracturing of the fuel pellets. Indeed, the thermal expansion of the center of the pellets, at a temperature of about 1,500 K in normal operation, is higher than that of the periphery. This difference in expansion induces internal compressive stresses at the center and tensile stresses at the periphery, whose maximum values $\sigma_{\theta M}$ can be calculated from equation [3.4].

$$\sigma_{\theta M} = \frac{E.\alpha}{2(1-\nu)} \Delta T_M \qquad [3.4]$$

with α being the coefficient of thermal expansion, E being Young's modulus (whose value decreases from 220 GPa to 190 GPa between 300 K and 1,300 K for UO_2), and ν being Poisson's modulus (whose value is 0.3 for UO_2).

For an average linear power of 200 $W.cm^{-1}$, the calculated value is 600 MPa, for a rupture value of the order of 100–150 MPa involving the creation of radial cracks, then the fracturing into several fragments of the fuel pellet. Moreover, thermomechanical calculations, which take into account all the deformation phenomena of the pellet (effect of temperature, stresses and resulting deformations), show that the initially cylindrical shape of the pellet changes towards a specific diabolo-like shape, for which the two planes located at the ends of the pellet are more displaced than its median plane. When the gap between the cladding and the pellet is progressively filled, the cladding follows the shape of the pellet and the

5. The variations in tenacity, for the same material, are mainly due to differences in microstructures (particularly grain size).

6. The Hugoniot elastic limit is the pressure that corresponds to the shock point at which a material changes from a purely elastic state to an elastoplastic state.

stresses are concentrated in particular points (triple point) at the intersection between the planes of the pellet ends and of the fracture, for which there is maximal stress on the cladding and can potentially lead to its rupture.

3.3. Aging of ceramics under irradiation

Boron carbide and uranium dioxide have similar thermomechanical properties, but different behavior under irradiation. Thus, it is remarkable that uranium dioxide remains crystallized, whereas boron carbide amorphizes like most ceramics. A detailed understanding of the mechanisms responsible for the aging of these materials under irradiation comes up against the complexity of the physical phenomena, which take place under normal operating conditions or under accidental conditions. The strategy consists of conducting research at all scales, from the macroscopic scale on the real material to the atomic scale, and from multiparametric behavior to separate parametric effect studies.

After being in a PWR, spent fuel rods undergo non-destructive examinations on site or destructive examinations in hot cells, for more detailed examinations, in order to quantify quantities of interest (such as internal pressure, fraction of released fission gas, density, etc.), to characterize microstructural changes, to carry out mechanical tests on the irradiated cladding or to carry out thermal annealing, so as to obtain global data on the release of fission products by simulating specific accident situations. In research reactors, irradiations of rods with thermocouples, pressure sensors or elongation sensors are carried out, in order to study the kinetics of the phenomena in conditions that are representative of PWRs. In these research reactors, it is also possible to carry out integral tests which can go as far as cladding rupture, in order to determine the technological limits of the products, whether this be in power transients or in accidental conditions. However, in addition to the fact that there are few test reactors for neutron irradiation in the world, their use presents significant burdens: a relatively low damage rate (20 dpa/year), implying excessively long irradiation times, and irradiated materials, which are highly radioactive, requiring restrictive handling precautions during post-irradiation analyses.

In contrast, “laboratory” experiments with particle accelerators allow a higher damage rate to be achieved (50 dpa/day) and with radiation protection problems that are often much more moderate. In particular, the damage of the fuel by fission fragments and their diffusion in the material are aspects that are widely studied, thanks to these particle accelerators, which allow for the selection of the appropriate ion and energy. The energy loss of an FP is mainly due to electronic excitations of the matrix. The simulation of this effect is therefore generally carried out by the use of fast and heavy ions (also called Swift Heavy Ions (SHI)). At the end of its course, the FP mainly gives up its residual energy through elastic collisions. To simulate this effect, it is then relevant to use low energy and massive ions to generate a maximum dpa. Thanks to adapted irradiation chambers, it is also possible to test the impact of a specific parameter such as the irradiation temperature or the composition of the atmosphere. Particle accelerators are also used to introduce an element of interest in a controlled concentration (which is particularly important for studying the effect of the solubility limit), but which is close to the surface (~1 μm or less implantation depth). It is then possible to heat and/or irradiate the samples to study the behavior of these elements which, according to their nature and their concentration, will be able to diffuse quickly or be immobilized in precipitates.

The question of the representativeness of ion beams to simulate neutron irradiation is still relevant. Indeed, the spatial distribution of primary defects that are created by neutron irradiation is different from that produced by ion beams, so the microstructure evolves differently. Mansur developed a model postulating that the change of an irradiation parameter can be compensated by a shift in the value of another parameter (in this case the irradiation temperature), so as to guarantee a comparable creation of point defects (Mansur 1978). Depending on this model and the material used, it will be necessary to use a higher ion irradiation temperature (between 50 and 200 K) than in the reactor. A commonly accepted rule for simulating neutron irradiation is to favor irradiation with ions of mass that are at least equal to that of the target atoms, in order to maximize the collision cascades (see Chapter 9).

3.3.1. *Evolution of the properties of uranium dioxide under irradiation*

The fission of uranium induced by thermal neutrons is asymmetrical: the atomic numbers of the fission products are between 29 and 67 (from copper to holmium), with a fission yield showing pronounced maxima around masses of 95 and 140 (light and heavy fission fragments, respectively). The fission products, which have a high neutron surplus, stabilize by beta(-) decay. In parallel, successive neutron captures on actinide nuclei generate transuranic nuclei (from neptunium $Z = 93$ to curium $Z = 96$), some of which will decay by alpha emission or fission, thus contributing to the energy of the reactor. Irradiation in a reactor is thus responsible for physico-chemical modifications of the nuclear fuel: (i) modification of the chemical composition by generation of fission products and transuranic actinides; (ii) modification of the atomic organization through creation of irradiation defects. These two often-coupled contributions, to which the temperature parameter should be added, are ultimately responsible for microstructural and structural changes.

3.3.1.1. *Behavior of fission products*

The fission products belong to different families of chemical elements. In addition, the two oxygen atoms released by the fission of a uranium nucleus are the driving force behind the chemical reactivity of the system. Oxygen regulates the chemical activity of all the fission products present in different degrees of oxidation. The state of the system and the speciation of the oxidizable fission products result, at each moment, from the competition between the different solid, liquid or gaseous phases in which these fission products are engaged, under different degrees of oxidation. Some elements therefore precipitate in metallic form (Mo, Tc, Ru, Rh, Pd, Ag, Cd, In, Sn, Sb, Te) or in oxide form (Rb, Ba, Zr, Nb, Mo, Te) in the solid, others are soluble (Sr, Zr, Nb, Y, and the lanthanides La, Ce, Pr, Nd, Pm, Sm) or are volatile or semi-volatile (Kr, I, Xe, Cs). Table 3.1 details the distribution of fission products and their concentrations (Sercombe et al. 2016).

Group	Elements	Concentration (mmol/mol UO_2/at%)	Category
Xe	Xe+Kr	3.06	Inert fission gases and volatile fission products
Cs	Cs+Rb	1.7	
Te	Te+Se	0.302	
I	I+Br	0.131	
Ba	Ba+Sr	1.35	Stable oxides
Zr	Zr+Nb	2.62	
Mo	Mo	2.33	Mixed (oxide and metal)
Ru	Ru+Tc+Rh	2.44	Metallic fission products
Pd	Pd	1.02	
Ce	Ce	1.24	Fission products and actinides in solid solution in UO_2
Eu	Eu+Sm	0.381	
La	La+Y	0.931	
Gd	Gd+Nd+Pm	1.92	
Pu	Pu+Pr	3.4	

Table 3.1. *Inventory and concentration of fission products and plutonium in irradiated UO_2 (by simplification, the number of FPs calculated by the CESAR code has been reduced and grouped into 14 elements of similar physico-chemical behavior)*

Although the presence of soluble species in the lattice leads to a global contraction of the cell parameter of the oxide, the creation of irradiation defects and the presence of fission products in the form of metallic precipitates, oxides and gas bubbles are responsible for a swelling of the fuel (increase of its volume) of the order of 0.7% for a burnup rate of 10 $GW.j.t^{-1}$, divided between a solid contribution and a gaseous contribution. However, at the very beginning of irradiation, the swelling is in competition with a densification phase (maximum variation of density of the order of 0.5%). This densification is consecutive to the progressive disappearance of the small pores (of the order of the micrometer), because of their dissolution in the oxide in the form of vacancies, then of their diffusion towards the defect sinks (grain boundary, free surfaces). Dissolution and diffusion are induced by the different sources of irradiation undergone by the fuel.

In accident scenarios such as Chernobyl in 1986 or Fukushima-Daiichi in 2011, the chain of events can lead to the rupture of various containment barriers, and thus to the release of fission products and other radioactive materials into the environment. It is therefore essential to determine the so-called "source term", that is, the speciation, the release kinetics and the overall release fraction of fission products.

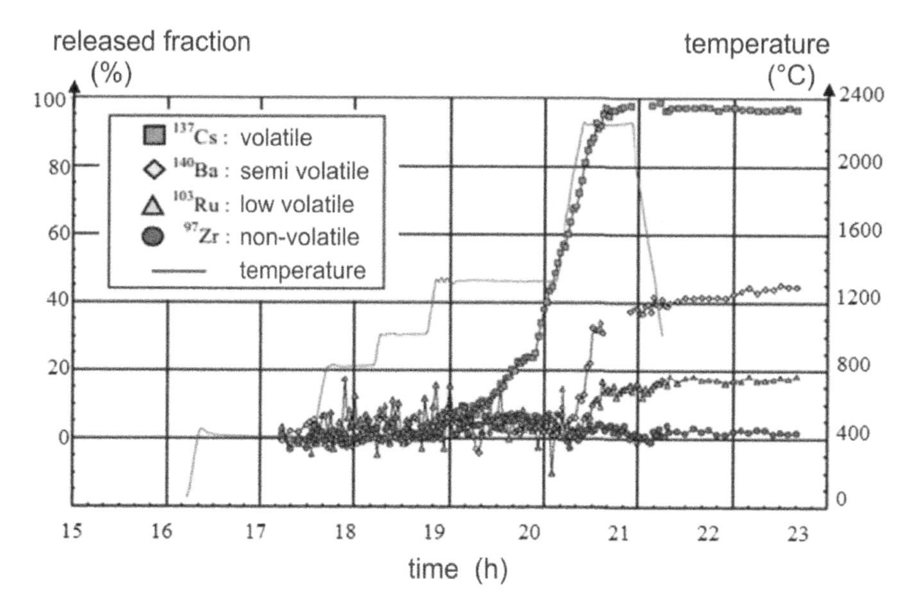

Figure 3.4. *Illustration of the volatility of fission products released (online measurement) during a VERCORS analytical test (from (CEA 2008)). For a color version of this figure, see www.iste.co.uk/bouffard/nuclear.zip*

Under these extreme conditions, the temperature greatly exceeds that of normal operation and very strong variations of the oxygen potential are likely to occur due to the oxidation of the fuel rod cladding (generating large quantities of H_2), followed by an infiltration of steam or air into the fuel. In order to quantify these phenomena, experimental programs are conducted such as the PHEBUS-PF integral tests or the VERCORS and VERDON separate effects (or analytical) tests. The experimental results thus generated have made it possible to draw up an inventory of the fission products according to their released fractions, as well as their specific behavior, that is, volatility (see Table 3.1). An example of the behavior of fission products according to their degree of volatility is given in Figure 3.4. Analysis of these data shows that the main radionuclides released in the event of an accident are from the iodine and cesium families, two highly volatile elements.

3.3.1.2. *Specific impact of fission gas release*

The noble gases resulting from fission (Kr and Xe) play a particular role in solids, because of their high production rate (0.3 atoms per fission, which corresponds to 2 cm^3 of gas produced per cm^3 of fuel, measured under normal temperature–pressure conditions) and their insolubility in the solid lattice. This insolubility leads to the formation of rare gas bubbles, which are responsible for the swelling of the fuel and the degradation of its thermal conductivity. Under the effect of its release in the plenum, it also leads the composition of the filling gas to be modified, which is harmful because of the lower thermal conductivity of xenon compared to helium. In addition, the control of the release of gases in the plenum is essential, as they contribute to the increase in the internal pressure of the rod. The fraction released remains in the range of a few percent, and the examination of the microstructure of the fuel shows the presence of nanosized bubbles that are strongly over-pressurized (pressure of the order of a GPa), which are localized in the grains, and whose equilibrium size is defined by the ratio between the aggregation of new gas atoms and their ejection by the atomic collisions induced by the fission fragments (resolution of the Xe atoms in the oxide, which leads to a supersaturation of the gas in the solid).

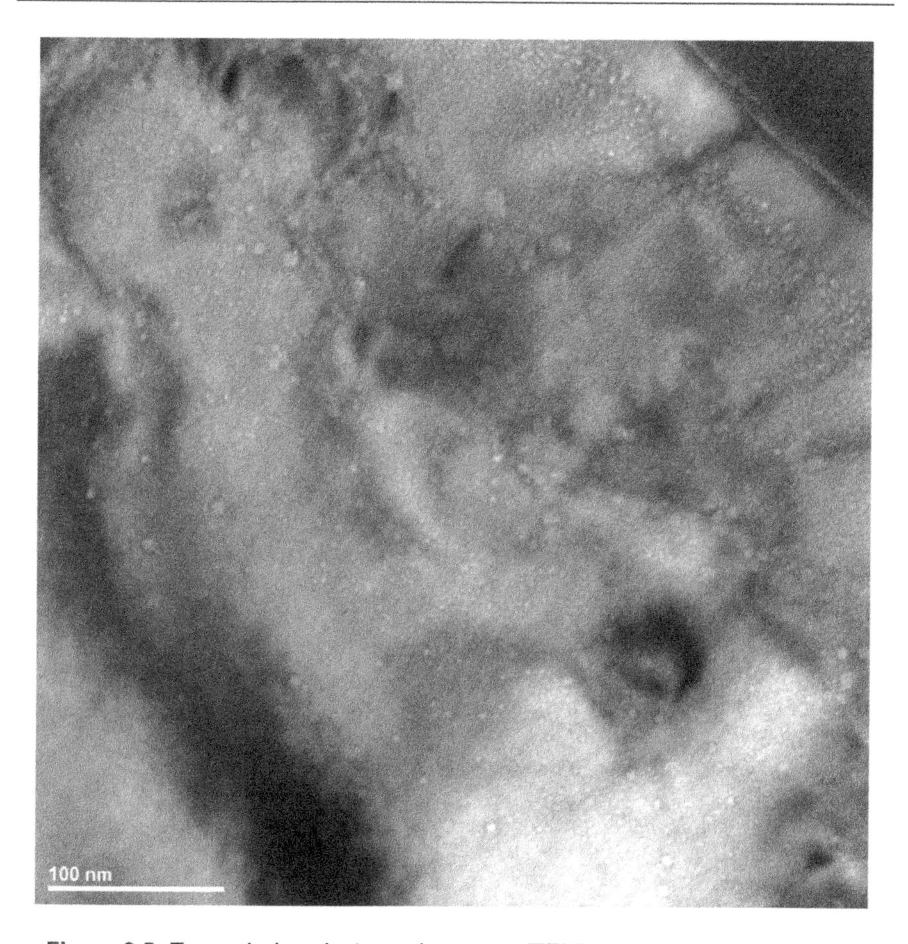

Figure 3.5. *Transmission electron microscopy (TEM) image recorded on a UO_2 crystal implanted with Xe ions (fluence of 10^{16} ions.cm^{-2}) and then annealed at 1,873 K (Xe bubbles in white) (photo credit: T. Epicier (CLYM, Villeurbanne))*

Figure 3.5 shows nanosized bubbles in a UO_2 crystal implanted in Xe. The gas also migrates in atomic form to the grain boundaries, as well as to hotter regions where it forms large, typically micrometer-sized bubbles. The gas accumulates in the hot zones, coalescing and forming channels through which it escapes by percolation (free surfaces, fractures, pores). The release of fission gases remains limited with a value of about 0.5% until a threshold burnup rate of about 35 GW.j.t^{-1}, and then increases sharply with this rate to reach values of 3%–10% at 60 GW.j.t^{-1}. The release results from athermal and thermally activated processes. At low burnup rates, irradiation plays a dominant role by several mechanisms: a gaseous fission fragment produced at the extreme periphery of the pellet (less than 8 μm from the surface, in other words, the maximum path of a fission fragment) can exit directly, or a gas atom located near the surface can be ejected by a succession of collisions or simply undergo an accelerated diffusion towards the surface thanks to the defects, in particular the vacancies, created by the irradiation. The quantity released remains small, but does increase regularly with the burnup rate and the fission products content. Beyond a burnup threshold (35 $GW.j.t^{-1}$), thermally activated processes become predominant. The central region of the pellets then presents a microstructure, characterized by the interconnection of bubbles at the grain boundaries and the formation of bubbles in the grains, the origin of which is linked to the crossing of a threshold temperature (about 1,500 K). This central region is depleted in fission gas, and the accelerated release comes from the release of gas from intergranular bubbles through fractures and pores. The thermal degradation of the fuel with the increase in the burnup explains the crossing of this threshold temperature.

3.3.1.3. *Fission product speciation*

The thermodynamic variables that govern chemical transformations are temperature, the elemental composition of the chemical system, imposed by the burnup rate, and the partial pressure of oxygen (pO_2), which is related to the oxygen potential[7] $\Delta G(O_2)$ by the relationship in equation [3.5].

$$\Delta G(O_2) = RTln(pO_2) \qquad [3.5]$$

with R being the ideal gas constant (8.314 $J.mol^{-1}.K^{-1}$) and T being the temperature (K).

During normal operation and regardless of the burnup rate reached, the oxygen potential of the fuel does not change very much, that is, about -450 $kJ.mol^{-1}$. On the contrary, during a power transient, the fuel undergoes significant heating at the pellet center, which results in a high-amplitude reduction disruption, causing the speciation of the volatile fission products to change significantly, some of which are corrosive for the cladding. Under certain conditions, the combination of the additional stresses imposed on the cladding at the triple point by the pellet expansion, the temperature and the presence of corrosive fission products allows the initiation of cracks in the fuel rod cladding, which can lead to a loss of integrity in the first containment barrier. This phenomenology is called the pellet–clad interaction (PCI). The cladding fracture surfaces are characteristic of an

7. The oxygen potential is the partial molar free enthalpy of oxygen at a given temperature in an atmosphere containing a certain amount of oxygen. It reflects the balance between the oxygen in the crystallographic lattice and the oxygen in the gas phase.

iodine-induced stress corrosion cracking (ISCC) mechanism (Cox 1990). The chemical form of iodine determines its corrosivity towards zirconium; its speciation being influenced by temperature, oxygen partial pressure, and also by the presence of certain fission products. The Cs-I-Te-Mo system in UO_2 is the most representative thermochemical system to study ISCC.

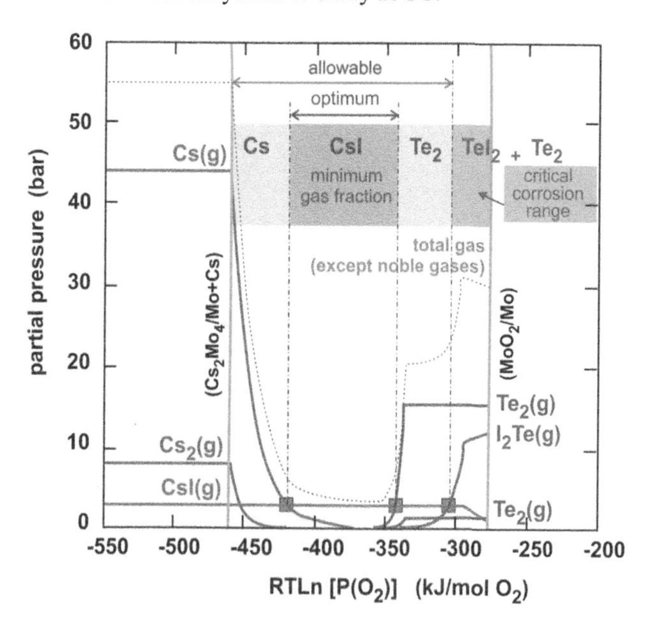

Figure 3.6. *Partial pressure and speciation of major gases (excluding He and Xe) in standard UO_2 fuel irradiated at 30 GW.j.t_U^{-1} at 1,770 K, as a function of operating oxygen potential (from Pennisi (2015)). For a color version of this figure, see www.iste.co.uk/bouffard/nuclear.zip*

The range of variation of the O_2 potential in a power ramp, from -550 to -200 kJ.mol^{-1}O_2, is divided into four areas of predominance of $Cs_{(g)}$, $CsI_{(g)}$, $Te_{2(g)}$ and TeI_2 + $Te_{2(g)}$ (see Figure 3.6) (Pennisi 2015). The issue is staying in the stabilizing range of non-corrosive chemical species, such as CsI(g), and avoiding the most corrosive form of iodine for the cladding, TeI_2. At approximately 1,770 K, the optimum range for limiting cladding corrosion corresponds to an oxygen potential between: $-425 < \Delta G(O_2) < -350$ kJ.mol^{-1}O_2 in which $CsI_{(g)}$ dominates, TeI_2 is absent and free cesium is still in small amounts. The degree of immobilization of the fission products is then maximal: Te can be found in the condensed form $Cs_2Te_{(s)}$ and Cs in the forms $Cs_2MoO_{4(s,l)}$ and $Cs_2Te_{(s)}$.

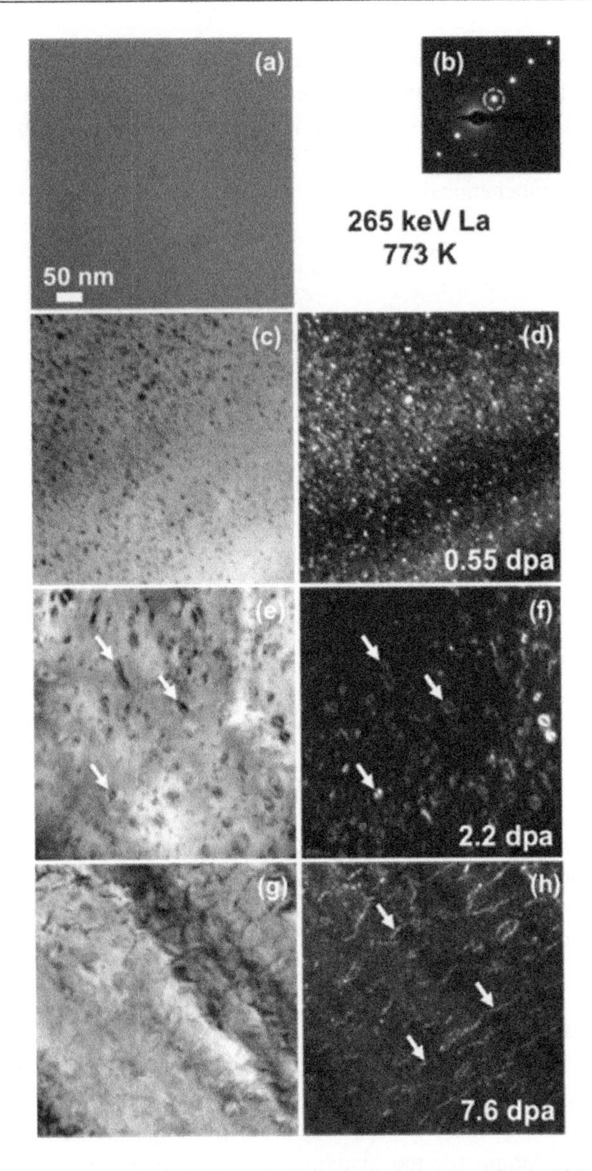

Figure 3.7. *Microstructure evolution of UO_2 irradiated with 265 keV La ions at 773K and followed in situ by TEM: (a) before irradiation; (c) and (d) 5×10^{13} cm^{-2} (0.55 dpa); (e) and (f) 2×10^{14} cm^{-2} (2.2 dpa); (g) and (h) 7×10^{14} cm^{-2} (7.6 dpa). The first column (a, c, e, g) shows bright field images; the second column (d, f, h) shows dark field images. In (e) and (f), arrows indicate dislocation loops and in (h), they show fusion of dislocation loops. The scale is identical for all images (photo credit: A. Gentils (IJCLab, Orsay))*

Finally, there is a threshold in diode partial pressure (P_{I_2}), above which the rupture time of a zirconium alloy cladding is no longer affected by the increase in this concentration. Below this threshold value, the lower the iodine content, the higher the rupture time. This threshold is approximately 60 MPa (600 bar) (Baurens 2014).

Controlling the amount of iodine or iodine chemistry in a fuel is a major challenge to prevent cladding failures by ISCC during power transients.

3.3.1.4. *Defects incorporated during irradiation*

The fuel is subjected to many sources of irradiation: fragments from the fission of ^{235}U (and odd isotopes of Pu), alpha decay of actinides (uranium and transuranics produced by successive neutron captures) leading to the formation of an alpha particle and a heavy recoil nucleus, neutrons from fission and beta(-) decay of fission products (emission of electrons, antineutrinos). As a result of their high kinetic energy and high charge state, fission fragments are the dominant source of fuel damage. At high velocity, they mostly lose their energy by electronic processes (electronic excitation and ionization) and by atomic collisions at the end of their path (95% of the kinetic energy is deposited on the electronic system, and the remaining 5% is in the form of atomic collisions). In spite of these extreme conditions, the structure of the oxide always remains crystallized, and the ultimate stage of damage observed is a structural transformation from a single crystal to a polycrystal known as polygonization.

The very high electronic stopping power of the fission fragments (in the range of 20 keV.nm^{-1}) leads to the formation of damaged material cylinders (tracks) along their trajectory. Model experiments carried out with accelerated ion beams bombarding virgin polycrystalline pellets (or single crystals) have shown that the cylinders remain crystalline, even for high electronic excitation rate values (60 keV.nm^{-1}) (Wiss et al. 1997). The creation of disordered tracks is explained by the localized abrupt increase in space and time (radius of a few nanometers, with a duration in the range of 10^{-13} s) of the temperature along the trajectory of the projectile (thermal spike model (Lifshits et al. 1960; Toulemonde et al. 1993)). The energy deposited on the electronic system is transferred to the atomic lattice via electron–phonon coupling, and this transfer creates a heating that can locally exceed the melting temperature, followed by an ultra-rapid tempering that prevents any perfect recrystallization and stabilizes structural defects. An ejection of material from the core of the tracks appears on the surface in the form of protrusions, which are similar to nanoeruptions (height of a few nanometers and diameter of a few dozen nanometers) (Bouffard et al. 1998; Garrido et al. 2009). This deposition of electronic energy, and the resulting increase in temperature, are also responsible for the accelerated diffusion of uranium atoms (the diffusion coefficient is constant and independent of temperature until about 1,300 K, and then follows an Arrhenius law) and for the resolution of atoms belonging to the gas bubbles. When the irradiation fluence increases, the damaged material cylinders overlap, and the structure of the oxide changes to the formation of crystalline domains of submicrometer size, with very low angular disorientation (size in the range of a few tenths of nanometers and angular disorientation of a few tenths of a degree).

Atomic collisions (dominant at low energy) produce point defects, with their accumulation leading to the formation of small loops of interstitial dislocations (typical sizes of 5–8 nm). They progressively evolve into lines of dislocations (size of a few tenths of nm) and the final high-fluence stage is the formation of an extensive lattice of entangled dislocations. Figure 3.7 shows the typical evolution of the microstructure that can be observed by transmission electron microscopy, during irradiation by ions that are representative of low-energy fission fragments (at the end of their range). The damage, in the formation of a mosaic structure, is all the more significant as the irradiation defects are created in the presence of a high concentration of impurities in the crystal lattice. This structure is only observed in the presence of insoluble impurities in the lattice; it does not appear when incorporating soluble species.

3.3.1.5. *Rim effect*

In the case of fuel irradiated in a reactor, a particular microstructure is observed at the periphery of the fuel pellets (a so-called "rim effect", known as a High Burnup Structure (HBS)), due to the local increase in the burnup rate. The periphery of the fuel pellets is subjected to irradiation by neutrons, whose energy exceeds that of thermal neutrons (epithermal neutrons). These neutrons, which are partially captured by the ^{238}U nuclei, produce various actinides, including fissile ^{239}Pu, which locally increase the burnup rate to values of about 150 $GW.j.t^{-1}$. The fuel then undergoes major microstructural transformations over a zone of thickness that is limited to the first 200 μm below the surface. The original grains, which have a size of about 10 μm, are then at the heart of a phenomenon of crystalline subdivision (polygonization, see Figure 3.8): a mosaic structure appears, whose crystallites have a size of about 100 nm with a small angle of disorientation (a few degrees). At the same time, the structure of the fuel becomes porous with a "cauliflower" shape, under the effect of the grouping of gas bubbles in intergranular pores, whose size reaches 1 μm (see Figure 3.9).

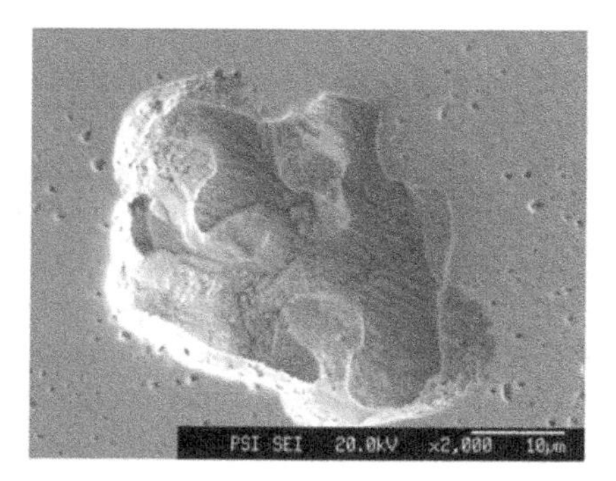

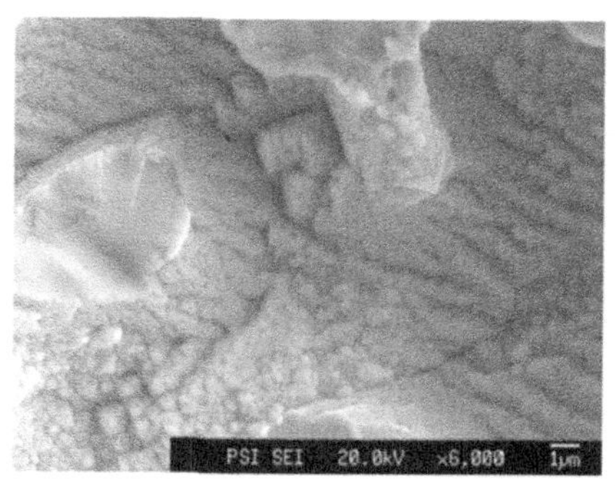

Figure 3.8. *Scanning electron microscopy observations of the polygonization phenomenon at the periphery of a* Cr_2O_3*-doped* UO_2 *fuel irradiated at 69* $GW.j.t^{-1}$ *(average rod burnup). On the left, a microstructural detail at 0.5 mm from the edge of the pellet, and on the right, the observation of the recrystallization (photo credit: Kernkraftwerk Leibstadt AG – Paul Scherrer Institute – Framatome)*

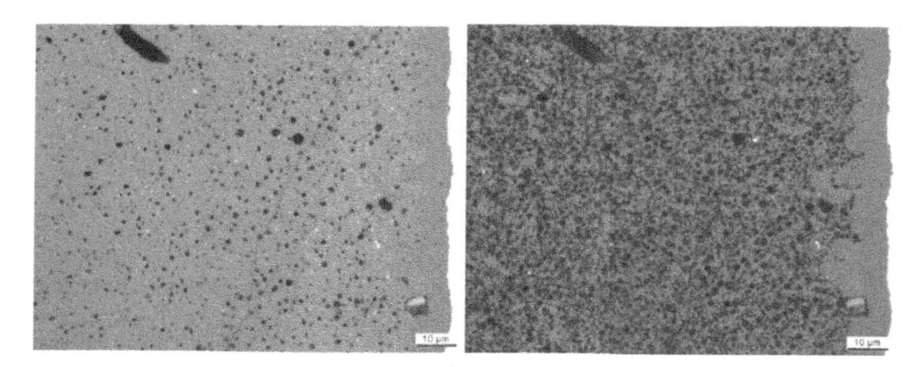

Figure 3.9. *Porous structure at the periphery of an irradiated* UO_2 *fuel with an average rod burnup of 63* $GW.j.t_U^{-1}$ *(on the left: polished state, on the right: after a chemical attack) (photo credits: Vattenfall – Studsvik AB)*

Under normal operating conditions, the release of fission gases remains limited from this peripheral zone. Indeed, the gaseous atoms that leave the grains are trapped by the enclosed porosity. On the contrary, a sudden and significant release of fission gases can appear in accidental conditions: the increase of pressure in the pores with temperature is ultimately responsible for the decohesion of the grains along the original grain boundaries.

In addition to the local increase in the burnup rate, the atomic-scale mechanisms responsible for this peculiar structure still remain highly debated (Nogita and Une

1995; Kinoshita 1997; Rondinella and Wiss 2010). Model experiments on UO_2 single crystals have shown that crystal subdivision can be induced by electronic energy deposition alone, but only for electronic excitation densities that are higher than those of fission fragments (Garrido et al. 2009). Furthermore, it has been shown that atomic collision cascades in the presence of insoluble impurities in concentrations beyond a few percent can also be at the origin of this structure (Matzke et al. 1994; Garrido et al. 2008). In the case of fuel irradiated in a reactor, it is likely that the temperature (lower at the periphery of the pellet), the concentrations of fission products, especially insoluble ones, as well as the irradiation defects of electronic and elastic origin, act in synergy to lead to the final microstructure, but their respective roles are not clarified. The different atomistic models proposed for the formation of this structure notably focus on the evolution of dislocations and their role in the formation of new grains, on the participation of a recrystallization of the grains in the final microstructure and on the nucleation of fission gases in the core of the tracks formed by electronic excitation that are rich in vacancies. In particular, the exact nature of the observed microstructural transformation has not yet fully been clarified and is divided between a *polygonization* scenario, that is, subdivision of the original grains that are weakly disoriented by the formation of subgrain boundaries induced by the movement of dislocations, and a *recrystallization* scenario, characterized by the formation of new grains that grow at the expense of the original grains and completely erase the initial microstructure.

3.3.2. *Evolution of boron carbide properties under irradiation*

In the reactor, boron carbide is mostly irradiated by fission neutrons. Thus, the evolution of the boron carbide properties is induced by two main processes: on the one hand, the damage resulting from elastic collisions between neutrons and boron or carbon nuclei and, on the other hand, the majority neutron absorption reaction[8] presented in equation [3.6], which can be written as $^{10}B(n,\alpha)$:

$$^{1}n + \, ^{10}B \rightarrow \, ^{4}He + \, ^{7}Li \qquad [3.6]$$

The predominant area of each of these processes is governed by neutron energy, as can be seen in Figure 3.10 (Siméone et al. 1997).

In a PWR, the neutron spectrum presents energies lower than 10 eV. Figure 3.10 therefore shows that the displacements due to the $B(n,\alpha)$ reaction are overwhelmingly dominant in this type of reactor. The boron carbide thus undergoes

8. Another reaction $^{10}B(n,2\alpha)^{3}H$ exists, but with a thermal neutron cross-section that is one million times smaller.

a change of composition, depleting its boron content and incorporating a quantity of helium and lithium that is proportional to the neutron fluence. Although the behavior of helium has been intensively studied, that of lithium is little known. Nevertheless, it is assumed to have little impact on the evolution of the properties of boron carbide. However, the Li and He species are created with a kinetic energy of about 1 MeV, which is enough energy to disorganize the structure, by ejecting boron and carbon atoms from their site onto their path of a few micrometers in the material. The creation of these numerous point defects can indirectly impact the properties of boron carbide, by facilitating the formation of helium bubbles.

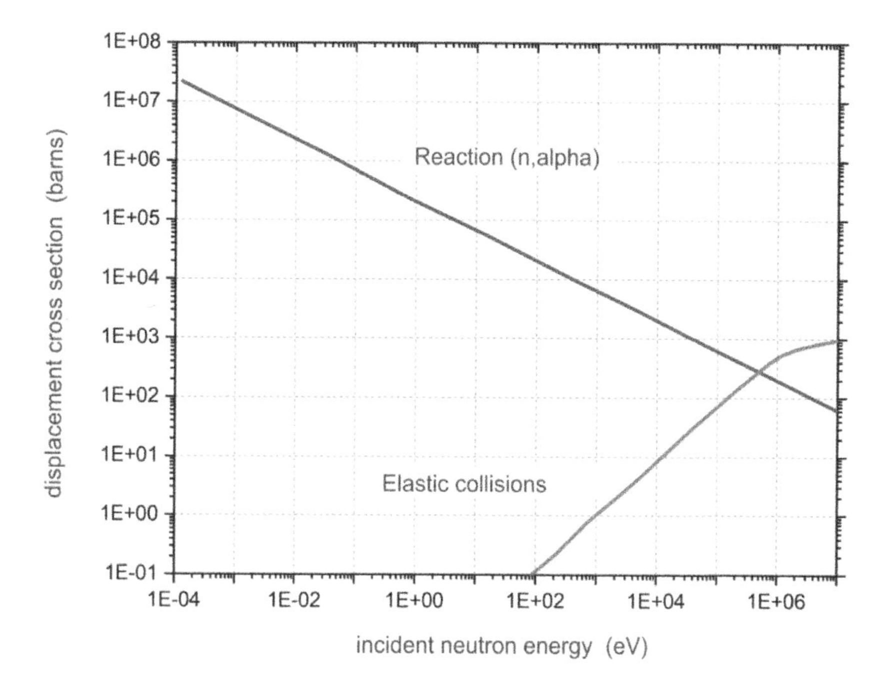

Figure 3.10. *Displacement cross-sections in boron carbide as a function of neutron energy (from the original figure of Siméone et al. (1997)). For a color version of this figure, see www.iste.co.uk/bouffard/nuclear.zip*

3.3.2.1. *Impact of helium on the aging of boron carbide*

Helium is easily trapped as bubbles in irradiated boron carbide under normal conditions. When created close to the grain boundaries, these bubbles take a lenticular shape, whereas in the grain, the bubbles are spherical and nanosized. In all cases, these bubbles contain highly pressurized helium gas (> 5 GPa), which will degrade the thermomechanical properties and create a microfracturing of the material.

The total quantity of helium created (about 10^{22} helium atoms per cm^3 or ~10% at.) during the presence of the material in the reactor is so significant that it is the limiting factor in the lifespan of boron carbide in PWRs. The macroscopic swelling linked to the appearance of these bubbles mainly occurs at the periphery of the boron carbide pellet, in other words, where the neutron absorption is the most significant. The stress gradient between the swollen periphery and the center of the pellet, which is almost free of helium, causes cracks to appear, mostly through the grain boundaries. We can simultaneously observe a release of helium located near the grain boundaries, as well as an important swelling of each grain. Ultimately, the combination of swelling, microfracturing and cracking of the material can cause a mechanical interaction between the absorbing material and the stainless-steel cladding, which can lead to the latter's rupture.

The cracks also result in a significant decrease in the thermal conductivity of the boron carbide. From the beginning of irradiation, from $\sim 5 \times 10^{20}$ neutron captures per cm^3, the thermal conductivity decreases. It is between 3 and 6 $W.m^{-1}.K^{-1}$ for values between 2×10^{21} and 10^{22} captures per cm^{-3}, and hardly changes with temperature. Nevertheless, for a PWR (contrary to fast neutron reactors), the impact on the physical properties remains limited, since the temperature gradient between the center and the periphery of the pellet remains low.

3.3.2.2. *Impact of the ballistic damage induced by the slowing down of He and Li*

During the slowing down of helium and lithium in boron carbide, atomic collisions lead to collision cascades that generate about 250 displacements per neutron capture. Furthermore, the energy of these light ions induces electronic excitations that correspond to an electronic stopping power slightly lower than 1 $keV.nm^{-1}$. Raman microspectroscopy and transmission electron microscopy are the two techniques that have allowed us to better understand the mechanisms of ballistic and electronic excitation damage of boron carbide. Indeed, its particular structure induces 12 modes (A_{1g} and E_g) that are active in Raman spectroscopy, and initial atomistic simulation calculations have made it possible to identify each vibrational mode of the $(B_{11}C^p)CBC$ structure, as can be seen in Figure 3.11 (the "Ico" modes indicate inter- or intra-icosahedral vibrations, and the "Ch" modes indicate chain vibration modes). The D and G modes indicate the presence of free carbon incorporated in the material during its development.

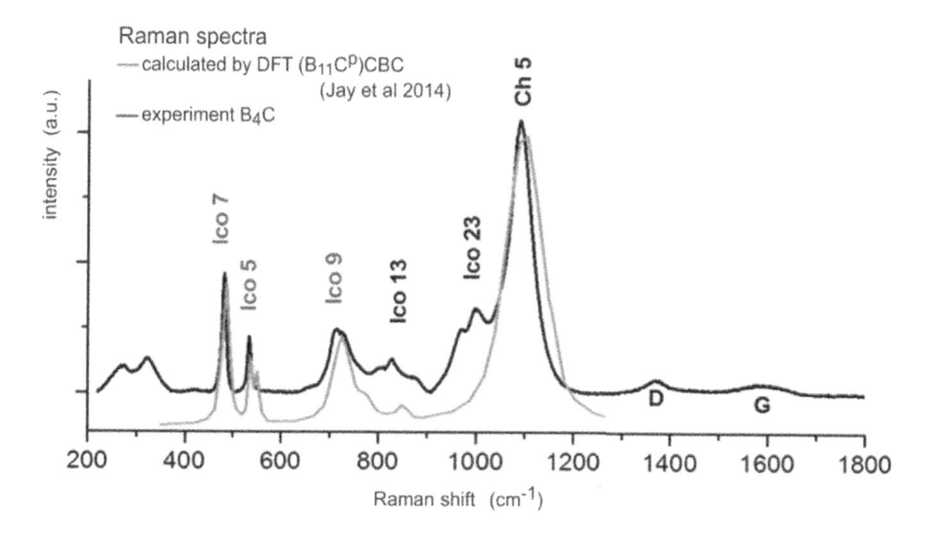

Figure 3.11. *Experimental Raman spectrum of B_4C compared to a theoretical spectrum calculated by Jay et al. (2014). D and G represent the graphite bands (presence of carbon). For a color version of this figure, see www.iste.co.uk/bouffard/nuclear.zip*

The elastic collision damage mechanisms of boron carbide were first clarified by Gosset et al. (2016) and then by Victor et al. (2019), through structural analysis by Raman microspectroscopy and TEM after irradiation with Au ions at room temperature. For low ballistic damage (~0.1 dpa), the structural changes correspond to point defects. Between 0.5 and 2.5 dpa, the concentration of point defect clusters and their size increase significantly. From 2.5 dpa onwards, extensive defects appear, particularly dislocation loops. Beyond that, amorphous zones start to form and then coalesce, before leading to a quasi-total amorphization between 4 and 7.5 dpa. This threshold of amorphization, which is higher than for most ceramics (e.g. that of silicon carbide is around 0.3 dpa), demonstrates this ceramic's good resistance to amorphization, and thus to damage. The recombination of point defects that can lead to restructuring largely explains this resistance to amorphization. In particular, the icosahedra remain globally conserved in spite of the disappearance of a boron atom (during neutron absorption, for example), managing to restructure themselves by taking advantage of the atoms of the linear chain whose rigid bonds cannot, on the contrary, reform once broken.

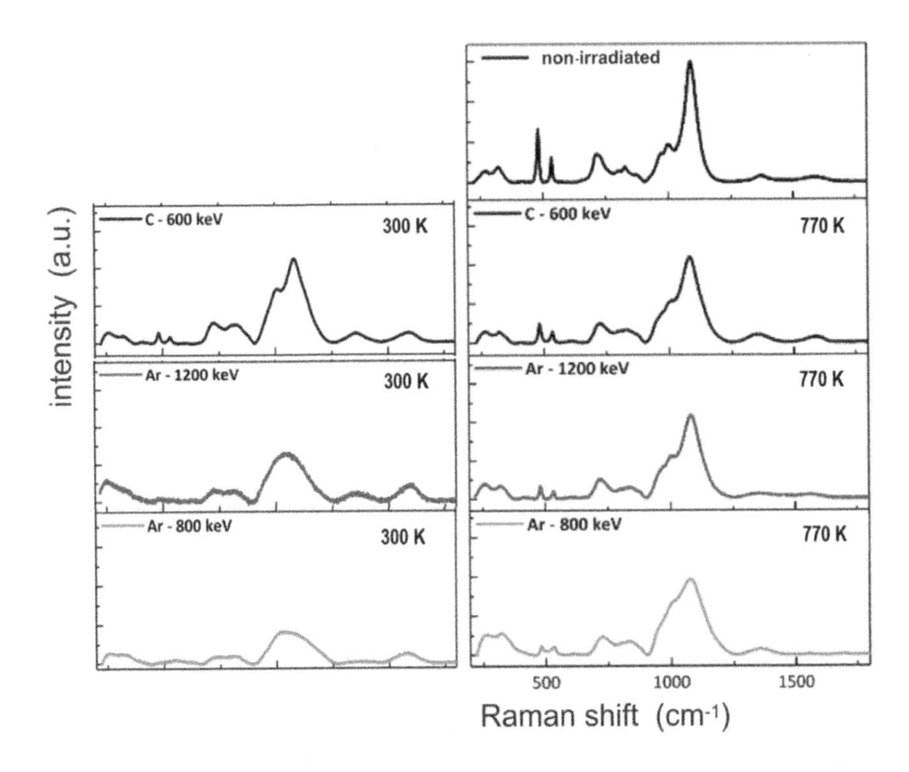

Figure 3.12. *Raman spectra performed after different irradiation conditions (600 keV C, 800 keV Ar and 1.2 MeV) at room temperature (RT) and at 770 K (Victor 2016). For a color version of this figure, see www.iste.co.uk/bouffard/nuclear.zip*

The temperature amplifies this effect, as can be seen in Figure 3.12, which shows, for different irradiation conditions, a partial self-healing of the structure at 770 K. The temperature in a PWR does not exceed 800 K in boron carbide (under normal operating conditions), and the damage to the structure remains globally low. Moreover, the electronic excitations induced by the recoil of Li or He ions do not induce damage but further enhance the healing effect, as shown by Pipon et al. (2021), when irradiating boron carbide with 100 MeV sulfur ions.

3.4. Future challenges

The irradiation behavior of UO_2 fuel and B_4C absorbing material is composed of a complex set of phenomena with numerous interactions more or less marked, combining effects of temperature, irradiation, mechanics, chemistry, etc. Although questions still remain today, research conducted over the past 60 years has made it possible to understand the phenomena involved and to identify the mechanisms that limit the performance of these materials in service. Consequently, this research has led to the development of remedies to improve, for example, the retention of fission gases or resistance to the PCI for UO_2 or to the development of new materials in order to overcome the B_4C problem. In both cases, the aim of these developments is to preserve, if not increase, the safety margins.

The improvement of the performance of the current PWR fuel is aligned according to the following three objectives:

– increased maneuverability of the reactors, in order to adjust their production to the demand of the network. In concrete terms, the rods must be able to withstand power variations of greater intensity than those authorized today, without damage;

– cycle economy with particularly better exploitation of the energy potential of the fissile material, by increasing the time spent in the reactor;

– increased fuel tolerance to accidental conditions.

In order to answer this question, two timeframes can be considered: (i) controlled evolution of the UO_2 fuel and (ii) development of technological breakthrough solutions.

The development of UO_2 fuel doped with chromium oxide (Cr_2O_3) falls into the first category. The R&D studies carried out since the 1990s within the framework of the three-party CEA-EDF-Framatome institute have led to the definition of a UO_2 pellet with a coarse-grained optimized viscoplastic microstructure (see Figure 3.13), obtained by doping with Cr_2O_3.

The feedback acquired with this fuel under irradiation highlights an increased retention of fission gases in the intragranular region, as well as a better resistance to the PCI phenomenon. This overall behavior generates additional performance margins that are in line with the first two objectives listed above (Delafoy Arimescu 2016). Concerning the PCI, post-irradiation observations show that, beyond the expected beneficial effects of the PCI on the mechanical behavior (i.e. reduction of the mechanical loading imposed on the cladding), the dopant also contributes to the decrease of the oxygen potential in the center of the fuel during a power transient. The thermomigration of oxygen released by Cr_2O_3 reduction from the hot central regions of the fuel to its periphery then mitigates the risk of ISCC failure (Sercombe et al. 2016).

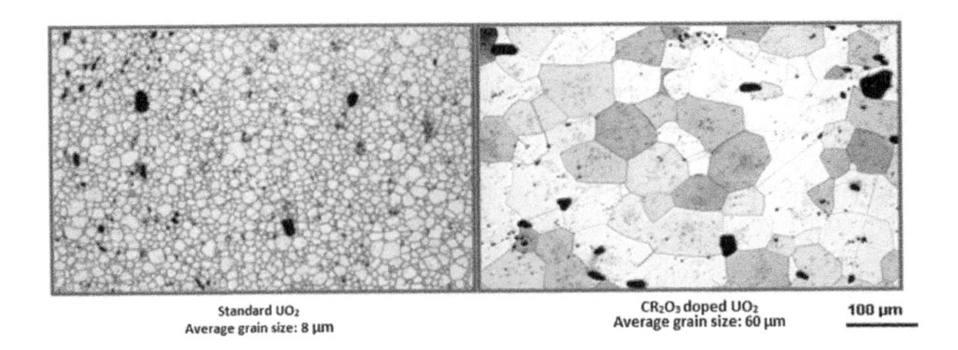

Figure 3.13. *Comparison of microstructures of non-doped and chromium oxide-doped UO_2 pellets*

Doped fuel could also contribute to increasing safety margins during a depressurization accident such as a LOCA (Loss of Coolant Accident). By lowering the pressure of the fuel rods through better gas retention, the risk of ballooning and bursting of the cladding is reduced. Furthermore, reducing the quantity of fission gases accumulated at the grain boundaries in the form of bubbles reduces the risk of grain decohesion, and thus the risk of fuel fragmentation, relocation and dispersion in the event of cladding rupture. Ultimately, a reduction of the source term of the broken rods is considered thanks to a more favorable kinetics of transient fission gas release with the doped fuel. Dedicated tests will make it possible to quantify these different phenomena (Delafoy Arimescu 2016).

The Fukushima-Daiichi accident in March 2011 has highlighted the need to continuously improve the safety of nuclear facilities, especially with Enhanced Accident Tolerant Fuel (E-ATF). The first involves delaying the start of the degradation process following an LOCA-type accident. Concerning the zirconium alloy cladding, the priority is to minimize its oxidation at high temperature in a steam environment and, consequently, the production of hydrogen gas likely to cause an explosion in the reactor building. In this context, several solutions are considered, such as the deposition of a 10–20 µm-thick chromium coating on the current fuel rod cladding or a radical change of material, for example, with a SiC/SiC ceramic cladding.

For the fuel pellet, improved tolerance to accidental conditions can be summarized in two major themes (OECD/NEA 2018):

– *Decreasing the temperature and the thermal source term*: during an accidental transient, the specific heat of the assembly (pellet + cladding) significantly affects the increasing temperature rate of the rod and, in the case of short accident scenarios, the energy stored in the fuel, which is linked in particular to its thermal conductivity, can influence the progress of the accident. Current developments consist of adding elements with high thermal conductivity to UO_2, such as chromium metal in the form of a continuous network of microcells for improved efficiency, silicon carbide in the form of filaments or beryllium oxide. In addition to oxide fuels, there are other variants that are part of an approach that would break in the current situation. These fuels – often categorized as dense fuels have a significantly higher thermal conductivity, up to 10 times that of UO_2. They are silicides (in particular U_3Si_2), nitride (UN), uranium carbide (UC) and metal alloys (in particular UZr). However, the deployment of these fuels in commercial light water reactors presents major technological challenges. Indeed, they all exhibit degraded behavior in the presence of water and steam compared to UO_2, rapidly leading to fuel fragmentation and hydrogen generation in the event of an uncontrolled accident. International work is underway to find concrete solutions to this problem. In addition, the large-scale production of these fuels is not without difficulties, notably because of the pyrophoric properties of the metal powders.

– *Improving the retention of fission products* (FP), in order to limit the dissemination of radioactive species in the environment much as possible. Beyond evolving solutions such as Cr_2O_3-doped UO_2, a concept of microencapsulated fuel has been proposed. For this, the principle of TRISO (TRi-structural ISOtropic) particles developed as fuels for high and very high temperature reactors (HTR and VHTR) is again being studied in this context. These particles have excellent fission product retention capabilities, due to the presence of multiple layers of ceramic coating (C and SiC) that are chemically stable, with high mechanical strength at high temperatures and very high burnup rates. The low density of fissile material is the main drawback of this concept for PWR application. In order to overcome this, developments include a combination of uranium enrichment up to the practical upper limit of non-proliferation (~ 19.7% ^{235}U), the use of a high-density UN core, an increase in the volume fraction of the core in relation to the particles and particle compaction fraction.

After the deployment of E-ATF, the control rod elements will become potentially limiting in the reactor under severe and major accident conditions. Under these conditions, the temperatures reached by the absorber rods are expected to be high enough for the current materials to melt, with vaporization of part of the bundle in the case of SIC. For B_4C, a eutectic is expected to form with the iron in the stainless-steel cladding of the absorber rods at about 1,420 K. This will result in a loss of integrity of the absorber materials and an acceleration of the degradation of the reactivity control elements. To address these issues, developments are moving towards absorber materials with more than three times the melting point of current materials, as well as high stability under accidental conditions (OECD/NEA 2018).

In this context, the new materials that are considered as neutron absorbers are either based on rare earth sesquioxides (RE_2O_3 with [RE] = Sm, Eu, Gd, Dy or Er) with additions of ZrO_2 and HfO_2, which improve the chemical stability of RE_2O_3 in the presence of water or steam or a combination of hafnium carbide (HfC) and samarium hafnate (Sm_2HfO_5) to replace the current SIC, or europium hafnate (Eu_2HfO_5) chosen as a replacement for B_4C, because it does not emit α particles, in addition to having favorable properties with respect to swelling under flux.

Research and development of ceramic materials for PWRs is therefore a very active theme, fueled by ever more advanced observations at the heart of the material, thanks to increasingly sophisticated characterization tools and numerical simulations that make it possible to obtain information that would otherwise be inaccessible through experience. It is now possible to better understand the effects of irradiation on materials, as well as the mechanisms that govern the evolution of these materials in service.

3.5. References

Baer, Y. and Schoenes, J. (1980). Electronic structure and Coulomb correlation energy in UO_2 single crystal. *Solid State Communications*, 33(8), 885–888.

Bailly, H., Menessier, D., Prunier, C. (1998). *Le combustible nucléaire des réacteurs à eau sous pression et des réacteurs à neutrons rapides. Conception et comportement.* Collection du Commissariat à l'Énergie Atomique, Édition Eyrolles, Paris.

Baurens, B. (2014). Couplage thermo-chimie-mécanique dans l'UO_2 – Application à l'interaction pastille-gaine. PhD Thesis, Université d'Aix-Marseille.

Bouchacourt, M. and Thevenot, F. (1981). The properties and structure of the boron carbide phase. *Journal of the Less Common Metals*, 82, 227–235.

Bouffard, S., Duraud, J.P., Mosbah, M., Schlutig, S. (1998). Angular distribution of the sputtered atoms from UO_2 under high electronic stopping power irradiation. *Nuclear Instruments and Methods in Physics Research Section B: Beam Interactions with Materials and Atoms*, 141(1), 372–377.

CEA (2008). *Les combustibles nucléaires*. CEA/Direction de l'Énergie Nucléaire, Éditions du Moniteur, Gif-sur-Yvette.

Cox, B. (1990). Pellet-clad interaction (PCI) failures of zirconium alloy fuel cladding – A review. *Journal of Nuclear Materials*, 172(3), 249–292.

Delafoy, C. and Arimescu, I. (2016). Developments in fuel design and manufacturing in order to enhance the PCI performance of AREVA NP's fuel. In *OECD/NEA Workshop on Pellet-Cladding Interaction (PCI) in Water-Cooled Reactors. Proceeding of a WGFS Worshop NEA/CSNI Report*, Lucca.

Fink, J.K. (2000). Thermophysical properties of uranium dioxide. *Journal of Nuclear Materials*, 279(1), 1–18.

Garrido, F., Vincent, L., Nowicki, L., Sattonnay, G., Thomé, L. (2008). Radiation stability of fluorite-type nuclear oxides. *Nuclear Instruments and Methods in Physics Research Section B: Beam Interactions with Materials and Atoms*, 266(12), 2842–2847.

Garrido, F., Moll, S., Sattonnay, G., Thomé, L., Vincent, L. (2009). Radiation tolerance of fluorite-structured oxides subjected to swift heavy ion irradiation. *Nuclear Instruments and Methods in Physics Research Section B: Beam Interactions with Materials and Atoms*, 267(8), 1451–1455.

Gosset, D. and Colin, M. (1991). Boron carbides of various compositions: An improved method for X-rays characterisation. *Journal of Nuclear Materials*, 183(3), 161–173.

Gosset, D. and Provot, B. (2001). Boron carbide as a potential inert matrix: An evaluation. *Progress in Nuclear Energy*, 38(3), 263–266.

Gosset, D., Miro, S., Doriot, S., Moncoffre, N. (2016). Amorphisation of boron carbide under slow heavy ion irradiation. *Journal of Nuclear Materials*, 476, 198–204.

IAEA (1997). *Thermophysical Properties of Materials for Water Cooled Reactors*. International Atomic Energy Agency – TECDOC-949, Vienna.

IAEA (2006). *Thermophysical Properties Database of Materials for Light Water Reactors and Heavy Water Reactors*. International Atomic Energy Agency – TECDOC-1496, Vienna.

Jay, A., Vast, N., Sjakste, J., Duparc, O.H. (2014). Carbon-rich icosahedral boron carbide designed from first principles. *Applied Physics Letters*, 105(3), 031914.

Kinoshita, M. (1997). Towards the mathematical model of rim structure formation. *Journal of Nuclear Materials*, 248, 185–190.

Lankford, J. (1982). Indentation microfracture in the Palmqvist crack regime: Implications for fracture toughness evaluation by the indentation method. *Journal of Materials Science Letters*, 1(11), 493–495.

Lazzari, R., Vast, N., Besson, J.M., Baroni, S., Dal Corso, A. (1999). Atomic structure and vibrational properties of icosahedral B_4C boron carbide. *Physical Review Letters*, 83(16), 3230–3233.

Leask, M.J.M., Roberts, L.E.J., Walter, A.J., Wolf, W.P. (1963). Low-temperature magnetic properties of some uranium oxides. *Journal of the Chemical Society*, 4788–4794.

Lifshits, I.M., Kaganov, M.I., Tanatarov, L.V. (1960). On the theory of radiation-induced changes in metals. *Journal of Nuclear Energy. Part A: Reactor Science*, 12(1), 69–78.

Mansur, L.K. (1978). Correlation of neutron and heavy-ion damage II: The predicted temperature shift if swelling with changes in radiation dose rate. *Journal of Nuclear Materials*, 78(1), 156–160.

Matzke, H., Turos, A., Linker, G. (1994). Polygonization of single crystals of the fluorite-type oxide UO_2 due to high dose ion implantation +. *Nuclear Instruments and Methods in Physics Research Section B: Beam Interactions with Materials and Atoms*, 91(1), 294–300.

Nogita, K. and Une, K. (1995). Irradiation-induced recrystallization in high burnup UO_2 fuel. *Journal of Nuclear Materials*, 226(3), 302–310.

OECD/NEA (2018). State-of-the-art report on light water reactor accident-tolerant fuels. NEA report, 7317.

Patarin, L. (2002). *Le cycle du combustible nucléaire*. EDP Sciences, Paris.

Pennisi, V. (2015). Contribution à l'identification et à l'évaluation d'un combustible UO2 dopé à potentiel oxygène maîtrisé. PhD Thesis, Université de Bordeaux.

Pipon, Y., Victor, G., Moncoffre, N., Gutierrez, G., Miro, S., Douillard, T., Rapaud, O., Pradeilles, N., Sainsot, P., Toulhoat, N. et al. (2021). Structural modifications of boron carbide irradiated by swift heavy ions. *Journal of Nuclear Materials*, 546, 152737.

Rondinella, V.V. and Wiss, T. (2010). The high burn-up structure in nuclear fuel. *Materials Today*, 13(12), 24–32.

Sercombe, J., Riglet-Martial, C., Baurens, B. (2016). Simulations of power ramps with ALCYONE including fission products chemistry and oxygen thermo-diffusion. In *OECD/NEA Workshop on Pellet-Clad Interaction (PCI) in Water-Cooled Reactors: Proceedings of a WGFS Workshop*, NEA/CSNI report, Lucca.

Siméone, D., Hablot, O., Micalet, V., Bellon, P., Serruys, Y. (1997). Contribution of recoil atoms to irradiation damage in absorber materials. *Journal of Nuclear Materials*, 246(2), 206–214.

Toulemonde, M., Paumier, E., Dufour, C. (1993). Thermal spike model in the electronic stopping power regime. *Radiation Effects and Defects in Solids*, 126(1–4), 201–206.

Victor, G. (2016). Étude des modifications structurales induites dans le carbure de bore B_4C par irradiation aux ions dans différents domaines d'énergie. PhD Thesis, Université de Lyon.

Victor, G., Pipon, Y., Moncoffre, N., Bérerd, N., Esnouf, C., Douillard, T., Gentils, A. (2019). In situ TEM observations of ion irradiation damage in boron carbide. *Journal of the European Ceramic Society*, 39(4), 726–734.

Wiss, T., Matzke, H., Trautmann, C., Toulemonde, M., Klaumünzer, S. (1997). Radiation damage in UO_2 by swift heavy ions. *Nuclear Instruments and Methods in Physics Research Section B: Beam Interactions with Materials and Atoms*, 122(3), 583–588.

Yakel, H. (1975). The crystal structure of a boron-rich boron carbide. *Acta Crystallographica Section B*, 31(7), 1797–1806.

4

Nuclear Graphite

Nicolas BÉRERD[1] and Laurent PETIT[2]
[1] *Institut de Physique des 2 Infinis de Lyon, Université Claude Bernard Lyon 1, IUT Lyon 1, CNRS/IN2P3, Villeurbanne, France*
[2] *EDF, Laboratoire Les Renardières, Moret-Loing-et-Orvanne, France*

4.1. What is nuclear graphite?

In its natural state, single-crystal graphite only exists in millimetric form. For industrial applications such as those of the nuclear industry requiring machinable blocks of metric dimensions, synthetic graphites are used, which are manufactured from sources of carbon that can be graphitized[1] at high temperatures (petroleum cokes), and a carbon binder (coal tar pitch) provides mechanical cohesion of the finished product (Cornuault 1981). Graphite is one of the two stable crystalline forms of carbon at room temperature. The hexagonal and lamellar-structured, hybridized sp^2 carbon atoms are arranged according to regular hexagons within poly-aromatic layers (or graphene layers), in which every other layer is shifted (see Figure 4.1a). The stacking of the layers forms the crystallites[2] (b and c). Imperfections exist within these crystallites. Three of them can be mentioned: carbon vacancies, stacking defects (the ABA sequence is not respected) and the non-parallelism of some graphene layers (a breakdown) (Pierson 1994). Finally, the coke grains are made up of clusters of crystallites (c and d). These are separated by

---

1. They have the capacity to be transformed into graphite through an anoxic treatment at 3,000°C.

2. Small graphite crystals of diameter L_a and height L_c, also known as coherent domains or basic structural units (BSU).

Nuclear Materials under Irradiation,
coordinated by Serge BOUFFARD and Nathalie MONCOFFRE.

inter-crystallite porosity, sometimes called Mrozowski cracks (Mrozowski 1954; Jones et al. 2008).

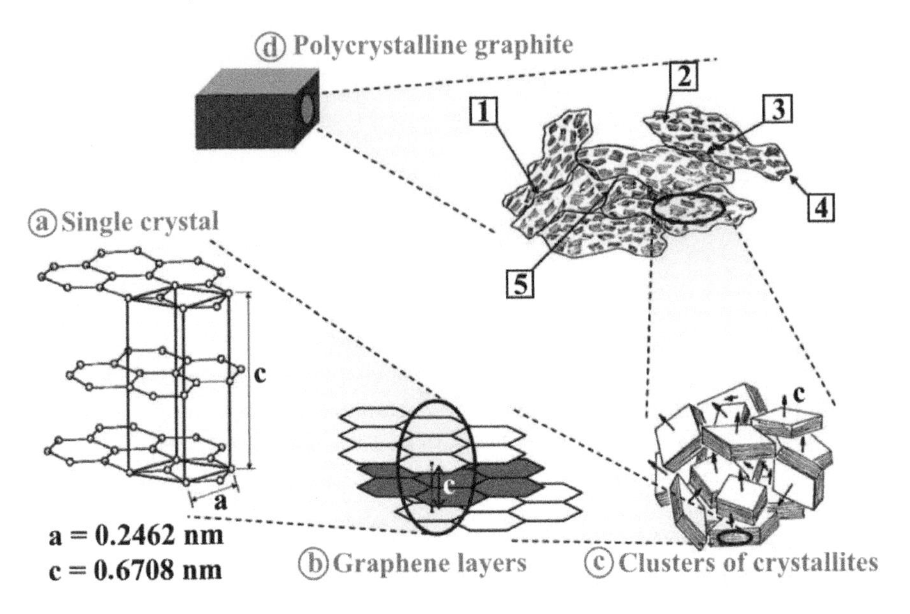

Figure 4.1. *Nuclear graphite structure – comparison with a single crystal structure from a figure in Bonal and Robin (2006). The details in part d correspond to the inter-granular porosity (1), the inter-crystallite porosity (2), the impregnated carbon (3), a coke grain (4) and the binder carbon (5). Reprinted with permission from J.P. Bonal. For a color version of this figure, see www.iste.co.uk/bouffard/nuclear.zip*

Strictly speaking, nuclear graphite is not a graphite. It can be described as a carbon–carbon composite, made up of a multi-scale assembly of crystallite grains that are more or less randomly oriented, bound by a carbon binder that is more or less well graphitized. Nuclear graphite differs from single-crystal graphite by a lower density (close to 1.7 vs. 2.265 g/cm^3), linked to a significant total porosity which can reach 25% in volume. This is due to the manufacturing process that is notably responsible for the formation of bubbles of volatile species, which are present as impurities in the petroleum cokes and binders. When they remain trapped in the graphite material, they create the so-called closed porosity (3% of the volume, the remaining 22% being the open porosity) (Comte et al. 2020). The photograph in Figure 4.2a illustrates the textural heterogeneity of a nuclear graphite with a high degree of graphitization in some places, as shown in the central area of Figure 4.2b, on which the stacking of graphene layers is observed along the <c> axis. The graphene layers thus appear on the photograph as alternating black and white "lines", with each line corresponding to a layer observed in the cross-section.

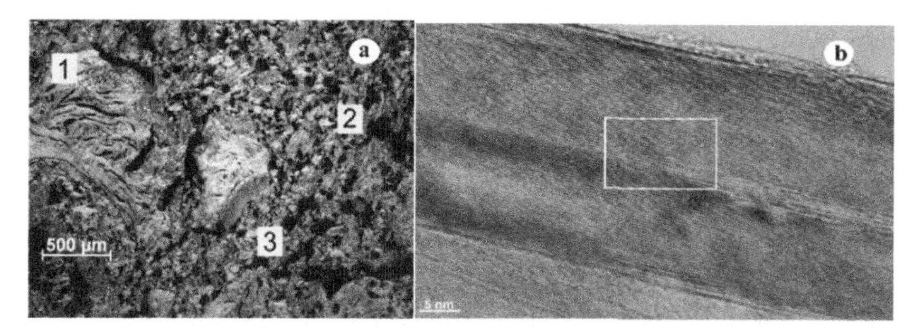

Figure 4.2. *(a) Optical microscopy (reflection mode) of a non-irradiated graphite sample from the St. Laurent A2 reactor stack (Vaudey et al. 2009). Zone 1: coke grain, 2: binder and 3: porosity. (b) "High-resolution" transmission electron microscopy of near-perfect stacks of graphene layers, attesting to the high degree of graphitization of certain zones such as, for example, the white-framed central zone of this non-irradiated graphite. Photo credit: J.N. Rouzaud (Silbermann 2013). For a color version of these figures, see www.iste.co.uk/bouffard/nuclear.zip*

Depending on the nature of the coke used for its manufacture and its mode of shaping before graphitization at high temperature, nuclear graphite more or less shows the marked anisotropic behavior characteristic of graphite crystal. Some cokes have a spherical structure (gilsonites), while others are in the form of needles. In the first case, the graphite obtained will have an isotropic nature, whereas the graphite made from needle cokes will be anisotropic. This anisotropic nature will be more pronounced for shaping by spinning of a needle coke (as is the case of graphites in Natural Uranium Graphite Gas (UNGG) reactors) than for shaping by hot isostatic pressing (modern nuclear graphite of high or very high temperature reactors (HTR or VHTR)). The physical and mechanical properties of nuclear graphite can depend on the angle (Heintz 1985) with respect to the general alignment of the crystallites (for an example, see Table II of Bonal and Gosmain (2006)). This is the behavior that is generally encountered for historical nuclear graphites.

4.2. Why use graphite in nuclear reactors?

Graphite is a material of choice for the nuclear industry because of its chemical inertia and thermal stability. Its use since the beginning of the nuclear industry is mainly related to its neutron properties and its availability on an industrial scale. Once purified, it is a relatively efficient neutron moderator. The role of the moderator is to slow down neutrons by a succession of collisions, so that their energy becomes thermal, and the area with the effective fission cross-section of ^{235}U is at its maximum. It takes on an average of 105 collisions to slow down a neutron from 1 MeV to 1/40 eV (see Chapter 1 of this book). Furthermore, its neutron capture cross-section is small (3.37 10^{-3} barn), two orders of magnitude smaller than that of light water (3.32 10^{-1} barn), the moderator of pressurized water reactors (PWRs), which allows the use of unenriched nuclear fuel as fissile material. These are the reasons why graphite was essential for the first atomic piles and why graphite was massively used as a moderator in the core of gas-cooled[3] power reactors[4]. The first natural uranium and CO_2-cooled power reactors, which were still in operation until very recently, can be found in this field. This technology includes the eight French UNGG reactors, the last of which, Bugey 1, was definitively shut down in 1994, and the 26 Magnox reactors operated in Great Britain until 2015 for the last of them (Wylfa 1). A few of these types of reactors have also been built and are operating in other countries, notably two Magnox reactors in Italy (Latina) and Japan (Tokai), as well as a UNGG in Spain (Vandellos). A second-generation reactor using low-enriched uranium (2.5% ^{235}U) in oxide form has also been developed in Great Britain. Fourteen Advanced Gas-cooled Reactors were commissioned between 1976 and 1989. They are still in operation, but will likely be shut down by 2025–2030.

This generation of gas-cooled power reactors has been progressively superseded by PWRs or boiling water reactors (BWRs), mainly for economic reasons. Nevertheless, graphite is still considered as a moderator in fourth generation gas-cooled reactors, generally using helium. This reactor technology, HTR – VHTR, is

3. This method is characterized by the use of a gas, generally CO_2, to cool the reactor core and ensure the transfer of the fission energy, so as to heat exchangers that produce the steam needed to operate the electric turbine.

4. Power reactors are reactors that are coupled to the electrical grid, as opposed to historically used atomic cells, in order to demonstrate the feasibility and control of nuclear fission, and then the production of plutonium for military purposes.

at the prototype stage. Readers interested in gas-cooled reactor technology using graphite can refer to the monograph published by the CEA (2006) or the GIF[5] roadmaps (DOE 2002; OECD 2014) for more details. Graphite was also used as a moderator in the RBMK[6] reactor technology developed in the Soviet Union era. In this reactor technology, the graphite core that serves as the moderator is cooled with light water.

In addition to its role as a moderator for the core of gas-cooled reactors, graphite has also been used as biological protection, mechanical support (sleeves) for fuel elements and for many other functions. Depending on the case, each of these nuclear graphites has significantly different characteristics and properties. The illustrations in Figure 4.3 show the different uses of graphite in the case of a UNGG reactor. Table 4.1 shows the estimated amount of graphite used by the four major users, based on available data.

Country	Number of reactors	Estimated mass of graphite (tons)
United Kingdom	46	96,000
USA	19	55,000
Russia	37	50,000
France	9	23,000

Table 4.1. *Major users of nuclear graphite in the world (extracted from IAEA (2006))*

4.3. Evolution of nuclear graphite in reactors

In reactors, nuclear graphite is subjected to multiple stresses which, together, cause material ageing: neutron irradiation, temperature and coolant chemistry, etc. Graphite therefore changes, both in terms of its microstructure and its nanostructure. This part aims to explain these changes and where they originate from.

5. The Generation IV International Forum (GIF) is a scientific collaboration, whose goal is to coordinate the research needed to demonstrate the feasibility and performance of Generation IV reactors, as well as to enable their industrial deployment by 2030. By 2021, this organization brought together 13 organizations or countries.

6. Reaktor Bolshoy Moshchnosti Kanalnyi or high-powered pressure tube reactor.

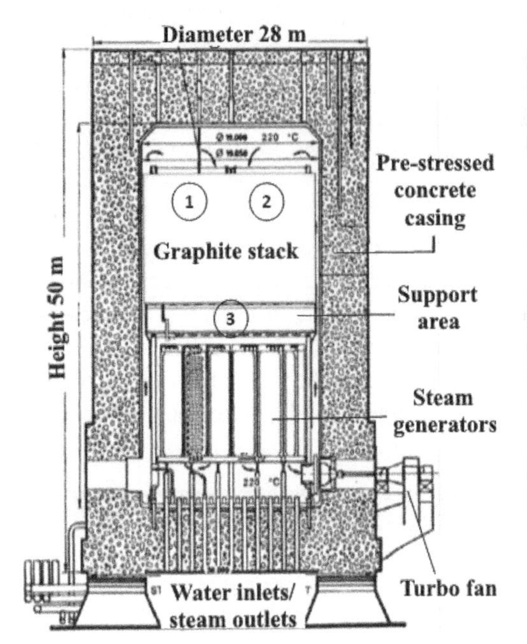

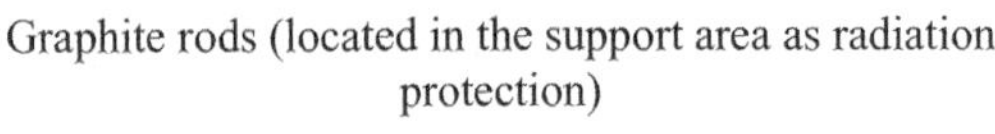
Graphite rods (located in the support area as radiation protection)

Graphite stack under construction

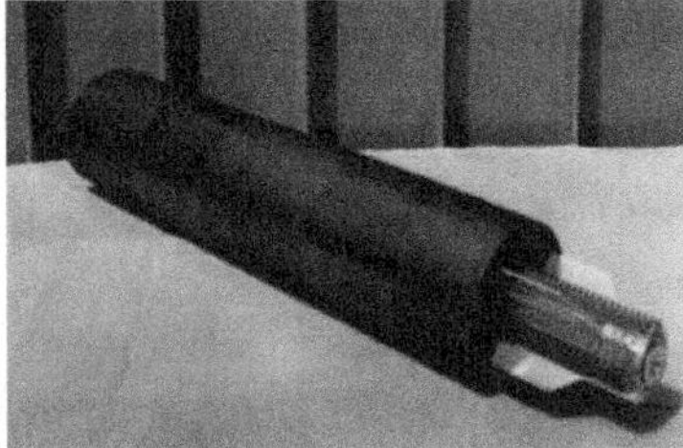
Graphite sleeve (with fuel rod) for the fuel channels of the graphite stack

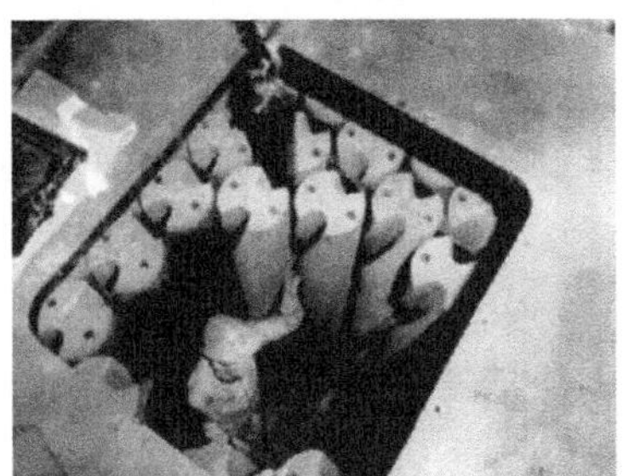

Figure 4.3. *Schematic cross-section of a UNGG reactor of the "integrated vessel" type with the location of the nuclear graphite used as a fuel element support (1), moderator (2) and biological protection (3). The moderator represents a total mass of about 2,000–3,000 tons of graphite, depending on the reactor (EDF photo library)*

4.3.1. *Neutron irradiation*

Neutrons are the main source of irradiation of nuclear graphite. The primary spectrum for neutrons with an energy higher than 100 keV is very hard, so graphite is thus the seat of many collision cascades. One way to assess the level of damage is to calculate the number of displacements per atom (dpa). For this calculation to be realistic, it is necessary to know the minimum energy that must be transferred to an atom to create a stable defect: the displacement threshold energy. The determination of the E_d displacement threshold value of carbon in a graphite material has been the subject of numerous studies since the 1960s. The values that have been obtained are between 12 and 60 eV (see Table 4.2). This difficulty in determining E_d accurately is related to the sensitivity of the methods and experimental techniques used, to the angle of incidence of the irradiating beam with respect to the surface of the sample and to the temperature of the graphite. Despite the fact that research began in 1964, this question has still not been resolved definitively. However, in their review of the effects of irradiation in graphite, Campbell and Burchell (2020) propose using a threshold displacement energy between 12 and 20 eV for irradiations whose direction is perpendicular to the <c> axis, and a value close to 60 eV for irradiations whose direction is parallel to the <c> axis.

E_d (eV)	References	E_d (eV)	References
60 //	Lucas and Mitchell (1964)	35.3 35.8	Steffen et al. (1992a, 1992 b)
28 ⊥ 42 //	Iwata and Nihira (1966)	47.3	Choi et al. (1993)
31–33 ⊥ > 60 //	Montet (1967) and Montet and Myers (1971)	34 – 34.5	Smith and Beardmore (1996)
24 ⊥	Ohr et al. (1972)	> 30 ⊥ 15-20 //	Banhart (1999)
34	Egerton (1977)	42	Hehr et al. (2007)
12	Nakai et al. (1991)	25 RT 30, 900 K	McKenna et al. (2016)

Table 4.2. *Carbon displacement threshold in a graphitized material for an incidence that is parallel (//) or perpendicular (⊥) to the <c> axis. Electrical resistivity measurements or TEM observations are the experimental methods used and, for the most recent ones, molecular dynamics calculation is used*

Nuclear reactions are also possible (such as the production of carbon-14 from carbon-13), but carbon remains insensitive to neutron capture reactions. On the contrary, impurities can happen.

Simmons (1965a) explains that the nature and level of damage of graphite depends in particular on the energy spectrum of the neutron flux in the reactor. The latter is heterogeneous in the reactor, which makes the quantification of the damage of irradiated graphite complex. Moreover, during the last 50 years, the methods for assessing the damage of materials have varied greatly, making direct comparisons difficult. Although the notion of dpa is currently used, this has not always been the case. Some methods were based on the measurement of the activity of a metal sheet placed in the reactor. This activity was compared to the activity of the same metal that remained in the core of a reference reactor (BEPO, PLUTO or DIDO pile in Great Britain, for example). The units derived from these methods were named in reference to the calibration: Equivalent DIDO Nickel Flux (EDNF), for example. Other methods were based on the measurement of the burn-up rate of the reactor and compared to a reference reactor (Calder Hall reactor). The Calder effective dose was expressed in megawatt days per ton of uranium.

In spite of these difficulties, several groups have proposed equivalence factors to calculate the damage of nuclear graphite used in a nuclear reactor. These factors can be applied to French UNGG reactors. As an example, the graphite of the A2 reactor at Saint Laurent des Eaux (SLA2) was subjected to a total fluence of 3.3 10^{21} $n.cm^{-2}$ ($E_{neutrons}$ > 0.1 MeV). For these conditions, the conversion factor between fluence and dpa is 8.9 10^{-22} dpa/($n.cm^{-2}$) (Burchell 1999). The nuclear graphite damage of SLA2 is about 3.4 dpa. Since this reactor operated 4,380 efdp[7], the damage rate is therefore about 6.4 10^{-9} $dpa.s^{-1}$ (for more on this topic, see Burchell (1999), Black et al. (2016) and Marsden et al. (2020)).

This damage is at the origin of a certain number of structural and physical property changes of nuclear graphite. By way of illustration, Table 4.3 compares the values of the mesh parameters *a* and *c*, as well as the density of a nuclear graphite before and after passage through the UNGG G2 reactor, located at the CEA center at Marcoule. The maximum neutron fluence of this reactor is of the same order of magnitude as that of SLA2. These values are also compared to those of a perfect single-crystal graphite. X-ray diffraction studies were used to obtain the *a* and *c* parameters, as well as helium or mercury pycnometer parameters for the densities (Comte et al. 2020).

Table 4.3 explicitly shows that neutron irradiation decreases the density of the graphite when it is already low, due to the presence of significant porosity. The value of parameter *a* barely changes, even though it decreases slightly. The value of parameter *c*, on the contrary, increases significantly.

---

7. efdp: effective full power days. The operating time is converted into an operating time in equivalent days or years at full power, in order to correct periods of shutdown or reduced power operation. Thus, during its 21 years of operation, the SLA2 reactor produced an amount of electricity that was equivalent to 12 years of operation at full power (515 MW).

	a (nm)	*c* (nm)	Density
Irradiated G2 graphite	0.2454–0.2463	0.6745–0.6805	1.61–1.67
Non-irradiated G2 graphite	0.2461	0.6725	1.71
Single-crystal graphite	0.2461	0.6708	2.265

Table 4.3. *Comparison between the values of the mesh parameters (a and c) (Gosmain et al. 2010) and the density of nuclear graphites (Comte et al. 2020) before and after irradiation in the G2 reactor of CEA Marcoule, as well as the density of a single-crystal graphite*

X-ray diffraction is not the only relevant structural characterization technique to study the effect of irradiation on graphite. Recent studies have shown the contribution of Raman spectroscopy associated with transmission electron microscopy in determining the average crystallite diameter, as well as in highlighting the heterogeneity of the distribution of structural defects in irradiated graphite from UNGG reactors (Ammar et al. 2015; Le Guillou et al. 2015; Pageot et al. 2015). For some areas of the G2 samples, the average L_a diameter of the crystallites decreases to values between 6 and 9 nm, while for some others, the L_a value varies very little (Gosmain et al. 2010). Before irradiation, the average L_a diameter of crystallites was about 49 nm.

All these experimental observations will make it possible to illustrate the influence of irradiation and temperature on the properties of nuclear graphite. As we will see in the following, this influence is linked to the appearance of defects in the material.

4.3.2. *Irradiation defects in nuclear graphite*

Vacancies and interstitials are mostly created in collision cascades, and are therefore close to one another. Their number will then quickly decrease by correlated recombinations for temperatures higher than 200°C, and their migration within the material will allow uncorrelated recombinations and their aggregation in the form of small clusters (see Figure 4.4).

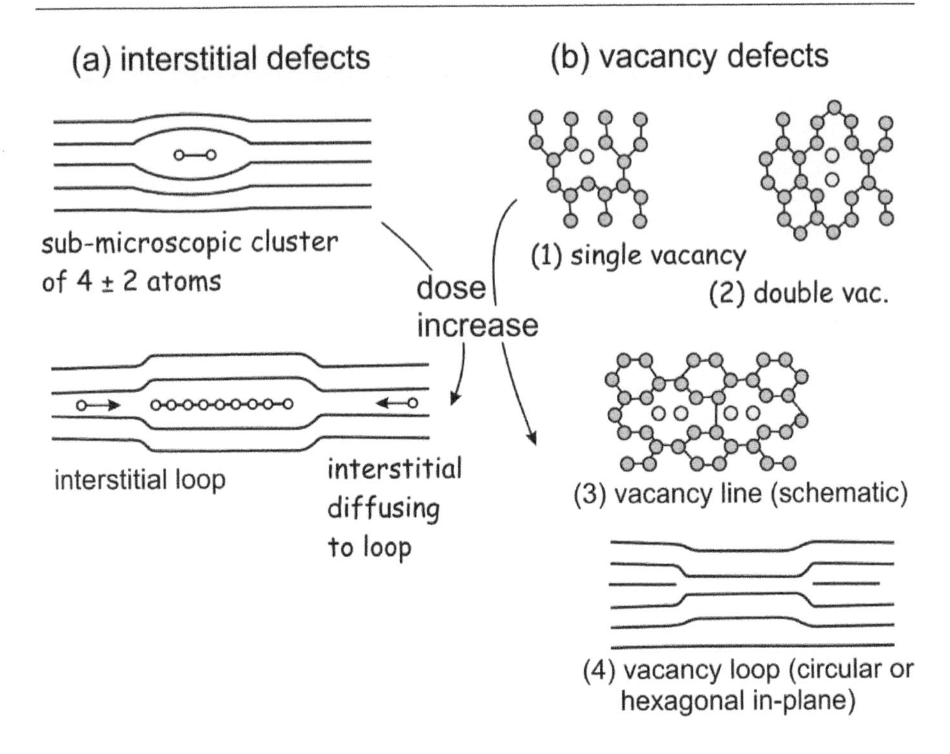

Figure 4.4. *Irradiation-induced structural defects in graphite as a function of dose (from Kelly (1982)). For a color version of this figure, see www.iste.co.uk/bouffard/nuclear.zip*

The carbon interstitials stabilize between the graphene planes, forming clusters. As they grow, these clusters become prismatic loops of interstitial dislocations (Amelinckx and Delavignette 1960). The vacancies, which are created in the graphene planes, produce dislocations in the base plane by agglomeration (Engle and Kelly 1984). In their review article, Telling and Heggie (2007) compare the results of the numerous studies on the determination of the formation and migration energies of point defects in graphite since 1946. In the discussion of their paper, they propose Table 4.4. Although these values are still being refined by other research teams, the suggested orders of magnitude are no longer discussed. For temperatures below -173°C, these authors report evidence of the existence of Frenkel pairs that are "not fully separated", which they call metastable Frenkel pairs. Moreover, concerning interstitials, they specify that their formation energy is quite well defined. However, uncertainties remain on the exact mechanisms of migration and, indeed, on the value of the migration energy of interstitials. Telling and Heggie propose a value of 1 eV for the latter. They also note that most authors who have worked on the subject have shown that interstitial diffusion between graphene planes is energetically favorable when compared to diffusion through the planes. Moreover, vacancies are less mobile than interstitials in graphite, because the first vacancy clusters are observed from about 650°C. Although the formation energy of the vacancies is well known, the migration energy value was re-assessed between 1980 and 2005 on the basis of labeling studies, Raman spectrometry and initial calculations. From about 650°C, the electrons of the dangling bonds at the edge of the vacancies will be able to associate, in order to create new bonds in the graphene plane, favoring the existence of bi-vacancies. The migration energy of the latter is high, so this type of defect is therefore not very mobile. However, it can be the site of nucleation of clusters of vacancies, and thus of dislocations. Ultimately, the simulations predict that bi-vacancies can exist through the association of two vacancies that are present on two different graphene planes. Two configurations are energetically possible (see Figure 4.5).

	Energy (eV)
Formation energy of a Frenkel pair	13.7
Formation energy of a metastable Frenkel pair	10.8
Formation energy of an interstitial	5.5
Migration energy of an interstitial	1.0
Formation energy of a vacancy	7.5
Migration energy of a vacancy	1.7
Formation energy of a bi-vacancy in the basic plane	8.7
Migration energy of a bi-vacancy in the basic plane	7.0
Formation energy of a bi-vacancy between two planes, in a "first inter-planar neighbor" configuration	14.6
Formation energy of a bi-vacancy between two planes, in a "second inter-planar neighbor" configuration	13.0

Table 4.4. *Summary of recent theoretical data on the formation and migration of point defects in graphite (simplified table from Telling and Heggie (2007))*

The generation of vacancies in the graphene planes thus changes the environment of the atoms. The new bonds created by the stabilization of the dangling bonds can induce new carbon hybridizations. In all cases, the local geometry is changed. Niwase (2005) mentions the existence of pentagonal, heptagonal and even octagonal rings. Recent molecular dynamics calculations show that the hybridization of the involved carbons can also evolve from sp^2 hybridization (planar geometry) to sp^3 hybridization (tetrahedral), sp (linear) or even to an isolated carbon, whose electronic procession is not involved in any bonding (Chartier et al. 2015). Using density functional theory (DFT), calculations highlight that the most energy-efficient carbon interstitial is a spiro-positioned carbon, hybridized sp^3, whose bonds create a bridge between two graphene planes (Li et al. 2005; Telling and Heggie 2007). Finally, other defects can be generated, such as the Stone-Wales defect, a 90° rotation of a pair of carbon atoms that transforms four hexagons into two pentagons and two heptagons (Stone and Wales 1986). These defects (changes in carbon hybridization, bridges between planes, etc.) deform the graphene planes, bending or even folding them.

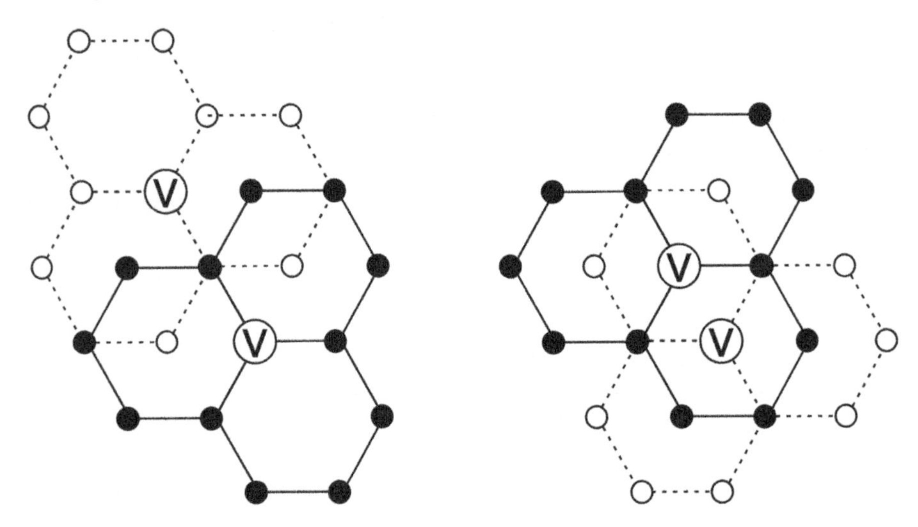

Figure 4.5. *Bi-vacancies between two planes. On the left, the slightly more energy-efficient "second inter-planar neighbor" configuration than the "first inter-planar neighbor" configuration (on the right) (from Telling and Heggie (2007))*

Defect production due to irradiation in graphite increases the internal energy of the lattice. This H_T increase, known as the Wigner energy, can be calculated using the relations [4.1] from Telling and Heggie (2007) or [4.2] from Gallego and Burchell (2011):

$$H_T = \sum_j N_j E f_j \quad [4.1]$$

$$H_T = \Delta H_i - \Delta H_{ni} \quad [4.2]$$

where N_j is the number of j-type defects, Ef_j is the value of the formation energy of defect j, and ΔH_i and ΔH_{ni}, are, respectively, the heats of combustion of an irradiated graphite and a similar but non-irradiated graphite.

The energy stored in irradiated graphite and its release have been the subject of numerous studies since the 1950s. Thus, the majority of experimental data was obtained between 1950 and 1960. Reference works exist on the subject, both old (Nightingale 1962; Simmons 1965b; Rappeneau et al. 1966) and recent (Burchell 1999; Bonal and Gosmain 2006; Bonal and Robin 2006; Telling and Heggie 2007), as well as very recent (Gallego and Burchell 2011; Marsden et al. 2020).

For a low irradiation temperature (typically below 115°C), and for a given neutron flux, the increase in H_T as a function of fluence is initially linear, before tending towards a saturation plateau (i.e. a horizontal asymptote) for high fluences. In reality, it is the differential enthalpy $\frac{dH_T}{d\theta}$ released during annealing at temperature θ that is a key parameter when defects are healing in graphite. Indeed, during a temperature increment, from θ_1 to θ_2, and if the heat $Q_{absorbed}$ given to the graphite is lower than the heat released $Q_{released}$ during defect healing (i.e. the transformation of the internal energy H_T into thermal energy during defect annihilation), then the excess heat leads to a significant local and spontaneous temperature rise up to a value θ_F: this is the Wigner effect. This effect is described by the following relation [4.3]:

$$Q_{absorbed} - Q_{released} = \int_{\theta_1}^{\theta_2} C_p(\theta).d\theta - \int_{\theta_1}^{\theta_2} \frac{dH_T}{d\theta}.d\theta < 0 \qquad [4.3]$$

where $C_p(\theta)$ is the heat capacity of graphite at temperature θ.

Knowledge of the variation of $\frac{dH_T}{d\theta}$ as a function of temperature is therefore of major importance, as the release of the Wigner energy could have catastrophic consequences for graphite and the nuclear reactor. Such an evolution is plotted in Figure 4.6. In this example, equation [4.3] is verified between θ_D and θ_A, that is, between about 150 and 250°C, with a maximum around 200°C. The energy then released (corresponding to the grey area 1) is transmitted adiabatically to the graphite. The temperature of the graphite increases spontaneously up to the value θ_F (about 400°C), which verifies the condition: area 1 = area 2.

A second peak, less known, is observed around 1,400°C (Rappeneau et al. 1966). These spontaneous energy releases are related to the mobility of defects. Urita et al. (2005) have shown that the critical temperature for which the sp^3 carbon interstitials annihilate on the vacancies is around 200°C. Below this value, it is accepted that the vacancies are immobile, or nearly so, and the mobility of the interstitials is quite low (Campbell and Burchell 2020). The most recent hypothesis to explain the peak at 1,400°C is, so far, unproven. Telling and Heggie (2007) report that the Wigner energy release may be related to interstitial emission from the most stable prismatic loops (Telling and Heggie 2007).

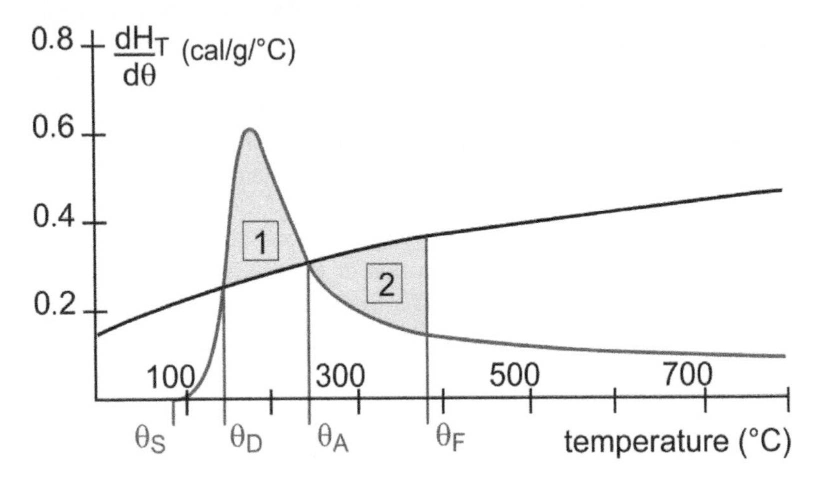

Figure 4.6. *Evolution of the differential enthalpy of an irradiated graphite (red curve) at 60°C, at a fluence of 1.74 10^{20} n.cm^{-2} (0.12 dpa), as a function of the annealing temperature, with θ_S being the temperature from which the stored energy starts to be released. θ_D and θ_A are the temperatures between which the stored energy is spontaneously released. The temperature θ_F corresponds to the final temperature of the graphite after release of the Wigner energy (Bonal and Robin 2006). The black curve represents the specific heat of non-irradiated graphite. Areas 1 and 2 correspond respectively to the Wigner energy released and absorbed by the graphite in an adiabatic way. For a color version of this figure, see www.iste.co.uk/bouffard/nuclear.zip*

In 1942, Eugene Paul Wigner realized (with Enrico Fermi) that the slowing down of neutrons by graphite could displace carbon atoms and that changes in the physical properties of the moderator had to be anticipated. Then, in charge of the theory group of the Metallurgical Laboratory of the University of Chicago, he proposed that this process could store energy in graphite. This energy could be released by heating the material (Wigner 1942; Burton and Neubert 1956). Wigner reaffirmed the importance of studying the effects of irradiation on matter in his introduction to the June 1946 American Physical Society conference in Chicago (Wigner 1946). Wigner (1902–1995) received the Nobel Prize in 1963 for his work on the theory of the atomic nucleus and elementary particles.

Box 4.1. *The Wigner effect*

Finally, Rapeneau has shown that for a spontaneous release of Wigner energy to occur, the irradiation temperature must be lower than 115°C and the neutron fluence must be higher than 1.6 10^{20} neutrons/cm². In order to avoid this risk, regular annealing operations of graphite, allowing the release of Wigner energy in a controlled way, have been carried out in graphite gas reactors with these characteristics. It was during one of these annealing operations, which was imperfectly carried out, that the first major accident in the nuclear field took place in Great Britain, at Windscale on October 10, 1957 (Arnold 1995).

4.3.3. *Evolution of lattice parameters and crystallite size in irradiated graphite*

Initially, the hypotheses explaining the evolution of the values of parameters a and c (see Table 4.3) were based on the creation of dislocations by aggregation of the same type of point defects. The presence of loops of interstitial dislocations, located between the graphene planes (see Figure 4.4), increases the value of parameter c. Their growth leads to the appearance of new graphene planes (Kelly 1977, 1978; Marsden et al. 2020). The decrease in the value of parameter a, on the contrary, is less well understood. However, Kelly (1965) proposed that the major contribution to the contraction of the basal planes is related to the modification of the inter-atomic bonds of the carbons in the vicinity of the vacancies. This hypothesis may find a link with the mechanism of dislocation formation proposed by Niwase (2005). In this reference, he proposes that a vacancy dislocation line leads to the reorganization of the crystal lattice in the form of deformed hexagons, a pentagon, a heptagon and a deformed ring of eight carbons.

While it is accepted by the scientific community that these proposals explain the evolution of the values of a and c for temperatures above 250°C, the dislocations formed in the basal planes are, on the contrary, "blocked" by the spiro-carbon at low temperatures. In particular, the improved resolution of transmission electron microscopes allowed Muto and Tanabe (1997) to: (i) reject the additional graphene plane hypothesis for room temperature irradiations; (ii) show that irradiation induces inter-plane bridges, as well as graphene plane breaks and bends (see the skeletons of the TEM images in Figure 4.7 from c to e) and (iii) attribute the apparent increase in the value of parameter c to these previous observations (Tanabe et al. 1992; Muto and Tanabe 1997). However, loops of interstitials at room temperature have been observed with superb TEM images (Karthik et al. 2011). That being said, they are too few and too unstable to explain the increase in parameter c. Heggie et al. (2011), proposed a mechanism based on graphene planes that would fold back on themselves. The seed of these folds would be the spiro-carbon which, below 250°C, would prevent the graphene planes from sliding over one another. The latter, under the effect of local stresses, would thus fold on themselves, causing an apparent increase in parameter *c*.

The breaking of graphene planes was observed by transmission electron microscopy (see Figure 4.7a) in samples implanted by carbon ions in Silbermann's (2013) thesis. Area 1 in Figure 4.7a shows the loss of crystal in a sample irradiated with carbon ions at 15°C. Ion beam irradiation has been used to simulate neutron irradiation in the laboratory. Galy et al. (2018) showed that the evolution of the graphite microstructure under ion beam was directly related to the ballistic damage of graphite, by nuclear interactions between the incident ions and the target nuclei, giving credence to this scientific approach.

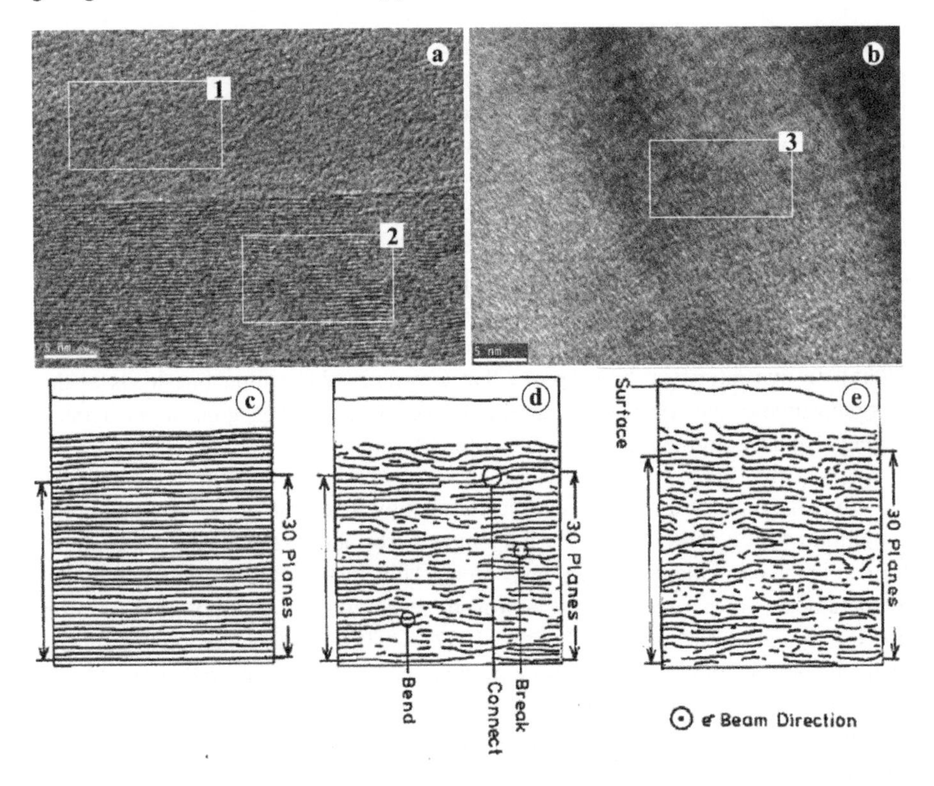

Figure 4.7. *TEM images of graphite irradiated with ^{13}C ions at 15°C (a, photo credit: J.N. Rouzaud) and 600°C (b), with the IMIO400 implanter of the Institut de Physique des Deux Infinis in Lyon (Silbermann 2013), to be compared with the skeletons of base planes for non-irradiated graphite (c) and electron-irradiated graphite (d and e) for increasing fluences. Diagrams according to Tanabe et al. (1992)*

The comparison with Figure 4.7c–e will help the reader to observe the TEM images. Figure 4.7c shows a non-irradiated graphite (similar to Figure 4.2b). In Figure 4.7d, the author shows how the plane breaks (break in the figure) appear on a TEM snapshot: a "black line" showing the rest of the broken plane and a "white line" showing an absence of a plane (Tanabe et al. 1992). In area 1, there is only alternation between black and white lines in different directions: only randomly oriented plane fragments remain. The diameter of the crystallites, estimated from this photograph, is $L_A < 2$ nm. On the same sample, area 2 corresponds to a less unstructured area, because the damage created is lower than in area 1. The irradiation dose gradient between areas 1 and 2 is explained by the irradiation technique, known as ion implantation, which generates Gaussian defect profiles. In this area, the graphene planes are wavy, sometimes broken up, but oriented globally in the same direction. The size of the L_A crystallites is about 2 nm. The lamellar structure is preserved.

Area 3, taken from a region equivalent to area 1, shows that the lamellar structure of graphite is preserved at 600°C, even though it is partially disordered. The crystallite size is about 2 nm for L_A. At 600°C, the migration of defects allows the partial recovery of the damage created by irradiation. Different authors have indeed shown experimentally that the degree of recovery of crystalline in graphite strongly depends on the initial disorder state and that, even with an annealing of 1,600°C, there is not a full recovery (Silbermann et al. 2014; Toulhoat et al. 2015; Galy 2016; Galy et al. 2018).

These results are consistent with previously mentioned observations on the decrease of the L_a radius of nuclear graphite crystallites irradiated by neutrons in UNGG reactors (Gosmain et al. 2010). The role of temperature, illustrated in this chapter by TEM results, could also be studied by Raman spectroscopy (e.g. Blondel et al. (2014) and Blondel (2015)).

Ultimately, the heterogeneity observed in the graphite nanostructure can be described as a set of lamellar area, separated by nanoscale pores.

4.3.4. *Density and porosity evolution by radiolytic corrosion of graphite*

In UNGG nuclear power plants, the radiolysis of carbon dioxide, which is the coolant, creates new radical, ionic or molecular species. Radical species with a strong oxidizing power will oxidize graphite and induce radiolytic corrosion of graphite (Wickham et al. 1977; Faircloth 1980; Norfolk et al. 1983). Overall, the chemical transformations can be summarized by the following equation (Silbermann

2013), the latter of which cannot be balanced because the elemental composition of the carboxidized solids is poorly known:

$$C_{graphite} + CO_2 \xrightarrow{irradiation} CO_2, CO, carboxidized\ solids$$

This equation expresses the fact that the solid transforms into carbon monoxide and dioxide gases, as well as carbonaceous deposits appearing on metal surfaces and on the graphite itself (Baird 1980; Blondel 2015).

Radiolytic corrosion occurs on all surfaces, especially on the surface of graphite pores. In this case, radical species can interact with the pore surface before they have time to recombine (Wood 1980). The diameter of the pores then increases (Standring and Ashton 1965; Wickham et al. 1977; Faircloth 1980; Comte et al. 2020). This also explains the decrease in the bulk density of the solid under reactor irradiation.

4.3.5. *What are the consequences at the macroscopic scale?*

These structural changes of nuclear graphite under irradiation are at the origin of the evolutions of its properties, which had to be taken into account during the design and operation of the reactors (Petit et al. 1991; Petit and Brié 1995):

– *Dimensional variations*: the variations of the lattice parameters observed on the graphite crystallites after irradiation are, in the case of the UNGG reactors, translated by dimensional variations of the core. The graphites used were, in fact, produced from needle cokes and shaped by extrusion, favoring a preferential orientation of the crystallites and thus conferring a strongly anisotropic nature to the final material[8]. After 12 effective full power years (efpy), the graphite core of the Bugey 1 reactor showed a shrinkage of up to 70 mm in some places for a total height of 10.20 m (7‰). In the radial direction (total core diameter of 15 m), almost no dimensional variation was found. These dimensional variations were taken into account from the design of these reactors, using a method of construction and assembly of the graphite bricks in the core that were sufficiently loose to tolerate this type of deformation (assembly by a system of tenons and mortises).

– *Loss of mass and drop in mechanical properties*: this phenomenon, which can be described as graphite wear, is the main issue encountered in the operation of

8. For example, for the graphite in the core of the Bugey 1 reactor, the compressive strength and elastic modulus measured along the axis that is parallel to the extrusion axis were 42 MPa and 11 GPa, respectively, compared to 38 MPa and 7.5 GPa in the direction that is perpendicular to the extrusion axis.

first-generation graphite-gas reactors and is what has led to their shutdown. Radiolytic corrosion leads to a significant loss of graphite mass during operation. For Bugey 1, some samples taken from the center of the graphite core showed a loss of mass that could reach 30–40% of the initial mass after 12 effective full power operation years. This loss of mass was coupled by a drastic drop in compressive strength to values of about 15 MPa, stretching to an initial value of about 42 MPa in extreme cases. In order to limit this phenomenon of wear (oxidation) of graphite by radiolytic corrosion, methane was injected into the coolant of certain reactors, up to 500 ppm vol. The methane's role was to react (consume) with the oxidizing species formed by the radiolysis of CO_2, thus preventing them from reacting with the graphite. However, the use of methane injection had to be limited because, as a result of its beneficial effect on the protection of the graphite, it also led to the formation of hydrocarbon deposits on the nuclear fuel cladding, disrupting the heat exchange and the neutron flux.

4.4. Conclusion

From the industrial point of view, for French nuclear graphite, the time has come to choose a strategy for dismantling UNGG power plants, as well as for the management of future irradiated graphite waste that will mostly come from the dismantling of the core of these reactors.

Unlike the current fleet of nuclear reactors (PWRs), for which dismantling could benefit from a series effect and a modular design, the dismantling of UNGG reactors is particularly complex. Each UNGG reactor can be considered a unique prototype, whose dismantling is made even more difficult by an internal design that is not adapted to dismantling constraints (small reactor vessel, difficult accessibility, multiple materials, etc.). EDF, which favors dry dismantling of its UNGG reactors, has set the goal of dismantling the Chinon A2 reactor as a pilot workshop by 2030. Engineering studies for this dismantling are underway and will be specifically based on an industrial demonstrator, which will make it possible to test the various processes and technologies that will be used on a 1:1 scale, under conventional conditions.

Regarding the long-term management of future irradiated graphite waste, there is currently no dedicated storage solution in France. According to French regulations, this waste is classified as long-lived low-level waste (LLW-LL), because of the presence of two long-lived radionuclides, called chlorine 36 and carbon 14 (ANDRA-a). Joint studies conducted by waste producers (EDF, CEA) and ANDRA between 2006 and 2014 have refined the knowledge of irradiated graphite, its radiological inventory, as well as its behavior in storage conditions (Le Guillou et al.

2014; Moncoffre et al. 2016; Comte et al. 2020). Alternative solutions to storage (treatment to extract certain radionuclides) have also been studied (Silbermann 2013; Pageot 2014). The reference scenario retained for the management of graphite waste is direct disposal without treatment. ANDRA is actively working on the definition and search for a storage site that is adapted to this type of waste (FA-VL) (ANDRA-b). The choice of the final solution will be made by the legislator, after examination of a technical dossier by the French Nuclear Safety Authority (*Autorité de Sûreté Nucléaire* – ASN).

From a scientific point of view, one of the key questions that remains to be addressed is the mechanism that is at the origin of the contraction of the *a*-parameter and the increase in the *c*-parameter of nuclear graphite, as well as the link that these changes have with the macroscopic dimensional variations of graphite. It is therefore necessary to obtain new data on the subject, whether it be experimentally or theoretically (DFT, molecular dynamics, etc.). Theoretical calculations are currently very popular, as they make it possible to test scenarios involving clusters composed of a very small number of defects or atoms (typically 2, 3 or 4). These dimensions are currently inaccessible through experimental techniques. Finally, at the beginning of this chapter, nuclear graphite has been described as a carbon–carbon composite material. **After irradiation in a nuclear reactor, graphite becomes a heterogeneous carbon–carbon composite, that is structurally and nanostructurally degraded**. Rouzaud compares monocrystalline graphite to a stack of paper sheets and irradiated nuclear graphite to crumpled paper sheets. Those who have ever lit a fire have noticed that crumpled paper ignites more easily and burns more fully than stacked paper, such as a non-crumpled newspaper, for example. From this comparison, it can be understood that irradiated nuclear graphite is chemically more reactive than single-crystal graphite, because the specific surface area of irradiated graphite is greater than that of non-irradiated graphite.

In conclusion, we would like to repeat this old ironic joke that Kelly (1982) used to begin his article on graphite: "It is an old and wry joke amongst people concerned with the design of graphite moderated gas-cooled reactors that **'the only property which does not change when it is exposed to neutron irradiation is its color'**, and some have added that **'that, appropriately, is black'**."

4.5. Acknowledgements

Research on graphite as a material of nuclear interest began more than 75 years ago. The authors would like to thank Sylvie Flores, from the IP2I documentation department, for her invaluable help in obtaining some of the oldest references, and also for the speed with which she was able to obtain the most recent ones.

4.6. References

Amelinckx, S. and Delavignette, P. (1960). Dislocation loops due to quenched-in point defects in graphite. *Physical Review Letters*, 5(2), 50–53.

Ammar, M.R., Galy, N., Rouzaud, J.N., Toulhoat, N., Vaudey, C.E., Simon, P., Moncoffre, N. (2015). Characterizing various types of defects in nuclear graphite using Raman scattering: Heat treatment, ion irradiation and polishing. *Carbon*, 95, 364–373.

ANDRA (2021a). Catégorie de déchets radioactifs [Online]. Available at: https://www.andra.fr/index.php/les-dechets-radioactifs/tout-comprendre-sur-la-radioactivite/classification [Accessed 6 May 2021].

ANDRA (2021b). Etudier des solutions de gestion pour les déchets de faible activité à vie longue [Online]. Available at: https://www.andra.fr/les-dechets-radioactifs/les-solutions-de-gestion/etudier-des-solutions-de-gestion-pour-les-dechets [Accessed 6 May 2021].

Arnold, L. (1995). *Windscale 1957: Anatomy of a Nuclear Accident*, 2nd edition. Palgrave Macmillan, London.

Baird, T. (1980). Carbon deposition on metals. *Proceeding 4 of Salford*, Salford.

Banhart, F. (1999). Irradiation effects in carbon nanostructures. *Report on Progress in Physics*, 62(8), 1181–1221.

Black, G., Marsden, B.J., Wright, G., Jones, A.N. (2016). Origin and validity of graphite dosimetry units and related conversion factors. *Annals of Nuclear Energy*, 94, 241–250.

Blondel, A. (2015). Effets de la température et de l'irradiation sur le comportement du chlore-37 dans le graphite nucléaire : conséquence sur la mobilité du chlore-36 dans les graphites irradiés. Thesis, Université Claude Bernard Lyon 1.

Blondel, A., Moncoffre, N., Toulhoat, N., Bérerd, N., Silbermann, G., Sainsot, P., Rouzaud, J.N., Deldique, D. (2014). New advances on the thermal behaviour of chlorine in nuclear graphite. *Carbon*, 73, 413–420.

Bonal, J.P. and Gosmain, L. (2006). Les graphites pour les applications nucléaires. *L'actualité chimique*, 295/296, 23–27.

Bonal, J.P. and Robin, J.C. (2006). Un matériau fascinant : le graphite. Les réacteurs nucléaires à caloporteur gaz. Monographie de la Direction de l'énergie nucléaire (CEA), 27–32.

Burchell, T.D. (1999). Fission reactor application of carbon. In *Carbon Materials for Advanced Technologies*, Burchell, T.D. (ed.). Pergamon Press, Oxford.

Burton, M. and Neubert, T.J. (1956). Effect of fast neutron bombardment on the physical properties of graphite: A review of early work at the Metallurgy Laboratory. *Journal of Applied Physics*, 27(6), 557–567.

Campbell, A.A. and Burchell, T.D. (2020). Radiation effects in graphite. In *Comprehensive Nuclear Materials*, 2nd edition, Konings, R.J.M. and Stoller, R.E. (eds). Elsevier, Oxford.

CEA (2006). *Les réacteurs nucléaires à caloporteur gaz*. Editions Le Moniteur – Monographie de la Direction de l'énergie nucléaire (CEA).

Chartier, A., Van Brutzel, L., Pannier, B., Baranel, P. (2015). Atomic scale mechanisms for the amorphisation of irradiated graphite. *Carbon*, 91, 395–407.

Choi, W., Kim, C., Kang, H. (1993). Interactions of low energy (lo-600 eV) noble gas ions with a graphite surface: Surface penetration, trapping and self-sputtering behaviors. *Surface Science*, 281, 323–335.

Comte, J., Guy, C., Gosmain, L., Parraud, S. (2020). Determining the porosity and water impregnation in irradiated graphite. *Journal of Nuclear Materials*, 528, 151816.

Cornuault, P. (1981). Modérateurs, graphite. *Techniques de l'Ingénieur – Génie nucléaire*, B3680, 1–16.

DOE (2002). A technology roadmap for generation IV nuclear energy system. U.S. DOE Nuclear Energy Research Advisory Commitee and the Generation IV International Forum.

Egerton, R.F. (1977). The threshold energy for electron irradiation damage in single crystal. *Philosophical Magazine*, 35(5), 1425–1428.

Engle, G.B. and Kelly, B.T. (1984). Radiation damamge of graphite in fission and fusion reactor systems. *Journal of Nuclear Materials*, 122–123, 122–129.

Faircloth, R.L. (1980). Coolant chemistry of the advanced carbon dioxide cooled reactor. *Specialists Meeting on Coolant Chemistry, Plate-out and Decontamination in Gas-cooled Reactors*, Jülich.

Gallego, N.C. and Burchell, T.D. (2011). A review of stored energy release of irradiated graphite. Report, ORNL/TM-2011/378, Oak Ridge National Laboratory.

Galy, N. (2016). Comportement du ^{14}C dans le graphite nucléaire : effets de l'irradiation et décontamination par vaporéformage. Thesis, Université Claude Bernard Lyon 1.

Galy, N., Toulhoat, N., Moncoffre, N., Pipon, Y., Bérerd, N., Ammar, M.R., Simon, P., Deldique, D., Sainsot, P. (2018). Ion irradiation used as surrogate of neutron irradiation in graphite: Consequences on 14C and 36Cl behavior and structural evolution. *Journal of Nuclear Materials*, 502, 20–29.

Gosmain, L., Comte, J., Ammar, M.R. (2010). Synthèse des caractérisations microstructurales réalisées sur les graphites vierges et irradiés des réacteurs G2 et SLA2. CEA report and technical document.

Heggie, M.I., Suarez-Martinez, I., Davidson, C., Haffenden, G. (2011). Buckle, ruck and tuck: A proposed new model for the response of graphite. *Journal of Nuclear Materials*, 413, 150–155.

Hehr, B.D., Hawari, A.I., Gilette, V.H. (2007). Molecular dynamics simulations of graphite at high temperatures. *Nuclear Technology*, 160(2), 251–256.

Heintz, E.A. (1985). Influence of coke structure on the properties of the carbon graphite artifact. *Fuel*, 64(9), 1192–1196.

IAEA (2006). *Characterization, Treatment and Conditioning of Radioactive Graphite from Decommissioning of Nuclear Reactors*. International Atomic Energy Agency, Vienna.

Iwata, T. and Nihira, T. (1966). Atomic displacements in pyrolitic graphite by electron bombardment. *Physics Letters*, 23, 631–632.

Jones, A.N., Hall, G.N., Joyce, M., Hodgkins, A., Wen, K., Marrow, T.J., Marsden, B.J. (2008). Microstructural characterisation of nuclear grade graphite. *Journal of Nuclear Materials*, 381, 152–157.

Karthik, C., Kane, J., Butt, D.P., Windes, W.E., Ubic, R. (2011). In situ transmission electron microscopy of electron-beam induced damage process in nuclear grade graphite. *Journal of Nuclear Materials*, 412, 321–326.

Kelly, B.T. (1965). Basal plane contraction in graphite due to mono-vacancies. *Nature*, 207, 257–259.

Kelly, B.T. (1977). The theory of irradiation damage in graphite. *Carbon*, 15, 117–127.

Kelly, B.T. (1978). Radiation damage in graphite and its relevance to reactor design. *Progress in Nuclear Energy*, 2, 219–269.

Kelly, B.T. (1982). Graphite – The most fascinating nuclear material. *Carbon*, 20(1), 3–11.

Le Guillou, M., Touhoat, N., Pipon, Y., Moncoffre, N., Bérerd, N., Perrat-Mabilon, A., Rapegno, R. (2014). Thermal behavior of deuterium implanted into nuclear graphite studied by NRA. *Nuclear Instruments and Methods in Physics Research B*, 332, 90–94.

Le Guillou, M., Rouzaud, J.N., Deldique, D., Toulhoat, N., Pipon, Y., Moncoffre, N. (2015). Structural and nanostructural behavior of deuterium implanted Highly Ordered Pyrolytic Graphite investigated by combined High Resolution Transmission Electron Microscopy, Scanning Electron Microscopy and Raman microspectrometry. *Carbon*, 94, 277–284.

Li, L., Reich, S., Robertson, J. (2005). Defect energies of graphite: Density-functional calculations. *Physical Review B*, 72, 184109.

Lucas, M.W. and Mitchell, E.W.J. (1964). The threshold curve for the displacement of atoms in graphite: Experiments on the resistivity changes produced in single crystals by fast electron irradiation at 15°K. *Carbon*, 1, 401–402.

Marsden, B.J., Hall, G.N., Jones, A.N. (2020). Graphite in gas-cooled reactors. In *Comprehensive Nuclear Materials*, 2nd edition, Konings, R.J.M. and Stoller, R.E. (eds). Elsevier, Oxford.

McKenna, A.J., Trevethan, T., Latham, C.D., Young, P.J., Heggie, M.I. (2016). Threshold displacement energy and damage function in graphite from molecular dynamics. *Carbon*, 99, 71–78.

Moncoffre, N., Toulhoat, N., Bérerd, N., Pipon, Y., Silbermann, G., Blondel, A., Galy, N., Sainsot, P., Rouzaud, J.N., Deldique, D. et al. (2016). Impact of radiolysis and radiolytic corrosion on the release of ^{13}C and ^{37}Cl implanted into nuclear graphite: Consequences for the behaviour of ^{14}C and ^{36}Cl in gas cooled graphite moderated reactors. *Journal of Nuclear Materials*, 472, 252–258.

Montet, G.L. (1967). Threshold energy for the displacement of atoms in graphite. *Carbon*, 5, 19–23.

Montet, G.L. and Myers, G.E. (1971). Threshold energy for displacement of surface atoms in graphite. *Carbon*, 9, 170–183.

Mrozowski, S. (1954). Mechanical strength, thermal expansion and structure of cokes and carbons. *Proceedings of the Conferences on Carbon*, 31–45.

Muto, S. and Tanabe, T. (1997). Damage process in electron-irradiated graphite studied by transmission electron microscopy I: High-resolution observation of highly graphitized carbon fibre. *Phylosophical Magazine A*, 76(3), 679–690.

Nakai, K., Kinoshita, C., Matsunaga, A. (1991). A study of amorphization and microstructural evolution of graphite under electron or ion irradiation. *Ultramicroscopy*, 39, 361–368.

Nightingale, R.E. (ed.) (1962). Stored energy. In *Nuclear Graphite*. Academic Press, Cambridge, MA.

Niwase, K. (2005). Formation of dislocation dipoles in irradiated graphite. *Materials Science and Engineering A*, 400–401, 101–104.

Norfolk, D.J., Skinner, R.F., Williams, W.J. (1983). Hydrocarbon chemistry in irradiated $CO_2/CO/CH_4/H_2O/H_2$ mixtures: A survey of the initial reactions. *Radiation Physics and Chemistry*, 21(3), 307–319.

OCDE (2014). Technology roadmap update for generation IV nuclear energy systems. OECD Nuclear Energy Agency for the Generation IV International Forum.

Ohr, S.M., Wolfenden, A., Noggle, T.S. (1972). Electron displacement damage in graphite and aluminium. In *Electron Microscopy and Structure of Materials*, Thomas, G., Fulrath, R.M., Fisher, R.M. (eds). University of California Press.

Pageot, J. (2014). Etude d'un procédé de décontamination du ^{14}C des déchets de graphite nucléaire par carboxy-gazéification. Thesis, Université Paris Sud.

Pageot, J., Rouzaud, J.N., Ali Ahmad, M., Deldique, D., Gadiou, R., Dentzer, J., Gosmain, L. (2015). Milled graphite as a pertinent analogue of French UNGG reactor graphite waste for a CO2 gasification based treatment. *Carbon*, 86, 174–187.

Petit, A. and Brié, M. (1995). Graphite stack corrosion of the Bugey 1 reactor. Specialists' meeting on graphite moderator lifecycle behavior, Bath. IAEA INIS RN:28008815.

Petit, A., Phalippou, C., Brié, M. (1991). Radiolytic corrosion of graphite surveillance and lessons drawn from the operation of Bugey 1 reactor. Specialists' meeting on the status of graphite development for gas cooled reactors, Tokai Ibaraki. IAEA INIS RN:24042595.

Pierson, H.O. (1994). *Handbook of Carbon, Graphite, Diamond, and Fullerenes: Properties*. Noyes Publications, Park Ridge, NJ.

Rappeneau, J., Taupin, J.L., Grehier, J. (1966). Energy release at high temperature by irradiated graphite. *Carbon*, 4, 115–124.

Silbermann, G. (2013). Effets de la température et de l'irradiation sur le comportement du ^{14}C et de son précurseur ^{14}N dans le graphite nucléaire. Etude de la décontamination thermique du graphite en présence de vapeur d'eau. PhD Thesis, Université Claude Bernard Lyon 1.

Silbermann, G., Moncoffre, N., Toulhoat, N., Bérerd, N., Perrat-Mabilon, A., Laurent, G., Raimbault, L., Sainsot, P., Rouzaud, J.N., Deldique, D. (2014). Temperature effects on the behavior of carbon 14 in nuclear graphite. *Nuclear Instruments and Methods in Physics Research B*, 332, 106–110.

Simmons, J.H.W. (1965a). *Radiation Damage in Graphite*. Pergamon Press, Oxford.

Simmons, J.H.W. (ed.) (1965b). Stored energy and annealing effects in reactor graphite. In *Radiation Damage in Graphite*. Pergamon Press, Oxford.

Smith, R. and Beardmore, K. (1996). Molecular dynamics studies of particle impacts with carbon-based materials. *Thin Solid Films*, 272, 255–270.

Standring, J. and Ashton, B.W. (1965). The effect of radiolytic oxidation by carbon dioxide on the porosity of graphite. *Carbon*, 3(2), 157–165.

Steffen, H.J., Marton, D., Rabalais, J.W. (1992a). Defect formation in graphite during low energy Ne bombardment. *Nuclear Instruments and Methods in Physics Research B*, 67, 308–311.

Steffen, H.J., Marton, D., Rabalais, J.W. (1992b). Displacement energy threshold for Ne+ irradiation of graphite. *Physical Review Letters*, 68(11), 1727–1729.

Stone, A.J. and Wales, D.J. (1986). Theoretical studies of icosahedral C60 and some related species. *Chemical Physics Letters*, 128, 501–503.

Tanabe, T., Muto, S., Niwase, K. (1992). On the mechanism of dimensional change of neutron irradiated graphite. *Applied Physic Letters*, 61(14), 1638–1640.

Telling, R.H. and Heggie, M.I. (2007). Radiation defects in graphite. *Philosophical Magazine*, 87(31), 4797–4846.

Toulhoat, N., Moncoffre, N., Bérerd, N., Pipon, Y., Blondel, A., Galy, N., Sainsot, P., Rouzaud, J.N., Deldique, D. (2015). Ion irradiation of ^{37}Cl implanted nuclear graphite: Effect of the energy deposition on the chlorine behavior and consequences for the mobility of ^{36}Cl in irradiated graphite. *Journal of Nuclear Materials*, 464, 405–410.

Urita, K., Suenaga, K., Sugai, T., Shinohara, H., Iijima, S. (2005). In situ observation of thermal relaxation of interstitial-vacancy pair defects in a graphite gap. *Physical Review Letters*, 155502.155501–155502.155504.

Vaudey, C.E., Toulhoat, N., Moncoffre, N., Bérerd, N., Raimbault, L., Sainsot, P., Rouzaud, J.N., Perrat-Mabilon, A. (2009). Thermal behavior of chlorine in nuclear graphite at a microscopic scale. *Journal of Nuclear Materials*, 395, 62–68.

Wickham, A.J., Best, J.V., Wood, C.J. (1977). Recent advances in the theories of carbon dioxide radiolysis and radiolytic graphite corrosion. *Radiation Physics and Chemistry*, 10(2), 107–117.

Wigner, E.P. (1942). U.S. Atomic Energy Report CP-387 (December 15), 4.

Wigner, E.P. (1946). Theoritical physics in the metallurgical laboratory of Chicago. *Journal of Applied Physics*, 17(11), 857–863.

Wood, C.J. (1980). Coolant chemistry in Cegb reactors. *Proceeding 1 of Salford*, Salford.

5

Nuclear Glasses

Magaly TRIBET
CEA, Centre de Marcoule, DES/ISEC/DE2D/SEVT/LMPA, Bagnols-sur-Cèze, France

Glass is a material that is largely part of our daily lives. For example, the first things that come to mind are window glass, stained glass windows in cathedrals, as well as the dishes we use every day (Pyrex glass) or for special occasions (lead-rich glass, known as "crystal" glass). This amorphous material first existed in the countryside, where it was produced by the rapid cooling of lava in volcanic regions. Then, since antiquity, it has been developed by humans from sands brought to high temperature (Richet 2000). Therefore, why was this material, which was seemingly common at first sight, chosen in the 20th century for an application in the nuclear field, within the framework of the radioactive waste confinement? And what impact can irradiation have on glass and its confinement properties?

5.1. Glass of nuclear interest: their role and their aging conditions under irradiation

5.1.1. *What is this kind of glass for?*

The glasses of nuclear interest – or nuclear glasses – are used to primarily manage two kinds of ultimate radioactive waste (Weber et al. 1997; Gin et al. 2017):

– Waste from the nuclear power industry and the reprocessing of spent fuel. In this case, the fuel is reprocessed to extract the recoverable elements (U and Pu),

Nuclear Materials under Irradiation,
coordinated by Serge BOUFFARD and Nathalie MONCOFFRE. © ISTE Ltd 2023.

while the ultimate waste is vitrified. For this application, we will speak hereafter of "nuclear glass for fission products and minor actinides confinement" (FPA).

– Waste of military origin, notably developed in the United States, such as weapon-grade plutonium.

Globally speaking, these nuclear glasses contain third-category radioactive waste according to the classification of ANDRA, which sorts radioactive waste according to their radiological activity and their half-life. This category includes long-lived intermediate-level waste and high-level waste (ANDRA). In France, long-lived intermediate-level waste represents 4.9% radioactivity for 3% of the volume of waste produced, while high-level waste represents 94.9% radioactivity for approximately 0.2% of the volume of radioactive waste produced (Débat Public 2020). This ultimate waste, which contains long-lived radionuclides, initially in solution or in the form of sludge, must be managed over periods of several hundred thousand years, hence the choice of some countries to stabilize it in a glass matrix (IAEA 2018). In this case, the resulting glass package is destined for deep geological disposal (see section 5.1.4).

5.1.2. *What does this glass actually contain and in what form?*

We will focus on the example of FPA conditioning glass. The composition of the fission product solution to be vitrified depends on factors such as the type of the reprocessed fuel, its history in the nuclear reactor and the reprocessing chemistry. The waste contains both long- and short-lived fission products, minor actinides (americium, neptunium and curium), a small fraction of unrecovered uranium and plutonium, residues or degradation products of the chemical compounds used during reprocessing and corrosion products of metals in contact with these reprocessing solutions. Thus, more than 30 chemical elements belonging to all the groups of the periodic table must be incorporated into the nuclear glass. **The composition of the glass must therefore be adapted to the composition of the waste to be immobilized, as well as to the proportion in which it is incorporated into the glass (i.e. the loading rate).** This compromise (see Figure 5.1) also considers the technological feasibility, the physico-chemical properties of the glass in the liquid and solid state, and its long-term confinement properties under irradiation (Gin et al. 2017).

The variety of wastes to be vitrified and the reprocessing/vitrification processes implemented on an industrial scale in different countries (France, UK, USA, Russia, Germany, Japan, India) lead to the existence of various compositions of nuclear glass around the world (Ojovan and Lee 2011; Gin et al. 2013; Kaushik 2014; IAEA 2018). Nevertheless, they can be classified into two main categories: the aluminophosphate glass used in Russia, and the borosilicate glass, which is universally selected in other countries because it gathers the properties described above (see Figure 5.1), allowing it to be produced on an industrial scale and in shielded cells due to the high radioactivity of the waste. We will focus on the latter in the rest of this chapter.

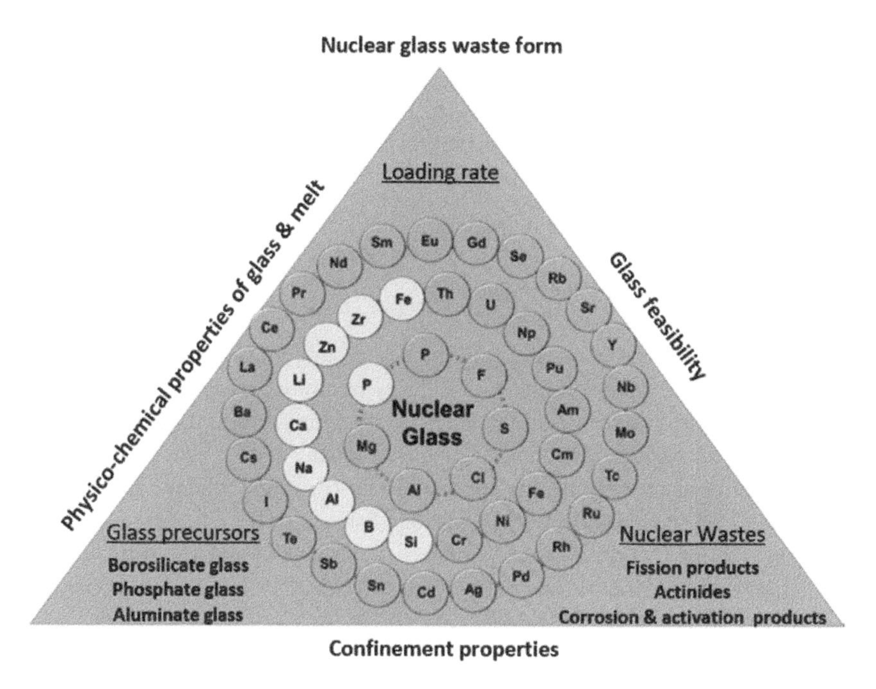

Figure 5.1. *Main chemical elements contained in a typical nuclear glass. Factors for its optimization (Gin et al. 2017). For a color version of this figure, see www.iste.co.uk/bouffard/nuclear.zip*

Finally, it should be pointed out that nuclear glass is not a coating matrix, like cements or bitumens, but that the **incorporated waste is an integral part of the glass structure**. Indeed, because of its amorphous nature, the glass is able to incorporate, in its structure, a great diversity of elements present in the waste. The incorporated atoms therefore form chemical bonds with the main constituents of the glass (see Figure 5.2).

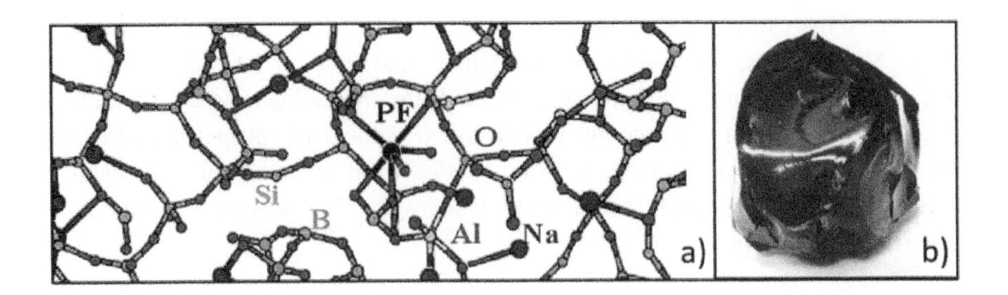

Figure 5.2. *(a) Principle of fission products (FP) incorporation in the glass matrix; (b) a block of borosilicate confinement glass, such as the French R7T7*[1] *glass. The glass appears black because it contains elements that absorb the light at variable wavelengths (Fe, Cr, Ni, Nd, etc.). For a color version of this figure, see www.iste.co.uk/bouffard/nuclear.zip*

5.1.3. *Nuclear glass radioactivity*

In the case of nuclear glass, there are numerous sources of irradiation. The glass is subjected to α decays resulting from the decay of minor actinides (or plutonium in the case of military-origin waste), to β decays resulting from the decay of fission products, as well as to γ transitions which accompany these decays. It should be noted that during α decays, there is emission of an α particle of about 5 MeV and a recoil nucleus of about 100 keV. Spontaneous fissions or (α, n)-type reactions can also occur and lead to the formation of fission fragments and neutrons, but their contribution to the total damage can be neglected because of their low probabilities of occurrence (Weber et al. 1997; Gin et al. 2017).

These different types of particles or radiation will lose their energy by interacting with the atoms of the glass through two types of processes (see Table 5.1): inelastic or electronic interactions (electronic excitations, ionizations) and elastic or nuclear interactions (ballistic collisions that can displace the atoms of the glass). These two types of interactions can lead to changes in the structure of the glass and, therefore, in the properties of the glass (see section 5.3). The β decays, the γ transitions and the helium nuclei from the α decays essentially interact electronically with the atoms of the glass network. The recoil nuclei from the α decays mostly lead to nuclear

1. French glass for conditioning FPA is produced at the Orano plant in La Hague (Normandy), more precisely in R7 and T7 facilities, hence the name R7T7.

interactions. This is also the case for the helium nuclei at the very end of its trajectory. They are both at the origin of the majority of the atomic displacements undergone by the glass during its aging (Mir and Peuget 2020).

In terms of the creation of defects, the γ rays can be assimilated to electrons. Indeed, the γ rays, with an energy of the order of MeV, interact with the matter by Compton diffusion, that is, by emission of an electron which will have a behavior identical to β in the matter.

Type of decay or radiation		Target		Depth into the glass
		Electrons	Nuclei	
α decay	α particle of 5 MeV	99.8% of the absorbed energy is 4.990 keV on 20 μm	0.2% of the absorbed energy and 200 atoms ejected from their site	~ 20 μm
	100 keV recoil nucleus	37% of the absorbed energy is 37 keV on 40 nm	63% of the energy absorbed and 1,500 atoms ejected from their site	~ 40 nm
β decay		A few hundred keV absorbed over several mm	~0	Several mm
γ transition		About 1 MeV absorbed over several cm	0	Several cm

Table 5.1. *Nature of the decay/radiation affecting the glass and the associated interactions with the electrons and nuclei of the target (i.e. the glass)*

From a more macroscopic point of view, these irradiation sources in the glass are described according to the following two parameters:

– the dose rate (or activity) at a given time t: this value, generally expressed in Gy/h, describes the intensity of the irradiation α β and/or γ;

– the dose: indeed, the aging of glass leads to an accumulation of these events over time (of electronic or ballistic origins).

As an illustration, the set of contributions related to irradiation for a typical French R7T7 glass is described in Figure 5.3.

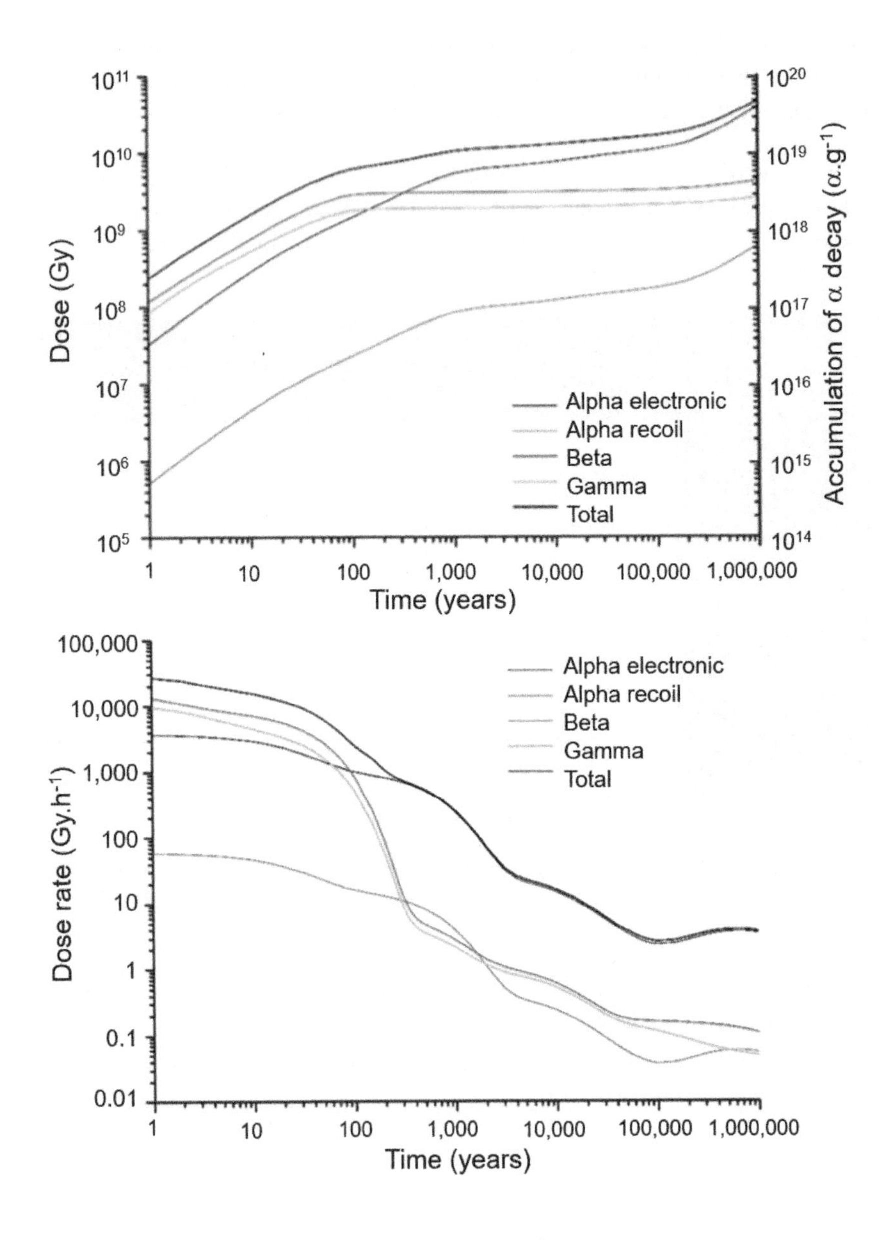

Figure 5.3. *Dose (top) and dose rate (bottom) evolutions of a typical R7T7 glass package, calculated by isotopic decay. Top: the right axis refers to the blue curve. For a color version of this figure, see www.iste.co.uk/bouffard/nuclear.zip*

5.1.4. *A complex scenario of glass aging under deep geological conditions*

After a cooling period of several decades, the nuclear glass is to be disposed of in a deep geological formation. In France, the CIGEO project aims to guarantee a radionuclides release much below the regulations, for several hundred thousand years (ANDRA). In order to do this, it is essential to understand how the glass would evolve in this scenario (see Figure 5.4).

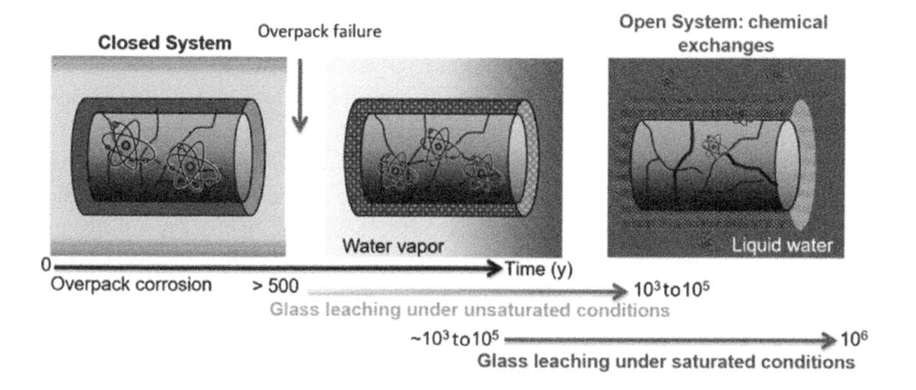

Figure 5.4. *Evolution of the glass package under disposal conditions as a function of time in years (denoted as y), according to the current French concept. On the left, the glass evolves in a closed system, subject to its own self-irradiation. After failure of the steel overpack, the glass will alter in the presence of water; first as a vapor and then as a liquid, and this water alteration may lead to the dissemination of radionuclides in the geosphere. We thus speak of an open system. For a color version of this figure, see www.iste.co.uk/bouffard/nuclear.zip*

Initially (over a period of around 500 years), the glass will evolve in a so-called closed system. Without interaction with the surrounding environment, it will be subjected to its own radioactivity, which will be very strong during the first phase of the life of the package, due to the short-lived radionuclides decay, which are mainly $\beta\gamma$ emitters (see Figure 5.3, bottom). The glass will therefore heat up under this stress during the first decades after its manufacture, before returning to a temperature below 100°C. In addition, over time, the glass will accumulate an increasing amount of damage, first by $\beta\gamma$ dose, then by α decays in a second phase (see Figure 5.3, top). Finally, each α particle becomes a helium atom when it thermalizes into the glass matrix. This helium atom, which stabilizes in the glass, could create helium clusters or bubbles that could exert local stresses in the material and risk cracking it. The maximum amount of helium to consider is in the range of 0.1 at.% to 1 at.% for FPA conditioning glass and military glass, respectively (Gin et al. 2017). All of these contributions (dose rate, heating, dose, helium accumulation) could lead to changes in the structure and properties of the glass, which need to be assessed, especially with regard to potential cracking or fracturing and consequently, the increase in reactive glass surface area that may later come into contact with water.

Over the centuries, water will reach the cells containing the waste packages. **The glass will then evolve into an open system, with the risk of radionuclides dissemination through the water**. In France, given the disposal concepts currently being considered by ANDRA, the glasses for high-level waste should first undergo an alteration stage by water vapor for a few hundred to a few thousand years, before they are leached into an aqueous medium by the groundwater of the disposal site. This period of glass alteration in the vapor phase could result from the slow re-saturation of the Callovo-Oxfordian clay[2] (the host rock of the French disposal site) due, among other things, to the H_2 release from the metallic overpacks and casings corrosion. However, this envisaged scenario by ANDRA, with a first alteration of the glass through a vapor phase, is very recent. As a result, most of the studies carried out to date – and in particular those concerning the impact of irradiation – have been concentrated on alteration by liquid water.

Currently in France, with almost all of the fuel reprocessed[3], R7T7 glass takes up the wide majority (> 95%) of the inventory of radionuclides that are present in the waste in the geological disposal site. Particular attention is therefore paid in our country to this material and its long-term behavior, the objective being to predict the source term, in other words, the quantity of radionuclides released by the glass as a function of time and environmental conditions (self-irradiation, presence of water and environmental materials in the deep geological site).

5.2. How are the effects of long-term irradiation studied at the laboratory scale?

The aim is to simulate the aging conditions that the waste packages will encounter during their life under disposal conditions, which are mainly self-irradiation (dose rate, dose, α and $\beta\gamma$) and interaction with the surrounding

2. During the excavation of these disposal zones, the air will have dried out the clay around their periphery. When the storage site is closed, the water will gradually take over: this is called resaturation.

3. With the exception of MOX fuel.

environment, particularly the groundwater, using a progressively complex approach, at the laboratory scale. It is therefore essential to find ways to accelerate the time scale, so as to simulate the potential changes that can occur over very long periods, typically from ten to several hundred thousand years. To this end, an approach based on four complementary axes has been implemented. This approach, which takes the impact of irradiation into account, can be used for studies on glass in a closed system, as well as studies concerning the alteration of glass by liquid water.

1) **Radionuclide-doped glass**. The choice of the radionuclide (mainly actinides) and its concentration impose the values of both the dose rate or the dose received by the glass over time (see Table 5.2). For example, to reach high dose values, the choice will be made for short-lived actinides such as ^{244}Cm ($T_{1/2}$ = 18 years). However, the two parameters (dose and dose rate) with relevant values for disposal conditions at a given aging time (e.g. low alpha dose rate and a consequent accumulation of alpha decays) cannot be directly combined on the same material for laboratory studies over a few years. In addition, alteration by water and irradiation can take place in a simultaneous way (see Chapter 6). In this specific case, irradiation does not just concern the solid, but the whole system, including water. We then talk about water radiolysis: under the effect of ionizing radiation, water will decompose into free radicals which, by successive recombinations, will form molecular (O_2, H_2, H_2O_2, etc.), radical (e^-_{aq}, $H^\bullet$, $OH^\bullet$, $HO_2^\bullet$, etc.) or ionic (H^+, OH^-, O_2^{2-}, etc.) species, whose proportions will depend on the nature of the ionizing radiation (α, βγ) and the chemical composition of the medium (solution, gaseous "sky") (Ferradini and Jay-Gerin 1999). Although the study of radioactive materials makes it possible to approach the most realistic conditions in terms of glass aging under irradiation, the use of these materials requires special precautions. They must be handled in nuclear facilities equipped with shielded cells and glove boxes, so as to protect against irradiation and contamination. Only a few laboratories in the world have the possibility of setting up such research.

2) **The use of external irradiation on non-radioactive glass** is widely used. In this case, non-radioactive glass is irradiated in particle accelerators or irradiators (γ sources). The resulting materials remain non-radioactive, so the panel of accessible characterization tools is therefore much more significant and the studies are much less expensive than those on radioactive materials.

The type of external irradiation, that is, the nature of the particle (heavy ions, electrons, neutrons, γ photons) and its energy, are chosen according to the particle–matter interactions being simulated (see Table 5.2). Nevertheless, these irradiation methods must be handled with care because they can induce artifacts. It is advisable to always take care of the representativeness of the studies carried out under external irradiation. For example, a transmission electron microscope allows high flux electron irradiations (nine orders of magnitude higher than the dose rates of real glass). However, under these conditions, the phenomena of alkali segregation linked with a change in the viscosity of the glass were observed, which is not the case for irradiations with lower flux nor in real-life cases.

The choice of the parameter (dose or dose rate) that we wish to consider determines the tool to be implemented for the irradiation. In general, irradiators with γ sources are favored to obtain dose rate values representative of the glass aging under disposal conditions (< 10 kGy/h). On the contrary, to reach dose values of interest with respect to disposal conditions (of the order of a few GGy), a more consequent dose rate (> 10^4 kGy/h) is required and, in this case, electron accelerators are used to reach said dose values in a few days.

The duration of the irradiation is also a crucial parameter in the case of experiments of glass alteration by water under irradiation: on particle accelerators, irradiation durations range from a few hours to a few days at most. The alteration experiments are then, most of the time, decoupled from the aging in water, or only so-called "perturbation" experiments are possible. On the contrary, on passive irradiation systems such as irradiators, experiments of the order of several months can be set up; the alteration and the irradiation then take place in a simultaneous way as in the case of the study of radioactive glasses.

	Dose rate	Dose
α decay	In relation to alpha particles interacting with glass and water → studies of glass doped with α emitters	In relation to ballistically interacting recoil nuclei → studies of α–doped glass and glass externally irradiated with few MeV heavy ions
β decay γ transitions	Electronic irradiation interacting with glass and water → studies of glass doped with β emitters or glass externally irradiated with γ rays	Electron irradiation of the solid (glass before or after alteration) → external electron irradiation

Table 5.2. *Schematic view of the main parameters considered in a decoupled way when studying glass under irradiation, in a closed system or in the presence of liquid water*

In addition, depending on the type of irradiation, the changes induced may affect the entire volume of the sample or just a layer near the surface. The thickness of this layer depends on the chosen particle and its energy (i.e. its range). The final choice will be a compromise between the thickness of the irradiated area (at least one to a few microns) and the representativeness of the chosen conditions with respect to the real-life case (ballistic or electronic interactions).

3) **In the particular case of the problem of helium accumulation** in the glass matrix, the use of nuclear neutron reactors or external He ion irradiations can be implemented, in order to produce high levels of helium respectively by nuclear reaction and by implantation in the material, while generating damage in the glass matrix by elastic and inelastic interactions.

4) **Modeling of the experimental observations**. For example, the modeling of the glass network by molecular dynamics allows a modeling at the atomic scale of the phenomena occurring during the alpha self-irradiation in glass (structural modifications, helium diffusion, etc.).

In summary, an approach based on several complementary axes, both experimental and computational, is implemented in the case of nuclear glass, so as to decouple the parameters related to irradiation and understand their impact on the glass behavior in a closed system or in the presence of liquid water. The complementarity of these different approaches aims to assess the impact of self-irradiation on the glass and its behavior under water, as well as to understand the mechanisms involved during the aging of this material, which is essential for the construction of long-term behavior models.

Finally, in terms of vocabulary, we will hereafter refer to "R7T7 glass" in the following results as *study glass*, whether it is radioactive or not, developed at the laboratory scale and which has a chemical composition close to that of a typical industrial R7T7 glass.

5.3. Closed system: evolution of glass subjected to its self-irradiation and to the accumulation of helium

5.3.1. *Impact of $\beta\gamma$ irradiation*

Irradiation with electrons or γ rays induces point defect formation in glass, by the homolytic bond breaks in the glass network, creating free electrons and holes, and then radicals (such as $\equiv Si^{\cdot}$; $\equiv Si - O^{\cdot}$) (Griscom 1985). Furthermore, in glass containing alkali, the capture of a hole in a site containing an ionic-type bond can induce the release of the corresponding alkali ion, which is then free to diffuse within the glass network (Weber et al. 1997).

The dose (i.e. the accumulation of events) will lead to a reorganization of the glass and to structural changes at the short (< 5 Å) and medium (5–20 Å) range levels: the coordination of boron partly changes from a tetrahedral (B^{IV}) to a trigonal (B^{III}) environment, causing changes in the polymerization degree of the glass and the value of the average angle between two silica tetrahedra (Si-O-Si angle) (Boizot et al. 1998, 2000, 2001, 2005).

These structural changes are limited in a glass that has a complex chemical composition, or close to real nuclear glass, due to two concomitant phenomena:

– The presence of mixed alkalis (Mixed Alkali Effect – MAE) compared to a single type of alkali in the glass composition. In this case, their migration is limited and the structural modifications observed in the glass are weaker (modification of the polymerization degree and of $B^{IV} \rightarrow B^{III}$). This indicates the importance of the role of alkalis in glass transformation (Ollier et al. 2004).

– The presence of transition metals or rare earth elements. Their ability to change their redox states acts as a trap for electrons or holes created during the interaction of the electron beam with the glass, thereby reducing or preventing the formation of point defects and the resulting transformation of the glass (Olivier et al. 2005; Malchukova and Boizot 2010).

In summary, the chemical complexity of glass is a favorable factor to limit the impact of these ionizing radiations.

From a more macroscopic point of view, these structural changes result in small modifications of some properties of the glass, which can be measured from doses in the range of 1 GGy. For example, hardness decreases by a few percent and the decrease in density is less than one percent (on a highly simplified glass composition, thus more impacted by this type of irradiation) (Mir et al. 2016).

It has been shown that a saturation phenomenon occurs for changes in the glass structure and properties from a dose between 1.5 and 3 GGy (Mir et al. 2016). These $\beta\gamma$-type doses will be reached for industrial nuclear glass after durations from a hundred to a few thousand years. Beyond this dose value, the glass structure and macroscopic properties stabilize. However, there are few studies carried out on glass with complex chemical compositions at a high irradiation dose, with the aim of quantifying its structural changes and the resulting property modifications.

Finally, the observation of the microstructure and the chemical composition of the irradiated complex glass at the SEM and TEM scale does not indicate any notable modifications after this type of irradiation. However, for high irradiation doses of the order of a few GGy performed by electron irradiation, a depletion of the sodium content on the surface of the glass by stimulated desorption, over a thickness of a few hundred nanometers, can be noted, which varies depending on the composition of the glass (Mir 2015). However, the representativeness of this depletion in reality currently remains unclear. It may be due to the desorption of a labile element that is favored in a vacuum and by beam heating. In this case, it is an experimental bias that should be avoided to study the behavior of the irradiated material. However, it could also come from a cumulative effect which could be observed in the case of high doses of electron irradiation: in this case, on the contrary, this modification could be expected under disposal conditions.

5.3.2. *Effects of α decays*

Valuable information on the impact of the α decays on the glass structure has been provided by the complementary study of radioactive glass that has been doped in actinides and glass that has been externally irradiated by different ions (Au, Kr, He, etc.) at different energy ranges, either by a single beam (in order to vary the part of the ballistic and electronic interactions) or by a multi-beam (to study a possible synergy between the recoil nuclei and the alpha particles). These experimental approaches, complemented by modeling, have shown that it is the ballistic interactions of the recoil nuclei that are responsible for the evolutions of the glass as a function of the α decay accumulation, by accumulation of ballistic damage (Peuget et al. 2014). A local fusion model followed by hyperquenching, named "ballistic supervitrification" has been proposed to explain the restructuring of the glass under the impact of recoil nuclei: the collision cascade generated by the recoil nuclei would lead to a thermal peak, in which the temperature within the collision cascade would exceed the melting temperature of the glass (see Figure 5.5 – stage 1). This "local fusion" would be followed by a hyperquenching, which would "freeze" the glass in a more disordered state (see Figure 5.5 – stage 2).

Recent results have also clarified the role of the α particle: it leads to a partial annealing (i.e. partial healing) of the damage generated by the recoil nuclei (Mir et al. 2015). Indeed, the electronic interactions induced by these α particles during their trajectory in the glass, with low electron stopping power, causes a local heating that can reach about 600 K (see Figure 5.5 – stage 3). This temperature, which is well below the melting temperature of the glass, nevertheless allows the glass structure to reorganize and thus to partially "heal" the damage induced by the recoil nuclei (see Figure 5.5 – stage 4). The final state of the glass damaged by the α decays is thus obtained by combining these two competing mechanisms: the generation of damage by the recoil nuclei, leading to a more disordered glassy state, and a partial self-healing phenomenon from the α particles.

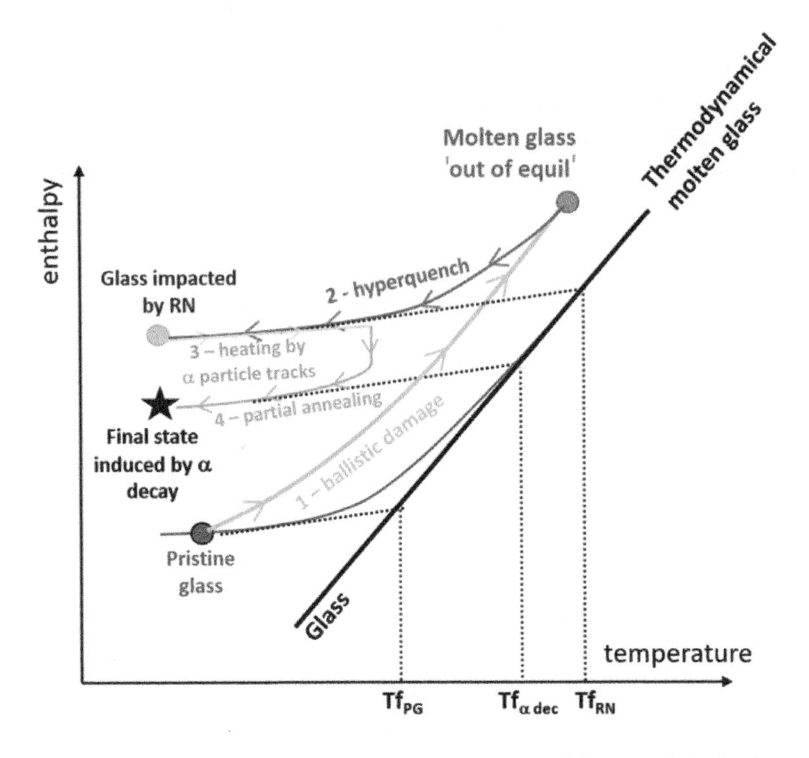

Figure 5.5. *Schematic view of the model making it possible to explain the impact of the α decay on the glass, starting from the "pristine glass" (PS), in other words, the glass before the α decay impact. The final state (black star) is an intermediate state between the pristine glass and the glass that is only subjected to ballistic interactions induced by the recoil nuclei (internal communication with S. Peuget, CEA). For a color version of this figure, see www.iste.co.uk/bouffard/nuclear.zip*

The damage to the glass increases with the accumulation of α decays until saturation: beyond a dose of the order of 40 MGy of nuclear origin (or of the order of 4 x 10^{18} α decays/g of glass and 0.3 dpa), the glass no longer changes. A direct impact model demonstrates this observation: a single event, whose elementary damage volume is about 300 nm^3 per recoil nucleus, is sufficient to definitely transform the glass. When the glass is 100% damaged, which is the case for 4 x 10^{18} α decays/g, its structure no longer changes under the impact of new ballistic particles (Peuget et al. 2014). This saturation dose will be reached by the glass after a few hundred years of aging (see Figure 5.3).

This α decay impact damage leads to structural and property changes in the glass that are quite similar to what is observed for glasses damaged by βγ-irradiation, although the underlying mechanisms are different. The resulting more disordered glass can be described as follows (Peuget et al. 2014, 2018; Mir 2015; Karakurt et al. 2016):

– at the short-range level (< 5 Å), there is an increase of the non-bridging oxygens, of the depolymerization degree of the glass network and of the proportion of trigonal boron B^{III} (at the expense of boron in tetragonal form, B^{IV});

– at the mid-range level (5–20 Å), there is an increase in the distribution of the ring sizes and in the angle values between two silica tetrahedra (Si-O-Si).

These structural modifications lead to changes in the macroscopic properties of the glass, such as hardness (which decreases by about 30%) or density (which decreases by about 0.5%). These changes are greater in magnitude than those generated by βγ-decay dose, but they remain fairly small and do not lead to a change in the confinement properties of the glass. Indeed, the microstructure of the damaged glass is not modified: neither phase separation nor crystallization has been observed at the TEM scale in any actinide-doped glasses, up to dose levels of the order of 10^{19} α decay/g of glass or 0.8 dpa (i.e. a dose reached after about 10,000 years).

5.3.3. *Accumulation of helium*

The helium content in the glass, at a given time, will depend on two parameters: on the one hand, its accumulation over time due to the accumulation of α decays, and on the other hand, its capacity to diffuse within the glass and thus to leave the glass matrix.

Several studies have investigated the maximum incorporation rate of helium in nuclear glass without modifying it, as well as the behavior of the helium contained in the glass when this value is exceeded. From an atomic point of view, helium will be incorporated in the free volumes present in the glass, that is, in the free spaces located between the network-forming polyhedra. However, helium will, for example, be in competition with sodium atoms, which are network modifiers. Therefore, the higher the sodium content in the glass, the lower the helium content that can be incorporated. The maximum content that can theoretically be incorporated in these free volumes is of the order of 2.5at.% for an R7T7 glass.

Experimentally, it is observed that nanobubbles appear beyond a certain limit. The appearance of these helium nanobubbles occurs at very variable helium contents and depends on parameters such as temperature, helium concentration, damage accumulation of the glass matrix and dose rate. As an illustration, at temperatures close to the real conditions of glass aging under disposal conditions, minimum concentrations of 0.03–0.1 at.% are reported in the literature for the observation of bubbles. It is difficult to be more precise on the conditions of helium bubble formation, but the various studies converge on one point: the absence of macroscopic effects of helium accumulation on the integrity of the glass, because no cracking has ever been observed in the zones where helium bubbles form.

Concerning the study of helium diffusivity, relatively consistent results are reported in the literature, regardless of the helium content and the state of glass damage. From room temperature, helium diffuses rapidly in the glass network, with a coefficient of about 10^{-12} $cm^2.s^{-1}$. An activation energy of the order of 60 $kJ.mol^{-1}$ was determined, which is consistent with the helium diffusion in the free volumes of the glass. The fact that the activation energy is not significantly impacted by the presence of damage in the glass suggests that the mid-range order, which controls helium diffusion in the glass, is not sufficiently modified to affect the topology of the glass, and thus the value of the helium diffusion coefficient.

Ultimately, considering the formation of helium in the glass and its ability to diffuse, it is possible to show that the diffusion of helium outside the glass could limit its concentration in it to a content level that is always lower than 0.02 at.%, for a borosilicate glass for FPA conditioning (R7T7). Therefore, it is unlikely that helium bubbles will form in such a glass. However, the question is more difficult to answer in the case of military-origin glass containing Pu, because the helium content that can form is about 1 at.%, as opposed to 0.1 at.% in the case of FPA conditioning glass.

5.3.4. *Summary of knowledge in closed system*

In summary, in a closed system, the structure of the glass will change under both the impact of a $\beta\gamma$ dose and the impact of an α dose. Mechanistic models exist to explain these observations.

It is observed that the modifications of the structure and macroscopic properties of the glass increase with the dose, until a saturation phenomenon is observed, starting from an electronic dose between 1.5 and 3 GGy for the $\beta\gamma$, and a nuclear dose of 40 MGy in the case of the α decays (i.e. 0.3 dpa or 4 x 10^{18} α decays/g). More pronounced modifications of the glass structure and properties can also be noted in the case of α decays. However, these modifications remain limited: no cracking of the glass matrix nor phase separation has ever been observed.

The generation of helium from α decays does not change the integrity of the nuclear glass either.

In summary, even under the impact of cumulative damage in its structure and helium accumulation, the glass remains intact and retains its radionuclide confinement capabilities. These data were acquired by considering the different irradiation parameters as independent. Further research should focus on the coupling parameters between the different irradiation sources and their impact(s) on the outcome of these glasses (see Figure 5.3).

5.4. Open system: alteration of glass by water under irradiation

5.4.1. *General information on the behavior of glass under water – methodology*

When glass is in contact with liquid water, four successive kinetic regimes characterize its alteration (see Figure 5.6): interdiffusion (grey), initial alteration rate (red), diminishing rate (green) and residual rate (yellow). These stages will follow each other as the alteration solution becomes loaded with elements released from the glass. The increase in the concentration of some of these elements, such as Si and Al in particular, which are relatively insoluble elements, will lead to the recondensation of these species on the surface of the glass and the formation of an alteration layer with protective properties (Vernaz et al. 2012; Gin et al. 2013).

Experimentally, the alteration of the glass is regularly monitored over time from the releases in solution of the various elements it contains. In addition to Si and Al, we are particularly interested in the behavior of the elements known as "alteration tracers" (B, Na and Li in general). Indeed, as these elements are not retained in the alteration layer and remain soluble in solution, their concentration in solution reflects the global alteration of the glass. Therefore, the alteration rate of the glass corresponds to the slope of the function representing the tracer element releases in solution as a function of time, and is expressed in $g.m^{-2}.d^{-1}$ or $nm.d^{-1}$. The "$g.m^{-2}.d^{-1}$" unit is used to normalize the rate to the glass surface (in m^2) that is in contact with the altering solution. At the end of the experiment, the altered solid can also be finely characterized, in order to provide additional information (chemistry and microstructure of the alteration layer).

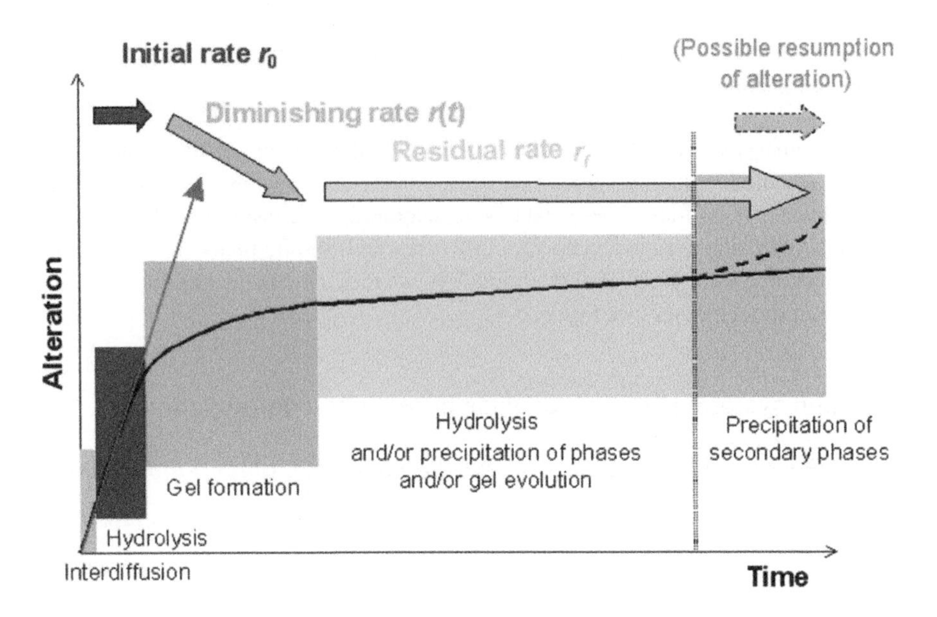

Figure 5.6. *Phenomenology of glass alteration over time (in liquid water). The regime of resumption of alteration, which can only take place in very particular cases of temperature and pH, will not be discussed here. For a color version of this figure, see www.iste.co.uk/bouffard/nuclear.zip*

Finally, it is assumed that the radionuclides contained in the glass will be altered at the same rate as these alteration tracers.

Regarding the impact of irradiation on the leaching behavior of glass, we will focus on:

– The initial alteration rate stage, which corresponds to the hydrolysis of the glass network forming bonds and, in terms of kinetics, to a maximum alteration rate value in a given medium. In this case, the alteration solution remains very diluted with respect to the recondensation of dissolved Si and Al. As an illustration, at 90°C and pH = 9, the initial rate of alteration of an R7T7 glass is about 2 $\mu m.d^{-1}$.

– The residual alteration rate stage, which is expected to predominate in the long term, and whose rate values are 3 to 4 orders of magnitude lower than the initial alteration rate values (< $nm.d^{-1}$) for an R7T7 glass. This slowing down is partly attributed to the formation of the protective alteration layer on the surface of the alteration glass (the protective part which is called "gel" or "passivating gel"), which thickens and restructures as the alteration occurs (see Figure 5.7). Several mechanisms may operate concomitantly and, to date, there is no international consensus on the mechanism(s) that control(s) this residual kinetic regime nor about its precise location. Is it passivation and the place of interdiffusion reactions throughout the thickness of the gel, exhibiting closed porosity and restructuring over time? Or is it an interfacial zone, located in the hydrated-dealkalized glass area, and corresponding to a confined zone which would be the place of hydrolysis/ precipitation reactions? At the experimental level, this alteration regime is studied over long alteration times (a few hundred days), under conditions that have favored the formation of this alteration film, by rapid saturation of the solution in elements that constitute it (Si, Al, etc.). In this case, the glass is altered in the form of powder or in a medium that is already pre-enriched in Si.

Ultimately, as the kinetics of glass alteration are generally very slow at room temperature, most studies are carried out at temperatures of up to 90–100°C, so as to increase the alteration kinetics (without modifying the predominant mechanisms). This change of temperature is also interesting for studying major cases under disposal conditions, where liquid water very prematurely comes into contact with a glass package that is still “warm” (50–70°C).

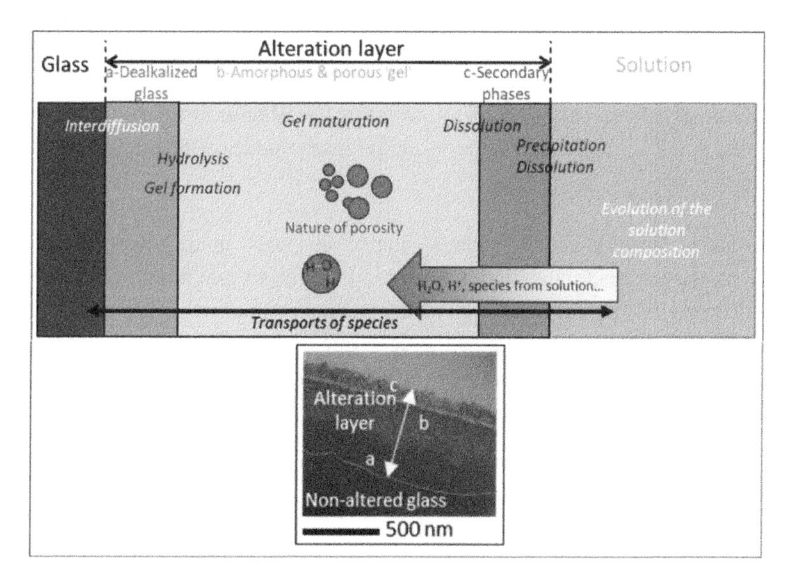

Figure 5.7. *Schematic view of the alteration profile at the glass/water interface. The thicknesses of the alteration layers are given as an indication, they can vary from one glass to another and from one area to another. For a color version of this figure, see www.iste.co.uk/bouffard/nuclear.zip*

5.4.2. *Taking irradiation into account in this multi-phase system*

The study of the radiation effects on the glass or the water, as well as the study of the alteration of the glass by water, if independently considered, are not enough to predict and understand the impact of irradiation on the alteration of the glass by water. The understanding of this complex alteration system requires the problem to be considered in its entirety. Taking irradiation into account when describing the long-term behavior of nuclear glass under water leads to the study of a very complex medium, with the implementation of two-phase systems (solid/liquid) under irradiation. This initially requires the implementation of a particular methodology, for example, through the development of specific experimental devices that are the most chemically inert, with respect to both water and irradiation.

For example, as a result of their very good chemical inertia, Teflon®-PFA or stainless steel are commonly used for leaching studies. However, irradiation in the presence of water makes the medium even more "aggressive". The degradation of these two types of materials under irradiation, and over time, releases species that can impact the alteration mechanisms of the glass under study (Rolland 2012; De Echave 2018). The choice of passivated titanium as a material for experimental devices, which has already been implemented in the early 2000s for studies about the behavior of spent fuel leached by groundwater (Jégou et al. 2005), is strongly recommended in the context of the simultaneous alteration of glass in water and under irradiation.

Ultimately, irradiation on glass alteration was mainly apprehended by considering the parameters as being decoupled and as simple alteration media: thus, the interactions between the effects of the various radiations were not taken into account. Only the parameter such as the dose rate and the dose were considered (see Table 5.2), and the alteration medium is initially pure water. The results obtained under these conditions give a descriptive view of the impact of irradiation on the alteration behavior of glass, rather than a mechanistic view.

5.4.3. *Irradiation and initial alteration rate*

The initial alteration rate of R7T7 glass under irradiation was initially measured in pure water over ranges of dose rate and dose values that were equivalent to several tens of thousands of years. These studies were carried out on radioactive glasses doped with alpha emitters and on externally irradiated glasses (see Figure 5.8). The set of results present in the literature on borosilicate glass, which has a complex chemical composition close to FPA conditioning glass, does not indicate that there is a significant difference in the measured rate on these systems, compared to a reference glass (non-radioactive or non-irradiated): the results remain within the measurement uncertainty (Wellman et al. 2005; Peuget et al. 2007; Tribet et al. 2014).

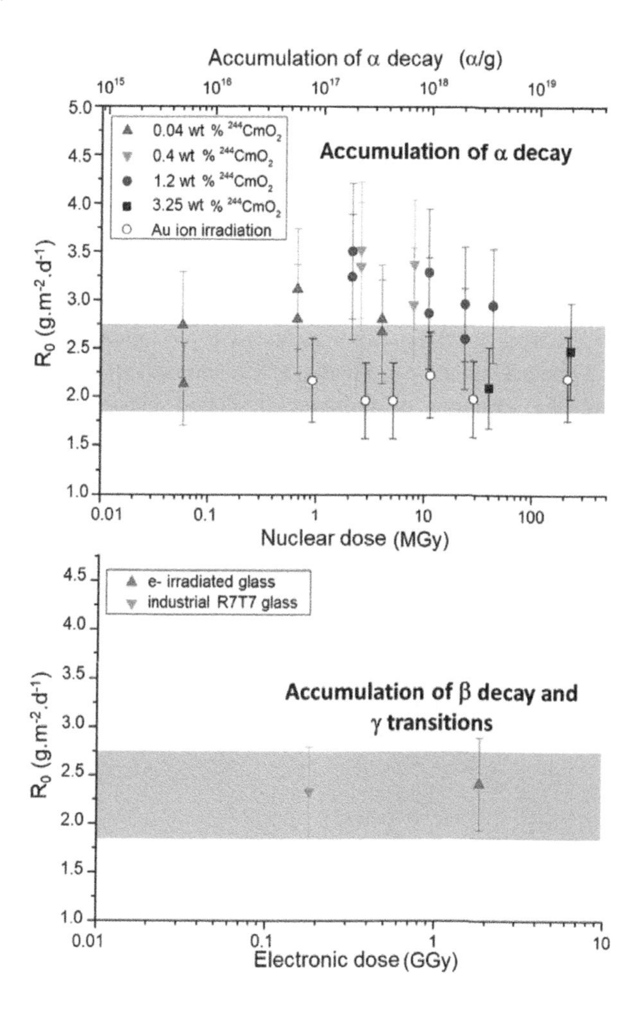

Figure 5.8. *Evolution of the initial alteration rate of R7T7 glass (R_0), measured at 100°C, as a function of the α decay accumulation (top figure), and of the electronic dose related to the β decays and the γ transitions (bottom figure). The shaded area represents the range of R_0 observed on similar non-radioactive glass. The error bar associated with each measurement is ± 20%. For a color version of this figure, see www.iste.co.uk/bouffard/nuclear.zip*

Over these short alteration times, it can be concluded that in pure water, neither radiolysis at the glass/solution interface nor the dose-related changes in glass structure and properties (see Figure 5.8) significantly impact the hydrolysis mechanism that operates during this kinetic alteration regime. For example, the impact of the cumulative dose on the structural modifications of the silicate network, such as the depolymerization degree or the increase in the distribution of Si-O-Si angles, is not enough to modify the reactivity towards water; reactivity which is measured at the macroscopic scale, via the tracer element releases in solution.

5.4.4. *Irradiation and residual alteration rate*

The influence of α and βγ dose rates on the establishment of the residual kinetic regime was studied from leaching experiments of R7T7 glass, in water that was initially pure and under external γ irradiation (^{60}Co sources) (Rolland et al. 2013b), and by studying the leaching of radionuclide-doped glasses (^{99}Tc β emitter, ^{238}Pu/^{239}Pu α emitters) (Rolland et al. 2013a; Tribet et al. 2021). In these various cases, alteration and irradiation take place simultaneously.

Figure 5.9. *TEM images of leached ^{239}Pu-doped glass grains. Phyllosilicates are the classical secondary phases that precipitate on the gel surface during the alteration of an R7T7 glass. The porous and dense areas may correspond to different degrees of gel maturation (Rolland et al. 2013a)*

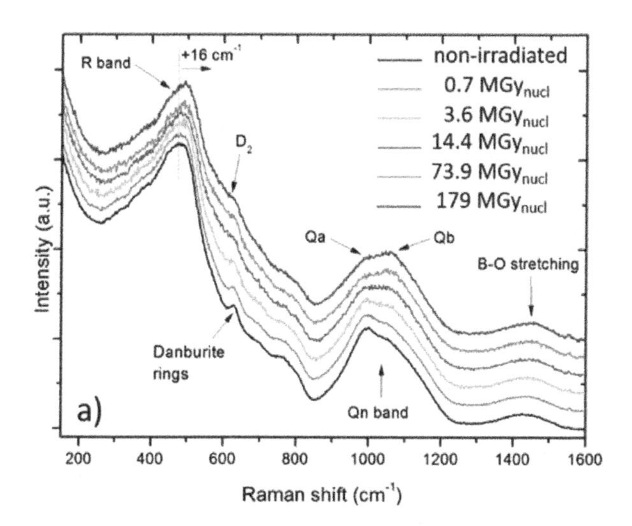

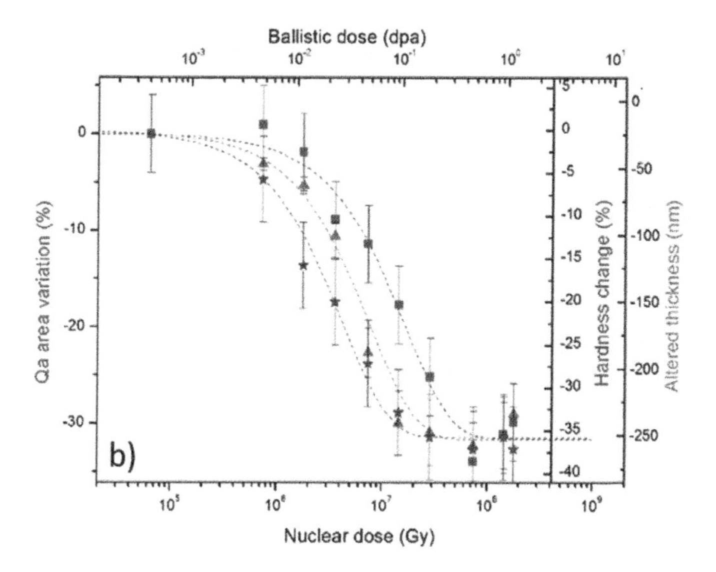

Figure 5.10. *(a) Raman spectra of simplified composition glass irradiated with Au ions of a few MeV, which generates ballistic damage. The decrease of the Qa band makes it possible to quantify the depolymerization of the glass network under the impact of the nuclear dose. (b) Evolution of structural parameters (Qa band), as well as glass properties (hardness and altered thickness), as a function of the nuclear dose (from Peuget et al. (2018)). For a color version of this figure, see www.iste.co.uk/bouffard/nuclear.zip*

In general, all the results obtained on alteration, both from leachate analyses and from the solid observations (see Figure 5.9), indicate that the residual rate of this borosilicate glass is not significantly affected by α and $\beta\gamma$ dose rate values that are typically expected under disposal conditions. Thus, the radiolysis phenomena that can occur at the interface between water and glass, both on the altered material surface and inside the water-filled porous gel, are not enough to macroscopically generate differences in the behavior of the glass leached by water.

An accumulation of α decays in the glass prior to the water arrival significantly increases the residual alteration rate value (by a factor of about 3 after a few hundred days): the glass alteration increases with the cumulative nuclear dose value, until it reaches saturation beyond a dose of 12 $MGy_{nuclear}$ (i.e. 10^{18} α decays/g) (Mougnaud et al. 2018; Tribet et al. 2021). Alteration and changes in the glass structure and properties follow the same evolution as a function of dose (see section 5.3.2 and Figure 5.10), demonstrating that there is a link between the structure of the glass and its alteration behavior. These data are congruent with recent results in the literature that confirm this link (Angeli et al. 2018; Stone-Weiss et al. 2018), although the underlying mechanism is not explained. The mechanistic hypotheses formulated, however, converge on the fact that the "interfacial zone" located between the non-altered glass and the "gel" (see Figure 5.7) constitutes a "key" area, in which the mechanisms (or part of the mechanisms) controlling alteration in the residual rate regime occur, and which are sensitive to the glass structure and thus to the damage caused by the accumulation of α decays (Mougnaud et al. 2018; Peuget et al. 2018; Tribet et al. 2020).

Recent results concerning the impact of a 3-GGy dose of $\beta\gamma$ electronic origin, which is the necessary dose to saturate the changes of structure and properties of borosilicate glass under this type of irradiation, indicate that the alteration of R7T7 glass is not significantly modified. The changes in structure and properties induced by the $\beta\gamma$ dose are of much smaller magnitude than the impact of the α decays. The same seems to be true for the alteration behavior, which supports the observations made in the case of α decays, on the link between the glass structure and properties, and its alteration behavior during the residual alteration rate regime.

5.4.5. *Summary on the behavior of glass under water and under irradiation*

Globally, under irradiation, glass deteriorates according to the same phenomenology as without irradiation.

The measured alteration kinetics are identical to those of the non-radioactive glass, regardless of the dose rate and the type of particle. Radiolytic phenomena at the interfaces (glass surface or alteration layer porosity) do not impact the overall alteration behavior of the glass.

On the contrary, an impact of the dose in relation to the α decays is observed: the long-term glass alteration kinetics increases by a factor of about 3 over the duration of the experiments. In the present state of knowledge, a link between glass structure and alteration behavior during the residual alteration rate regime is noticed. Current research is now focusing on the underlying mechanism(s). Given that an α dose equivalent to 10^{18} α decays/g would be reached by the glass about 50 years after its fabrication, the impact of the α decay accumulation is therefore a parameter that should be considered as soon as the water arrives in the deep repository microtunnels (see Figure 5.4).

5.5. Summary and prospects

In this chapter, we have discussed the impact of irradiation on the aging of glass in a closed system (glass subjected to its own irradiation only) and in an open system (glass subjected to alteration by water, under irradiation).

We have seen that the glass is a stable material under irradiation. Certainly, modifications of structure and macroscopic properties are observed, in particular the impact of doses of ballistic origin (α decays) has been noted, but the glass remains intact and it still has its capacities to confine radionuclides in a closed system.

Overall, the same is true for glass alteration in water and under irradiation. The same alteration phenomenology than without irradiation is observed, namely, an initial alteration rate regime similar to what is observed without irradiation, followed by a phase in which there is a drop in the kinetics, and then the residual alteration rate regime.

An impact of the ballistic dose caused by the α decays, that could induce an increase in the alteration kinetics at a given time compared to a non-radioactive glass, has been highlighted. Current research is focusing on the mechanism(s) at the origin of this impact, but this effect seems to be very limited and does not question the confinement capacities of the nuclear glass in an open system and under irradiation.

However, there are still two major pending questions regarding the aging of glass under irradiation.

Firstly, in the real-life case expected under disposal conditions, the parameters related to irradiation are not dissociated: real glass will accumulate increasing $\beta\gamma$ and α doses over time, while the dose rate values ($\beta\gamma$ and α) will decrease. Should we expect synergistic effects (positive or negative) between these different parameters, regardless of if they are in closed or open systems?

Moreover, in the current scenario of a deep repository, alteration of the glass by water vapor is to be initially expected. In addition, in the microtunnels, environmental materials will be present, such as the clay from the disposal site and steel, which will serve as an overpack for the glass package and as a casing for the microtunnel. The liquid water that will then leach the glass will therefore be loaded with all these elements from the surrounding environment. How will the glass leach under these conditions and under irradiation? It is therefore advisable, in the years to come, to question the impact of irradiation on this first phase of alteration by water vapor, and with a later alteration phase by liquid groundwater charged with all the elements from the environment.

5.6. Acknowledgements

The author would like to thank Sylvain Peuget for leading the studies on the evolution of glass under irradiation in a closed system, as well as the whole LMPA team (Active Materials and Processes Laboratory – *Laboratoire des Matériaux et Procédés Actifs*, CEA Marcoule) and the PhD students/post-doctoral fellows who collectively contributed to obtain many of the presented results.

5.7. References

ANDRA (n.d.). Site internet de l'Agence nationale pour la gestion des déchets radioactifs. Report [Online]. Available at: www.andra.fr.

Angeli, F., Charpentier, T., Jollivet, P., de Ligny, D., Bergler, M., Veber, A., Gin, S., Li, H. (2018). Effect of thermally induced structural disorder on the chemical durability of International Simple Glass. *Npj Materials Degradation*, 2(1), 31.

Boizot, B., Petite, G., Ghaleb, D., Calas, G. (1998). Radiation induced paramagnetic centres in nuclear glasses by EPR spectroscopy. *Nuclear Instruments & Methods in Physics Research Section B: Beam Interactions with Materials and Atoms*, 141(1–4), 580–584.

Boizot, B., Petite, G., Ghaleb, D., Pellerin, N., Fayon, F., Reynard, B., Calas, G. (2000). Migration and segregation of sodium under beta-irradiation in nuclear glasses. *Nuclear Instruments & Methods in Physics Research Section B: Beam Interactions with Materials and Atoms*, 166, 500–504.

Boizot, B., Petite, G., Ghaleb, D., Calas, G. (2001). Dose, dose rate and irradiation temperature effects in beta-irradiated simplified nuclear waste glasses by EPR spectroscopy. *Journal of Non-Crystalline Solids*, 283(1–3), 179–185.

Boizot, B., Ollier, N., Olivier, F., Petite, G., Ghaleb, D., Malchukova, E. (2005). Irradiation effects in simplified nuclear waste glasses. *Nuclear Instruments & Methods in Physics Research Section B: Beam Interactions with Materials and Atoms*, 240(1–2), 146–151.

De Echave, T. (2018). Étude des mécanismes d'altération des verres nucléaires sous radiolyse alpha et en conditions environnementales. PhD thesis, Université de Montpellier.

Débat Public (2020). Dossier du Maître d'Ouvrage pour le débat public sur le plan national de gestion des matières et des déchets radioactifs (5ème édition, PNGMDR). Report, Direction Générale de l'Energie et du Climat et Autorité de Sûreté Nucléaire.

Ferradini, C. and Jay-Gerin, J.-P. (1999). La radiolyse de l'eau et des solutions aqueuses : historique et actualité. *Canadian Journal of Chemistry*, 77(9), 1542–1575.

Gin, S., Abdelouas, A., Criscenti, L.J., Ebert, W.L., Ferrand, K., Geisler, T., Harrison, M.T., Inagaki, Y., Mitsui, S., Mueller, K.T. et al. (2013). An international initiative on long-term behavior of high-level nuclear waste glass. *Materials Today*, 16(6), 243–248.

Gin, S., Jollivet, P., Tribet, M., Peuget, S., Schuller, S. (2017). Radionuclides containment in nuclear glasses: An overview. *Radiochimica Acta*, 105(11), 927–959.

Griscom, D.L. (1985). Defect structure of glasses – Some outstanding questions in regard to vitreous silica. *Journal of Non-Crystalline Solids*, 73(1–3), 51–77.

IAEA (2018). *Status and Trends in Spent Fuel and Radioactive Waste Management*. International Atomic Energy Agency, Vienna.

Jégou, C., Muzeau, B., Broudic, V., Peuget, S., Poulesquen, A., Roudil, D., Corbel, C. (2005). Effect of external gamma irradiation on dissolution of the spent UO_2 fuel matrix. *Journal of Nuclear Materials*, 341(1), 62–82.

Karakurt, G., Abdelouas, A., Guin, J.-P., Nivard, M., Sauvage, T., Paris, M., Bardeau, J.-F. (2016). Understanding of the mechanical and structural changes induced by alpha particles and heavy ions in the French simulated nuclear waste glass. *Journal of Nuclear Materials*, 475, 243–254.

Kaushik, C.P. (2014). Indian program for vitrification of high level radioactive liquid waste. *Procedia Materials Science*, 7, 16–22.

Malchukova, E. and Boizot, B. (2010). Reduction of Eu^{3+} to Eu^{2+} in aluminoborosilicate glasses under ionizing radiation. *Materials Research Bulletin*, 45(9), 1299–1303.

Mir, A.H. (2015). Radiation effects on oxide glasses: Importance of energy deposition and relaxation processes. PhD Thesis, Université de Caen Normandie.

Mir, A.H. and Peuget, S. (2020). Using external ion irradiations for simulating self-irradiation damage in nuclear waste glasses: State of the art, recommendations and, prospects. *Journal of Nuclear Materials*, 539, 152246.

Mir, A.H., Peuget, S., Toulemonde, M., Bulot, P., Jégou, C., Miro, S., Bouffard, S. (2015). Defect recovery and damage reduction in borosilicate glasses under double ion beam irradiation. *Epl*, 112(3), 36002.

Mir, A.H., Boizot, B., Charpentier, T., Gennisson, M., Odorico, M., Podor, R., Jégou, C., Bouffard, S., Peuget, S. (2016). Surface and bulk electron irradiation effects in simple and complex glasses. *Journal of Non-Crystalline Solids*, 453, 141–149.

Mougnaud, S., Tribet, M., Renault, J.P., Gin, S., Peuget, S., Podor, R., Jégou, C. (2018). Heavy ion radiation ageing impact on long-term glass alteration behavior. *Journal of Nuclear Materials*, 510, 168–177.

Ojovan, M.I. and Lee, W.E. (2011). Glassy wasteforms for nuclear waste immobilization. *Metallurgical and Materials Transactions A*, 42(4), 837–851.

Olivier, F.Y., Boizot, B., Ghaleb, D., Petite, G. (2005). Raman and EPR studies of beta-irradiated oxide glasses: The effect of iron concentration. *Journal of Non-Crystalline Solids*, 351(12–13), 1061–1066.

Ollier, N., Boizot, B., Reynard, B., Ghaleb, D., Petite, G. (2004). Beta irradiation borosilicate glasses: The role of the mixed alkali effect. *Nuclear Instruments & Methods in Physics Research Section B: Beam Interactions with Materials and Atoms*, 218, 176–182.

Peuget, S., Broudic, V., Jégou, C., Frugier, P., Roudil, D., Deschanels, X., Rabiller, H., Noel, P.Y. (2007). Effect of alpha radiation on the leaching behaviour of nuclear glass. *Journal of Nuclear Materials*, 362(2–3), 474–479.

Peuget, S., Delaye, J.M., Jégou, C. (2014). Specific outcomes of the research on the radiation stability of the French nuclear glass towards alpha decay accumulation. *Journal of Nuclear Materials*, 444(1–3), 76–91.

Peuget, S., Tribet, M., Mougnaud, S., Miro, S., Jégou, C. (2018). Radiation effects in ISG glass: From structural changes to long term aqueous behavior. *Npj Materials Degradation*, 2, 23.

Richet, P. (2000). *L'âge du verre*. Découvertes Gallimard, Paris.

Rolland, S. (2012). La lixiviation des verres nucléaires de type R7T7 sous irradiation. Étude du régime de cinétique résiduelle d'altération. PhD Thesis, Université Claude Bernard Lyon I.

Rolland, S., Tribet, M., Jégou, C., Broudic, V., Magnin, M., Peuget, S., Wiss, T., Janssen, A., Blondel, A., Toulhoat, P. (2013a). Tc-99- and Pu-239-doped glass leaching experiments: Residual alteration rate and radionuclide behavior. *International Journal of Applied Glass Science*, 4(4), 295–306.

Rolland, S., Tribet, M., Jollivet, P., Jégou, C., Broudic, V., Marques, C., Ooms, H., Toulhoat, P. (2013b). Influence of gamma irradiation effects on the residual alteration rate of the French SON68 nuclear glass. *Journal of Nuclear Materials*, 433(1), 382–389.

Stone-Weiss, N., Pierce, E.M., Youngman, R.E., Gulbiten, O., Smith, N.J., Du, J.C., Goel, A. (2018). Understanding the structural drivers governing glass-water interactions in borosilicate based model bioactive glasses. *Acta Biomaterialia*, 65, 436–449.

Tribet, M., Rolland, S., Peuget, S., Broudic, V., Magnin, M., Wiss, T., Jégou, C. (2014). Irradiation impact on the leaching behavior of HLW glasses. *Procedia Materials Science*, 7, 209–215.

Tribet, M., Mir, A.H., Gillet, C., Jégou, C., Mougnaud, S., Hinks, J.A., Donnelly, S.E., Peuget, S. (2020). New insights about the importance of the alteration layer/glass interface. *Journal of Physical Chemistry C*, 124(18), 10032–10044.

Tribet, M., Marques, C., Mougnaud, S., Broudic, V., Jégou, C., Peuget, S. (2021). Alpha dose rate and decay dose impacts on the long-term alteration of HLW nuclear glasses. *Npj Materials Degradation*, 5(1), 36.

Vernaz, E., Gin, S., Veyer, C. (2012). Waste glass. In *Comprehensive Nuclear Materials*, Konings, R.J.M. (ed.). Elsevier, Oxford.

Weber, W.J., Ewing, R.C., Angell, C.A., Arnold, G.W., Cormack, A.N., Delaye, J.M., Griscom, D.L., Hobbs, L.W., Navrotsky, A., Price, D.L. et al. (1997). Radiation effects in glasses used for immobilization of high-level waste and plutonium disposition. *Journal of Materials Research*, 12(8), 1946–1978.

Wellman, D.M., Icenhower, J.P., Weber, W.J. (2005). Elemental dissolution study of Pu-bearing borosilicate glasses. *Journal of Nuclear Materials*, 340(2–3), 149–162.

6

Radiolysis of Porous Materials and Radiolysis at Interfaces

Sophie LE CAËR and Jean-Philippe RENAULT
CEA/Saclay, DRF/IRAMIS/NIMBE UMR 3685, Gif-sur-Yvette, France

6.1. Introduction

Although the radiation chemistry of aqueous solutions has been studied since the beginning of the 20th century, the radiation chemistry of heterogeneous systems has only been developed in the last 20 years, linked to the issue of nuclear waste storage. Indeed, in most confining matrices, large material/water interfaces are subjected to the action of ionizing radiation. Up until now, the questions addressed in this field have mainly concerned the production of dihydrogen, but the understanding of corrosion phenomena under irradiation is also an important and rapidly developing research area.

In this chapter, we will begin with general information on radiolysis and, more particularly, on the radiolysis of water, before quickly presenting the most studied porous materials. We will then look at the phenomena taking place on the surfaces and at the interfaces (production of dihydrogen, kinetic aspects, etc.).

Nuclear Materials under Irradiation,
coordinated by Serge BOUFFARD and Nathalie MONCOFFRE. © ISTE Ltd 2023.

6.2. General information on radiolysis

6.2.1. *A few definitions*

To quantify the energy deposited in matter by ionizing radiation, we have several concepts including the absorbed dose and the linear energy transfer (LET).

The absorbed dose is the amount of energy, expressed in joules, deposited in matter by ionizing radiation per unit of mass, expressed in kilograms. Its unit is the Gray (Gy) (1 Gy = 1 $J.kg^{-1}$). The dose rate corresponds to the dose transmitted to the material per unit of time.

The LET, which characterizes the energy deposit in the medium, is the average energy deposited (dE) per unit path (dx) in the absorbing medium, by a particle at a given energy:

$$\mathrm{LET} = -\frac{\mathrm{dE}}{\mathrm{dx}} \qquad [6.1]$$

It is generally expressed in $keV.\mu m^{-1}$. For particles of high energy that are greater than a hundred keV for the lightest ions and a hundred MeV for heavy ions (Xe, Pb, etc.), the Bethe formula allows the calculation of the LET. It is proportional to the charge number of the particle considered and to $\frac{\ln(v)}{v^2}$, with v being the velocity of the latter. Thus, with all things being equal, the more charged a particle is, the higher the value of its LET will be, and the stronger the interactions with the matter will be. Electrons of 10 MeV energy and gamma radiation have comparable and relatively low LET values, around 0.2 $keV.\mu m^{-1}$, whereas the LET of a 5.3 MeV alpha nucleus is 130 $keV.\mu m^{-1}$ (Spinks and Woods 1990). In the case of gamma radiation, the LET actually corresponds to the electrons ejected by gamma photons through the Compton effect.

In order to assess the reactivity induced by ionizing radiation, it is necessary to define a quantity that makes it possible to measure the number of a species produced (or destroyed), according to the quantity of energy absorbed by the material: this is the radiolytic yield. The radiolytic yield, usually denoted as G(X), where X is the species formed (or destroyed), is the quantity of X species formed (or destroyed, in mole) per unit of energy deposited in the matter (in joules).

$$G_{(species)} = \frac{\text{(Number of mole of species formed or destroyed)}}{\text{(Amount of matter receiving the dose)}*\text{Dose}} \qquad [6.2]$$

It is also expressed in molecules per 100 eV. The conversion factor between the two units is: 1 molecule/100 eV = 1.036 × 10^{-7} mol J^{-1}. Let us note that the measurement of the radiolytic yield of a species generated by radiolysis is very useful when assessing the reactivity of a material under irradiation. The radiolytic yields of the different species are well known in the case of water radiolysis: they depend on time (from initial yields at 1 ps to escape yields at 1 μs, after the initial energy deposit), the pH of the solution, the LET of the incident particles, etc.

In all that follows, we will mainly focus on low LET radiation.

6.2.2. *Radiolysis of liquid water*

The distribution of species as a function of time during water radiolysis is summarized in Figure 6.1. The mechanism strongly depends on the spatial distribution of the species formed by the ionizing radiation. A distinction is made between a heterogeneous and a homogeneous phase.

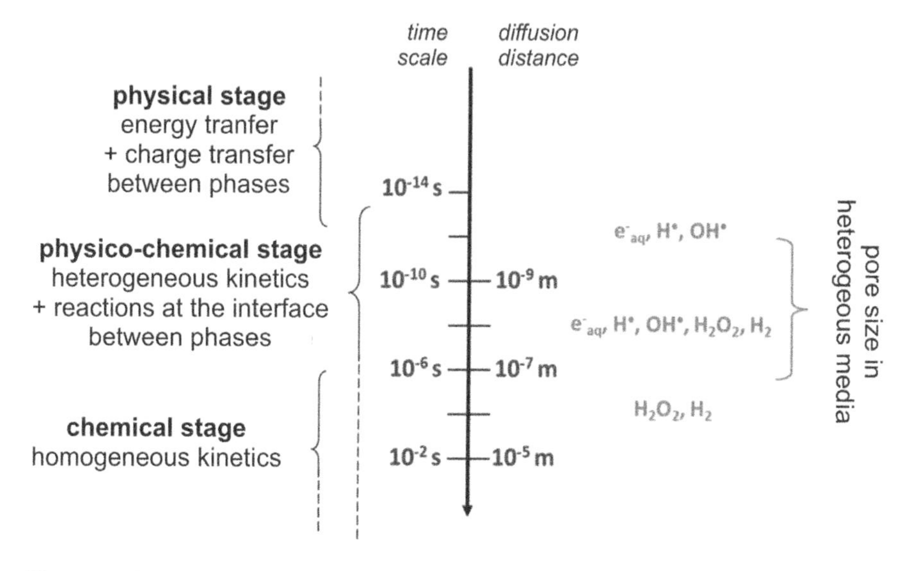

Figure 6.1. *Time/distance equivalence diagram for the different steps of the radiolysis of liquid water. On the left, we can see the additional phenomena occurring in heterogeneous media. For a color version of this figure, see www.iste.co.uk/bouffard/nuclear.zip*

During the heterogeneous phase, species are created locally on the radiation "track", resulting in an extremely high concentration of generated species near the track, but almost none elsewhere. This phase can be temporally separated into three stages: physical, physico-chemical and, lastly, a chemical stage (heterogeneous chemistry). Ultimately, at the homogeneous phase, the species generated during the heterogeneous phase have had time to diffuse into the solution and are therefore homogeneously distributed.

The reactions taking place during water radiolysis are well understood, both experimentally and theoretically (Spinks and Woods 1990; Le Caër 2011). In general, radiolytic events occur in three main stages and take place at different characteristic time scales:

1) *The physical stage*, which is carried out in about 1 fs after the initial ionizing radiation/matter interaction, consists of an energy deposit followed by fast relaxation processes. It leads to the formation of ionized water molecules (H_2O^+), excited water molecules (H_2O^*) and electrons (e^-).

2) During the *physico-chemical stage* (10^{-15}-10^{-12} s), many processes occur, including ion-molecule reactions (R1), dissociative relaxation phenomena (R2), electron solvation (R3), etc.

$$H_2O^+ + H_2O \rightarrow H_3O^+ + HO^\bullet \text{ (R1)}, H_2O^* \rightarrow HO^\bullet + H^\bullet \text{ (R2)}, e^- \rightarrow e^-_{aq} \text{ (R3)}$$

These two stages take place with species present in the tracks of the ionizing radiation. The phenomena taking place are therefore strongly heterogeneous.

3) During the *chemical stage* (10^{-12}–10^{-6} s), the species react in the tracks of the radiation and then diffuse into solution. They can thus react with one another, as well as with the molecules present in the medium (solute). The trace of the particles expands, due to the diffusion of the radicals and their subsequent chemical reactions.

The overall balance of water radiolysis leads to the formation of radicals, as well as stable molecules such as dihydrogen and hydrogen peroxide (R4 reaction). Let us mention that a radical is a chemical species with one (or more) unpaired electron(s) on its outer layer, denoted by a dot. Most of the time, the presence of a single electron gives these species a high reactivity. They react with many compounds in processes that are often non-specific, and their life span in solution is generally very short.

$$H_2O \xrightarrow{ionizing\ radiation} e_{aq}^-, H^\bullet, HO^\bullet, H_2, H_2O_2, H_3O^+ \quad \text{(R4)}$$

Note that dioxygen O_2 is not a primary species during water radiolysis, and that hydrated electrons and hydrogen atoms are strong reducing agents with respective standard potentials:

$$E°(H_2O/e_{aq}^-) = -2.9\ V_{NHE} \text{ and } E°(H^+/H^\bullet) = -2.3\ V_{NHE}.$$

These free radicals can then easily reduce dissolved metal ions to their lower oxidation states. On the other hand, hydroxyl radicals are very strong oxidizing species with a standard potential $E°(HO^\bullet/H_2O) = +2.7\ V_{NHE}$.

Figure 6.1 illustrates these three stages, as well as the changes brought about by the presence of a porous solid. In heterogeneous confinement systems, these three stages are modified from what happens in water. The energy deposit is different and various energy transfer processes can take place during the first stage. During the second stage, the rate constants of the reactions are modified when these reactions occur in confined environments; moreover, the species produced may also react with those present on the surface of the solid, for example. Radiolysis in heterogeneous media is much more complex than radiolysis in homogeneous media. Indeed, if we look at the distance scales in porous media with characteristic distances of less than a few hundred nanometers, we can see that it is difficult to reach the homogeneous chemistry stage (see Figure 6.1), since the diffusion distances of the produced species are of the same order of magnitude as the size of the porosities that are present. Moreover, it is not enough to simply know the effects of radiation on two or more phases and to sum up these effects, in an effort to deduce the radiolysis phenomena on the entire heterogeneous system.

Before showing how radiolysis phenomena can be modified in porous media, it is important to specify the nature of the porous materials of interest.

6.3. Main porous materials of interest

Figure 6.2 shows important porous materials, either for fundamental studies, due to their well-defined geometry, or for the nuclear industry.

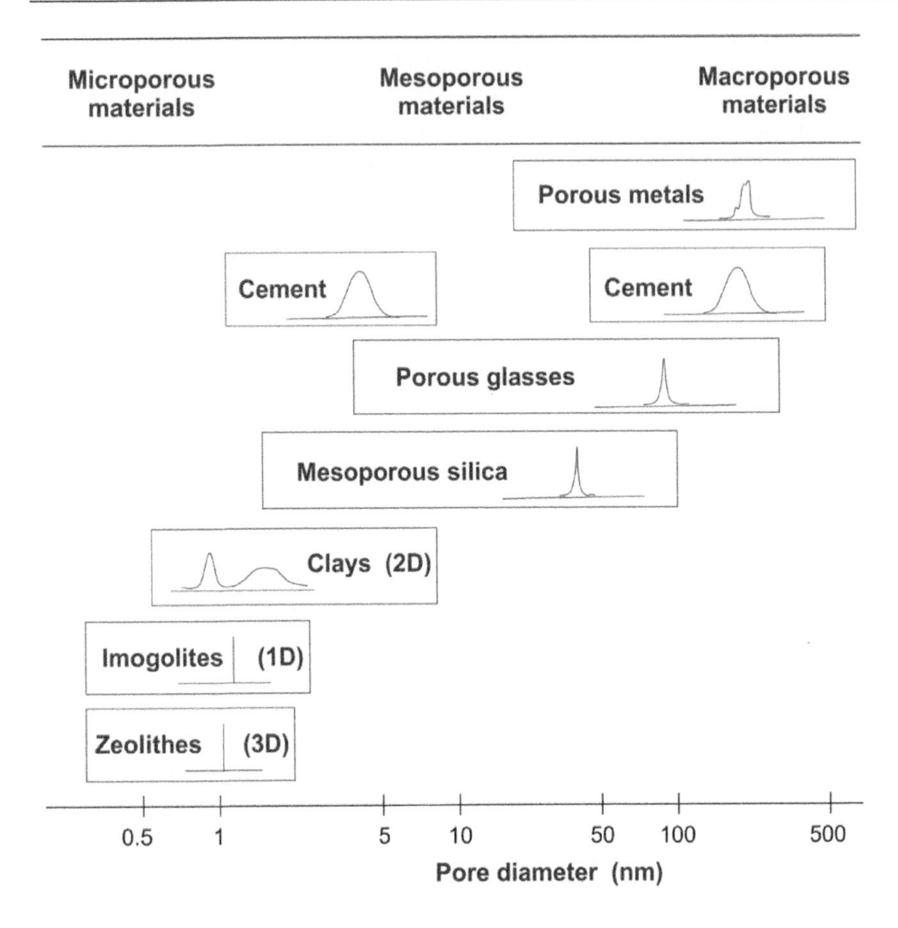

Figure 6.2. *Main porous materials of interest*

6.4. Dosimetry in heterogeneous media

Accurate dosimetry is a prerequisite for any radiolysis study. However, not all standard chemical dosimetry methods can be used in nanoporous media, as they rely on known radiolytic yields that would have to be modified by confinement. Therefore, an alternative is to use a Monte Carlo particle transport code. In this case, each particle is tracked individually and the type of interaction occurring is drawn at each step based on tabulated cross-sections, allowing the deposited dose to be calculated. The PENELOPE (positron and electron energy penetration and loss) code has thus proven to be particularly effective for performing micro-dosimetry calculations (Stewart et al. 2002; Mainardi et al. 2004). This code can handle electron transport up to low energies of 100 eV, and for any material.

As the case of heterogeneous silica/water systems has been widely studied, the results obtained with these systems will be detailed in the following. The dose received by water confined in nanometric to micrometric silica pores has thus been calculated to be very similar (within 15%) to the dose received by water that is not confined. These similarities arise from the fact that silica and water have comparable behavior, with respect to secondary particle production and absorption. Therefore, the usual approximation proposed to correct the dose by the average density of the system should not be used in porous media.

In contrast, when the confining material has a very different stopping power than water, as is the case for metal/solution interfaces, the increase of the dose in the solution can be much larger. For example, the calculated average dose in water near steels (Hastelloy type, SS316L) is increased by a factor of up to 1.5 (Moreau et al. 2014). Simulations suggest that the increase in dose received by water in porous metals is related to a greater interaction of secondary electrons with water than with metals. It generally occurs at a distance of about 10 nanometers, but with a limited impact on the interfacial dose. The same pattern can be identified for gold, but in this case, the difference in stopping power induces a dose increase up to a factor of 7 (Musat et al. 2010a).

6.5. Production of dihydrogen by radiolysis of water in a confined medium

Dihydrogen production from confined water has been studied in a wide variety of systems, ranging from borosilicate glass with controlled pore size (Le Caër et al. 2005; Rotureau et al. 2005), mesoporous silicas and zeolites (Frances et al. 2015) to clay materials (Fourdrin et al. 2013; Lainé et al. 2016, 2017) and cement materials (Le Caër et al. 2017), etc. These studies have shown that dihydrogen production comes from both the energy deposited in the confined water and the energy deposited in the confining material. These various works have allowed us to describe the phenomena of energy redistribution. Indeed, the energy of the ionizing radiation absorbed in one phase can be delivered to the interface, and induce physical and chemical processes in the other phase.

6.5.1. *Methods for calculating the yield of dihydrogen production*

In a heterogeneous system (material/water), in which most of the mass of the system is the mass of the material, the yield of dihydrogen production can be calculated in two ways: by considering that all the energy of the ionizing radiation is only deposited in the water, or that this energy is deposited in the entire water and material system. In the first case, it is the mass of the water that is considered in the denominator of equation [6.2] (see above), while the total mass of the sample is used in the second case. These differences in calculation are illustrated in Figure 6.3 for the case of a porous silica sample-containing water (Rotureau et al. 2005). By considering that all the radiation energy is only deposited in the water (black circles in the figure), the measured dihydrogen yield values are extremely high when compared to the yield measured in the water (4.7 10^{-8} mol.J^{-1}), especially when there is only a small amount of water in the sample. Let us note that yield values as high as 3 10^{-6} mol.J^{-1} have no physical meaning, and are simply due to the fact that the amount of water in the system is very low. Therefore, the yields were also calculated while taking the total mass of the system into account (empty circles in Figure 6.3). Evidently, the values obtained in this way are much lower than when the calculations are performed when considering water alone; they also vary very little with the amount of water present in the porosity. The only way to explain this near absence of evolution is to assume that the energy deposited in the silica is totally transferred to the (confined) water.

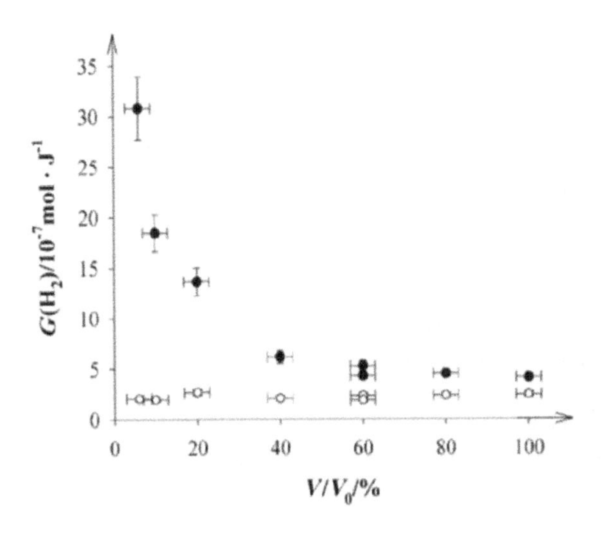

Figure 6.3. *Radiolytic yields of dihydrogen production as a function of the fraction of water present in the system, for glass with a 50 nm pore size under γ irradiation. Yields are calculated relative to the energy deposited in the water (•) or relative to the energy deposited in the entire glass-water system (○). V is the volume of water introduced into the sample and V_0 is the pore volume. Figure taken from Rotureau et al. (2005). Copyright © 2005 WILEY-VCH Verlag GmbH & Co KGaA, Weinheim*

It is therefore necessary to look at certain very important dihydrogen production yield values that are given in the literature with care, in particular when the method of calculation is not explicitly specified. Lastly, it is important to underline the fact that the calculation carried out by considering the total mass of the system is preferred, because it gives more realistic and directly exploitable values.

6.5.2. *Reaction mechanisms*

The most studied example is that of the silica/water interface under irradiation: we will therefore detail the behavior of this system. The mechanisms accounting for the production of dihydrogen under irradiation can be described as follows (Brodie-Linder et al. 2010). Initially, the ionizing radiation creates electron–hole pairs in silica (R5). These charge carriers can recombine through an excitonic state (R6). The excitons can react with the surface hydroxyl groups in the hydroxylated silica (R7 reaction, where "≡" represents the three bonds made by the silicon atom with oxygen atoms) (Shkrob and Trifunac 1997). However, the water molecules adsorbed on the silica surface also constitute traps for the excitons (R8). The hydrogen atoms thus formed (R7-R8) can then dimerize and generate dihydrogen H_2 (R9). This production of hydrogen atoms has been observed experimentally by EPR spectroscopy (Electron Paramagnetic Resonance, a technique used to detect radical species) during the radiolysis of fused silica (Griscom 1984) and MCM-41-type mesoporous silicas (Chemirisov et al. 2001), among others.

$$SiO_2 \xrightarrow{ionizing\ radiation} e^- + h^+ \qquad \text{(R5)}$$

$$e^- + h^+ \rightarrow {}^3exciton \qquad \text{(R6)}$$

$$^3exciton + \equiv SiOH \rightarrow \equiv SiO^\bullet + H^\bullet \qquad \text{(R7)}$$

$$^3exciton + H_2O \rightarrow HO^\bullet + H^\bullet \qquad \text{(R8)}$$

$$H^\bullet + H^\bullet \rightarrow H_2 \qquad \text{(R9)}$$

This model can be modified for other systems (alumina, zeolites, clay materials) (Thomas 1993), where the surface chemistry reflects a more ionic situation than in the case of silica, or for materials with very thin walls (a few nanometers thick, at the most) for which the recombination of electron–hole pairs cannot take place. In such cases, the reaction mechanisms are no longer based on the presence of excitons, but on the reactions induced by electrons and holes. Thus, hydrogen atoms can result from the dissociative attachment of electrons produced by ionization of the solid to the –OH groups, located on the surface of the material:

$$e^- + ROH \rightarrow RO^- + H^\bullet \qquad \text{(R10)}$$

6.5.3. *Different parameters influencing the production of dihydrogen under irradiation*

Several parameters can affect the yield of dihydrogen production: the value of the band gap energy of the material (Petrik et al. 2001) (this value determines the energy of the excitons and thus their ability to break the O–H groups), the surface density of the hydroxyl groups and the possible presence of water molecules adsorbed on the surfaces (Brodie-Linder et al. 2010), the migration distance of the excitons or charge carriers, etc.

The dose rate strongly influences the energy transfer from solid to liquid (see Figure 6.4). Studies of dihydrogen production by dried or hydrated borosilicate glass during high-dose-rate irradiation (pulses of about 2 $Gy.ns^{-1}$ of 10 MeV electrons) have shown that the dihydrogen mainly originates from the water contained in the pores, with the contribution from the interface being negligible (of the order of a few percent). Behavior under γ irradiation (experiments at a low dose rate, about 2 $Gy.min^{-1}$) is very different, given that the surface plays a preponderant role under these irradiation conditions. Indeed, the dried glass contributes to more than 50% of the dihydrogen production. At a low dose rate, interfacial chemistry is predominant, whereas radiolysis of nanoconfined water is predominant at a high dose rate (Le Caër et al. 2005). Moreover, in the case of dried glass, the dihydrogen production is about 100 times lower for high-dose-rate irradiations, which, during the physical phase, create a much higher density of species per unit time than for low-dose-rate irradiations. Therefore, during its diffusion, a given species has a much higher probability of reacting according to a recombination reaction between an electron and a hole within the material, instead of meeting the surface and reacting there, leading to the production of dihydrogen. Furthermore, high dose rates favor other reaction pathways (spraying of water molecules onto the surface, etc.) at the expense of dihydrogen production (Brodie-Linder et al. 2010). When compared to low-dose irradiations, all this explains why high-dose-rate irradiations (i) produce much less dihydrogen in the dried material and (ii) have a much lower energy transfer from the solid to the liquid, therefore leading the material to contributing less to the production of H_2. In real storage conditions (low dose rates), the interface plays a major role and its study is a key point for understanding the behavior of nuclear materials.

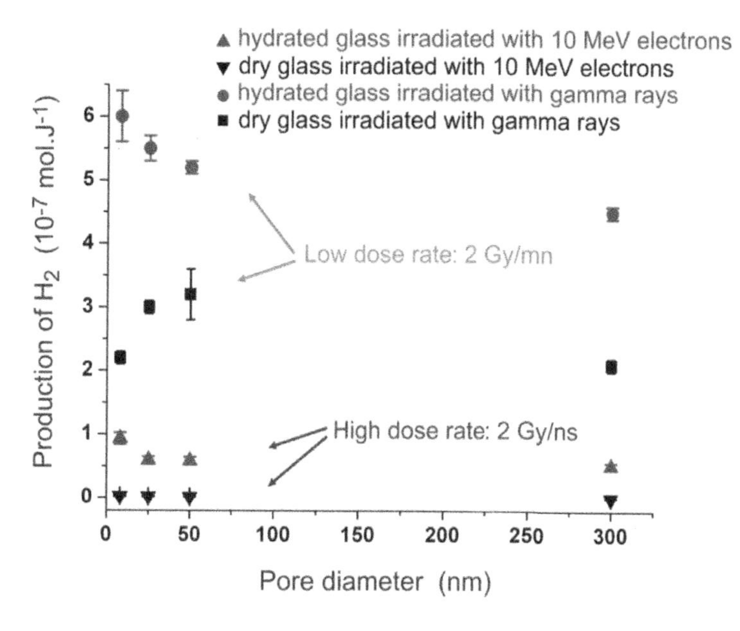

Figure 6.4. *Comparison of the radiolytic yields obtained for borosilicate glass with controlled porosity, both dried or totally hydrated, in the case of a low- or high-dose-rate irradiation (Le Caër et al. 2005; Rotureau et al. 2005). The yields are calculated by considering that the energy is deposited in the whole glass-water system (if present). For a color version of this figure, see www.iste.co.uk/bouffard/nuclear.zip*

6.6. Understanding transient phenomena

6.6.1. *Study of a short-lived species, the hydroxyl radical*

In order to understand the origin of radiolysis exaltation phenomena near interfaces, it is necessary to try to describe the underlying mechanisms of radical production and recombination.

The production of the hydroxyl radical ($HO^\bullet$) in confined water in zeolites was identified by Liu et al. (1997), through the trapping of organic compounds. The high value of the measured yield (about 6 10^{-7} $mol.J^{-1}$) is the sign of a hole transfer from matter to water. This work has been systematized to other porous materials, through the development of coumarin/benzoate hydroxyl radical capture systems (Louit et al. 2005; Musat et al. 2010a). These molecules, known as probes, produce fluorescent compounds by reaction with $HO^\bullet$ radicals that can be accurately measured (see Figure 6.5, with the example of benzoate). A characteristic time is associated with this reaction equation [6.3] It depends on the probe concentration and on the rate constant k associated with the capture reaction.

$$\tau = \frac{1}{k[probe]} \quad [6.3]$$

For example, an initial concentration of 100 mM of sodium benzoate makes it possible to scavenge hydroxyl radicals that are still present 1 to 2 ns after the ionizing radiation/matter interaction.

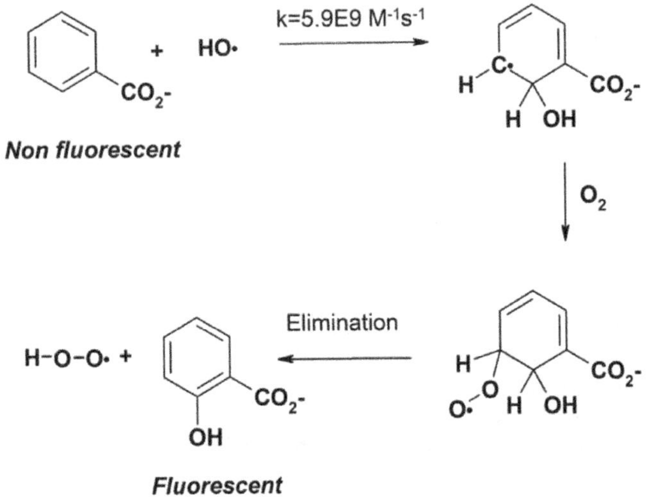

Non fluorescent

k=5.9E9 $M^{-1}s^{-1}$

O_2

Elimination

H-O-O• +

Fluorescent

Figure 6.5. *Benzoate hydroxylation mechanisms. The value of the rate constant for hydroxyl radical capture has been taken from Buxton et al. (1988)*

The production of hydroxyl radicals has also been measured in glass with controlled porosity, through selective capture by coumarin (Foley et al. 2005). In these systems, the energy transfer contribution to $HO^{\bullet}$ production is more limited than in zeolites. Indeed, it is at a maximum 30% for small pore sizes, and is negligible for large pore sizes. This behavior, which is significantly different from that of reducing species, has been associated with the lower mobility of holes in the material (Ouerdane et al. 2010).

The situation is completely different in the case of metal surfaces, which are susceptible to oxidation by hydroxyl radicals, especially since an overproduction of these radicals has been identified at short times and are mainly associated with a dose increase at the metal/solution interface (see Figure 6.6). However, this overproduction is transient and the radicals are, in most cases, rapidly consumed at the interface. In this case, comparison of the dose calculations with the measured yields also points to a second source of hydroxyl radicals, through activation of hydrogen peroxide, in a similar mechanism close to that of the Fenton reaction (Toijer and Jonsson 2019). This type of mechanism could also be implemented in the case of inorganic surfaces (Lousada et al. 2012), although probably in a less efficient way.

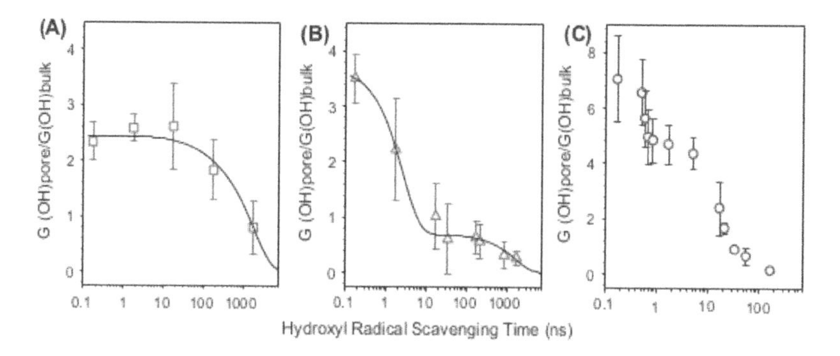

Figure 6.6. *Evolution of hydroxyl radical production relative to free water in different porous metals: (A) for SS316L steel; (B) for Hastelloy; (C) for porous gold. Figure taken from Moreau et al. (2014). Copyright © 2014 Elsevier Ltd. All rights reserved. For a color version of this figure, see www.iste.co.uk/bouffard/nuclear.zip*

6.6.2. *Confinement effect on the reactions taking place and their rate constants*

Beyond the measurement of the yields of transient species, a detailed mechanistic description requires the knowledge of the rate constants associated with the reactions of interest. However, the influence of surfaces or confinement phenomena on the values of the rate constants is, on the one hand, difficult to predict, and, on the other hand, difficult to determine experimentally. Indeed, in order to have kinetic data, the sample, which is excited by the radiation, must be sufficiently transparent to the analysis probe light that passes through it within the wavelength range of interest, in order for it to emerge. Moreover, in addition to this, this sample must have a certain thickness in order to obtain a measurable signal. It is clear that under these conditions, the number of porous materials that could be studied is very limited. The studies are restricted to relatively thin and optically transparent films, and the experiments are delicate to perform. A few pulse radiolysis studies have thus been able to be carried out (Musat et al. 2012; Lainé et al. 2017). They focus on the production of solvated electrons, which is a species whose absorption band is in the visible range, when the solvent is water. In addition, the molar extinction coefficient at the absorption band of solvated electrons is generally high (Torche and Marignier 2016).

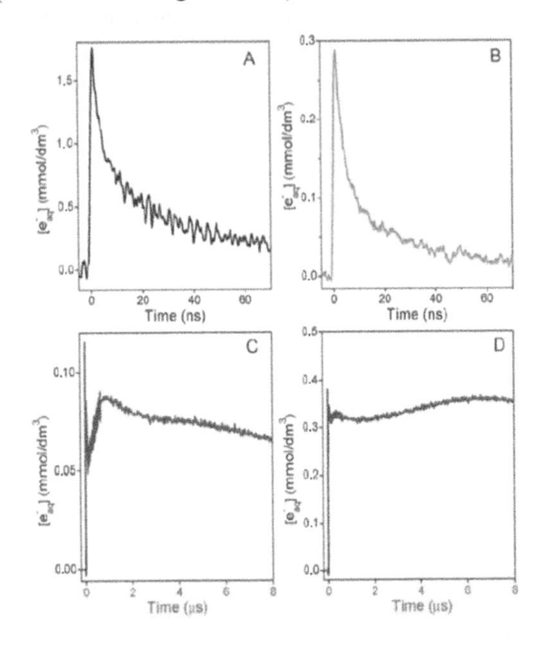

Figure 6.7. *Evolution of the hydrated electron concentration in different systems: (A) water; (B) water in silica pores of 1 nm diameter; (C) water in silica pores of 50 nm diameter; (D) water in alumina pores of 100 nm diameter. Figure from Musat et al. (2010b), reproduced with permission from the Royal Society of Chemistry. For a color version of this figure, see www.iste.co.uk/bouffard/nuclear.zip*

In the following, we have chosen to detail results obtained by exciting porous samples by laser photolysis (250 fs laser pulse at 266 nm). The production of the solvated electron is probed by a continuous laser. The samples are thin films of nanoporous silica (controlled porosity glass of 1 and 50 nm pore size) and alumina (100 nm pore size) (Musat et al. 2010b). By studying the evolution of the hydrated electron (see Figure 6.7) in free water and confined water, we note that small pores are particularly inefficient at producing and stabilizing the electron, and exhibit a behavior very similar to that of water in the homogeneous phase (see Figure 6.7(A) and (B)). Under these conditions, electron decay is governed by reactions with the material surface (Musat et al. 2012), and virtually no electrons are detected for times longer than 50 ns. In contrast, in large silica and alumina pores, the concentration of the solvated electron is stabilized around 10^{-5}–10^{-4} mol.dm^{-3} on times of several microseconds (see Figure 6.7(C) and (D)). The stabilization of this concentration is preceded by a delayed production of electrons (within 10 ns).

This delayed production is too slow to be attributed to direct ionization of the solid. However, it is consistent with energy transfer via excited species formed in the solid matrix. In silica under laser excitation, most of the excited species are self-trapped excitons (STEs) that are formed within 150 fs and have long lifetimes (R11, which is a combination of the R5 and R6 reactions above). Aqueous electrons generated by these excitons are consequently observed (R12):

$$SiO_2 \overset{h\nu}{\rightarrow} exciton \approx 150\, fs \qquad \text{(R11)}$$

$$exciton \xrightarrow{surface} e^-_{aq} + h^+_{silice} \qquad \text{(R12)}$$

This reaction (R12), with a rate constant of $10^7\ s^{-1}$, can account for the delayed electron production. These phenomena are related to the fact that the thickness of the silica walls increases as the pore size increases (this thickness is typically of the order of magnitude of the pore diameter). This also explains why this delayed production is not observed in the case of 1 nm pore size glass, since the diffusion of self-trapped excitons towards the surface takes place in less than a nanosecond.

Another question concerns the stability of the electron over long time scales. In large nanopores, the stability of the aqueous electron could only be explained by a slowing down of the disappearance reactions (reactions with itself or with oxidizing species), which would not be consistent with other experimental observations made in water and in pores that are 50 nm in diameter. The stability of the electron signal over long periods of time most likely comes from a compensation between the reactions that lead to its disappearance and the continuous injection from the solid matrix, generating a quasi-stationary state. The different stages are summarized in Figure 6.8.

This study thus highlights the fact that water in large pores (several tens of nanometers) exhibits an original behavior compared to neat water, which is not necessarily an intuitive result, but is related to the increase in wall thickness with pore size.

Lastly, let us note that up until now, excluding the phenomena of energy transfer and reaction on surfaces, no direct impact of porous media on reaction rates could be highlighted.

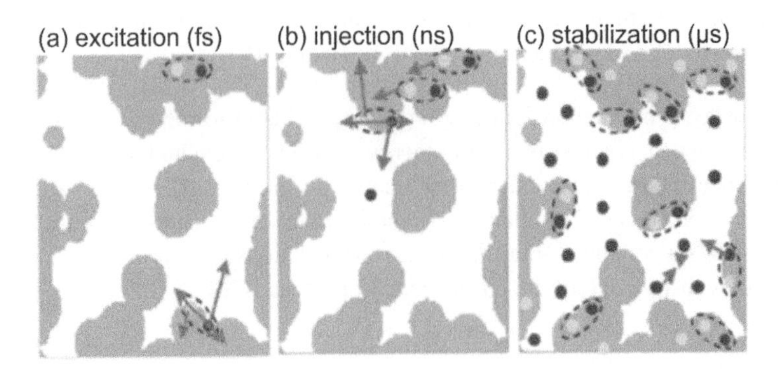

Figure 6.8. *Injection of electrons from the solid matrix to the confined liquid: (a) at the femtosecond time scale, the excitation of the solid and the formation of excited species within the solid take place. A direct injection from silica to water can be considered if the excited species are close to the interface; (b) the injection of electrons, resulting from the dissociation of excitons at the interface, takes place at the nanosecond time scale; (c) the concentration of electrons in the confined liquid is stabilized by the continuous injection from the solid. Figure from Musat et al. (2010b), reproduced with permission from the Royal Society of Chemistry. For a color version of this figure, see http://www.iste.co.uk/bouffard/nuclear.zip*

6.7. Conclusion: what about the effects of radiolytic species on materials?

In the case of heterogeneous solid/water systems, water chemistry is dominated by energy transfer. Evidence suggests a significant effect of species reactions on surfaces. This should therefore lead to questions about the impact of species created under radiolysis on the material. Hydrogen peroxide has long seemed to be the most important species. Its role has been particularly pointed out in the phenomena of exacerbated corrosion phenomena under radiation, because its long lifetime allows it to diffuse over large distances, from the solution to the surface. For metals of interest to the nuclear industry, even low concentrations of hydrogen peroxide can have a significant effect on corrosion potentials (Wang et al. 2020). However, oxidizing radicals such as $HO^•$ could also have a more local effect on material stability (Springell et al. 2015). The effect of reducing species is less documented, but the $H^•$ radical appears to absorb into copper, leading to the appearance of hydrogenated phases (Lousada et al. 2016). A large field of study is thus open to discovery, so as to better understand the effect of radiolytic species created on materials.

6.8. References

Brodie-Linder, N., Le Caër, S., Alam, M.S., Renault, J.P., Alba-Simionesco, C. (2010). H_2 formation by electron irradiation of SBA-15 materials and the effect of Cu^{II} grafting. *Physical Chemistry Chemical Physics*, 12, 14188–14195.

Buxton, G.V., Greenstock, C.L., Helman, W.P., Ross, A.B. (1988). Critical review of rate constants for reactions of hydrated electrons, hydrogen atoms and hydroxyl radicals (OH/O^-) in aqueous solution. *Journal of Physical and Chemical Reference Data*, 17, 513–531.

Chemirisov, S.C., Werst, D.W., Trifunac, A.D. (2001). Formation, trapping and kinetics of H atoms in wet zeolites and mesoporous silica. *Radiation Physics and Chemistry*, 60, 405–410.

Foley, S., Rotureau, P., Pin, S., Baldacchino, G., Renault, J.P., Mialocq, J.-C. (2005). Radiolysis of confined water: Production and reactivity of hydroxyl radical. *Angewandte Chemie International Edition*, 44(1), 110–112.

Fourdrin, C., Aarrachi, H., Latrille, C., Esnouf, S., Bergaya, F., Le Caër, S. (2013). Water radiolysis in exchanged-montmorillonites: The H_2 production mechanisms. *Environmental Science & Technology*, 47, 9530–9537.

Frances, L., Grivet, M., Renault, J.P., Groetz, J.-E., Ducret, D. (2015). Hydrogen radiolytic release from zeolite 4A/water systems under γ irradiations. *Radiation Physics and Chemistry*, 110, 6–11.

Griscom, D.L. (1984). Thermal bleaching of X-ray-induced defect centers in high purity fused silica by diffusion of radiolytic molecular hydrogen. *Journal of Non-Crystalline Solids*, 68, 301–325.

Lainé, M., Balan, E., Allard, T., Martin, F., von Bardeleben, H.-J., Robert, J.-L., Le Caër, S. (2016). Reaction mechanisms in talc under ionizing radiation: Evidence of a high stability of $H^\bullet$ atoms. *Journal of Physical Chemistry C*, 120, 2087–2095.

Lainé, M., Balan, E., Allard, T., Paineau, E., Jeunesse, P., Mostafavi, M., Robert, J.-L., Le Caër, S. (2017). Reaction mechanisms in swelling clays under ionizing radiation: Influence of the water amount and of the nature of the clay mineral. *RSC Advances*, 7, 526–534.

Le Caër, S. (2011). Water radiolysis: Influence of oxide surfaces on H2 production under ionizing radiation. *Water*, 3, 235–253.

Le Caër, S., Rotureau, P., Brunet, F., Charpentier, T., Blain, G., Renault, J.P., Mialocq, J.-C. (2005). Radiolysis of confined water: Hydrogen production at a high dose rate. *ChemPhysChem*, 6, 2585–2596.

Le Caër, S., Dezerald, L., Boukari, K., Lainé, M., Taupin, S., Kavanagh, R.M., Johnston, C.S.N., Foy, E., Charpentier, T., Krakowiak, K.J. et al. (2017). Production of H_2 by water radiolysis in cement paste under electron irradiation: A joint experimental and theoretical study. *Cement and Concrete Research*, 100, 110–118.

Liu, X., Zhang, G., Thomas, J.K. (1997). Spectroscopic studies of electron and hole trapping in Zeolites: Formation of hydrated electrons and hydroxyl radicals. *Journal of Physical Chemistry B*, 101, 2182–2194.

Louit, G., Foley, S., Cabillic, J., Coffigny, H., Taran, F., Valleix, A., Renault, J.P., Pin, S. (2005). The reaction of coumarin with the OH radical revisited: Hydroxylation product analysis determined by fluorescence and chromatography. *Radiation Physics and Chemistry*, 72(2–3), 119–124.

Lousada, C.M., Johansson, A.J., Brinck, T., Jonsson, M. (2012). Mechanism of H_2O_2 decomposition on transition metal oxide Surfaces. *Journal of Physical Chemistry C*, 116(17), 9533–9543.

Lousada, C.M., Soroka, I.L., Yagodzinskyy, Y., Tarakina, N.V., Todoshchenko, O., Hänninen, H., Korzhavyi, P.A., Jonsson, M. (2016). Gamma radiation induces hydrogen absorption by copper in water. *Scientific Reports*, 6, 24234.

Mainardi, E., Donahue, R.J., Wilson, W.E., Blakely, E.A. (2004). Comparison of microdosimetric simulations using PENELOPE and PITS for a 25 keV electron microbeam in water. *Radiation Research*, 162(3), 326–331.

Moreau, S., Fenart, M., Renault, J.P. (2014). Radiolysis of water in the vicinity of passive surfaces. *Corrosion Science*, 83, 255–260.

Musat, R., Moreau, S., Poidevin, F., Mathon, M.H., Pommeret, S., Renault, J.P. (2010a). Radiolysis of water in nanoporous gold. *Physical Chemistry Chemical Physics*, 12(39), 12868–12874.

Musat, R., Vigneron, G., Garzella, D., Le Caër, S., Hergott, J.F., Renault, J.P., Pommeret, S. (2010b). Water reduction by photoexcited silica and alumina. *Chemical Communications*, 46, 2394–2396.

Musat, R.M., Cook, A.R., Renault, J.-P., Crowell, R.A. (2012). Nanosecond pulse radiolysis of nanoconfined water. *Journal of Physical Chemistry C*, 116, 13104–13110.

Ouerdane, H., Gervais, B., Zhou, H., Beuve, M., Renault, J.P. (2010). Radiolysis of water confined in porous silica: A simulation study of the physicochemical yields. *Journal of Physical Chemistry C*, 114, 12667–12674.

Petrik, N.G., Alexandrov, A.B., Vall, A.I. (2001). Interfacial energy transfer during gamma radiolysis of water on the surface of ZrO_2 and some other oxides. *Journal of Physical Chemistry B*, 105(25), 5935–5944.

Rotureau, P., Renault, J.P., Lebeau, B., Patarin, J., Mialocq, J.C. (2005). Radiolysis of confined water: Molecular hydrogen formation. *ChemPhysChem*, 6(7), 1316–1323.

Shkrob, I.A. and Trifunac, A.D. (1997). Spin-polarized H/D atoms and radiation chemistry in amorphous silica. *Journal of Chemical Physics*, 107(7), 2374–2385.

Spinks, J.W.T. and Woods, R.J. (1990). *An Introduction to Radiation Chemistry*, 3rd edition. Wiley, New York.

Springell, R., Rennie, S., Costelle, L., Darnbrough, J., Stitt, C., Cocklin, E., Lucas, C., Burrows, R., Sims, H., Wermeille, D. et al. (2015). Water corrosion of spent nuclear fuel: Radiolysis driven dissolution at the UO2/water interface. *Faraday Discussions*, 180, 301–311.

Stewart, R.D., Wilson, W.E., McDonald, J.C., Strom, D.J. (2002). Microdosimetric properties of ionizing electrons in water: A test of the PENELOPE code system. *Physics in Medicine and Biology*, 47(1), 79–88.

Thomas, J.K. (1993). Physical aspects of photochemistry and radiation chemistry of molecules adsorbed on SiO_2, gamma-Al_2O_3, zeolites, and clays. *Chemical Reviews*, 93, 301–320.

Toijer, E. and Jonsson, M. (2019). H_2O_2 and γ-radiation induced corrosion of 304L stainless steel in aqueous systems. *Radiation Physics and Chemistry*, 159, 159–165.

Torche, F. and Marignier, J.L. (2016). Direct evaluation of the molar absorption coefficient of hydrated electron by the isosbestic point method. *Journal of Physical Chemistry B*, 120(29), 7201–7206.

Wang, P., Grdanovska, S., Bartels, D.M., Was, G.S. (2020). Effect of radiation damage and water radiolysis on corrosion of FeCrAl alloys in hydrogenated water. *Journal of Nuclear Materials*, 533, 152108.

7

Concrete and Cement Materials under Irradiation

Pascal BOUNIOL
CEA/Paris-Saclay, Service d'Étude du Comportement des Radionucléides, Gif-sur-Yvette, France

7.1. Introduction

Concrete and other cement-based materials (mortars, grouts, cement-coated waste) have a wide variety of compositions that cover a large number of applications in the nuclear field: civil engineering (reactor vessels, cooling towers), radiation shielding (casemates, irradiators, accelerators), packaging of low- and medium-level radioactive waste (containers, coating matrices). The materials that are most exposed to radiation and for which there are behavior and/or aging problems, also referred to as durability under irradiation, fall into two completely different fields of application in both their functioning and their issues:

– Reactor vessel well concretes with mixed gamma-neutron dose rates of the order of 3,000 Gy/h. With a multi-metric thickness and a high density of steel reinforcements, they have both a structural and a radiation attenuation role.

– Coating matrices for long-lived intermediate-level waste (up to 10^{14} Bq/pack), with mixed alpha–beta–gamma dose rates, depending on the case, of the order of 30 to 300 Gy/h. Their role is to physically immobilize the waste and limit the spread of radionuclides.

These two categories of materials are currently being researched in relation to their respective operating conditions and durations, with the modalities of the

Nuclear Materials under Irradiation,
coordinated by Serge BOUFFARD and Nathalie MONCOFFRE.

radiation-material interaction being specific in each case. In the first case, the desire to extend the operating life of nuclear power plants from 40 to 60 years (France), or even 80 years (USA), must also consider the problem of aging under irradiation, due to the more or less pronounced progressive swelling of certain concrete aggregates beyond neutron fluences of $2\ 10^{19}$ n/cm^2 (Rosseel et al. 2015). In the second case, the systematic presence of residual water in the porosity of cement materials is the cause of radiolytic dihydrogen production, once the waste activity is in direct contact with the cement matrix. From dose rates of the order of 10^{-3} and 10^{-5} Gy/s respectively in β, γ and α irradiation, the emission of gaseous H_2 becomes significant, and the exceeding 4% content in the air surrounding the cemented waste packages poses a safety problem from the first few years of operation. The particularities associated with these two examples make it possible to focus respectively on the physical phenomena appearing in the long term, or the chemical phenomena appearing in the short term.

7.2. Radiation shielding concrete

7.2.1. *Overview*

Whether they are ordinary or specially designed for radiation shielding, concretes are composite materials that are made of mineral aggregates, hydraulic binder (usually Portland cement) and water. With the intention of significantly reducing the thickness of the material while maintaining the same attenuation power, traditional aggregates (limestone, silica or a mix) of low density (about 2,600 kg/m^3) can be replaced by special aggregates, such as hematite (4,900 kg/m^3) for protection against X-rays and γ rays, or colemanite[1] (2,420 kg/m^3) for protection against neutrons (Bouniol 2001). After hardening (resulting from the reaction of water with cement to give hydrated products with binding properties), concretes are presented in the form of a granular framework, coated with a porous cement paste. Within this composite material, which is made up of solids with different physical properties (thermal conductivity, expansion coefficients, etc.), radiation induces necessarily differential effects (thermomechanical stresses) at the origin of aging. In addition to the direct effect of irradiation on the aggregates, the thermal cycling associated with periods of operation is the cause of fatigue, which itself contributes to aging. Overall, the concretes resist well to gamma and neutron irradiation during the first decades. The decrease in mechanical performance (compressive and tensile strength, elasticity) begins when the quantity of accumulated defects crosses a percolation threshold, corresponding to a fluence greater than 10^{19} n/cm^2. However, this is a

1. Colemanite: hydrated calcium borate, $Ca_2B_6O_{11}.5H_2O$.

trend (see Figure 7.1), as the composition of the materials and their operating conditions are very varied.

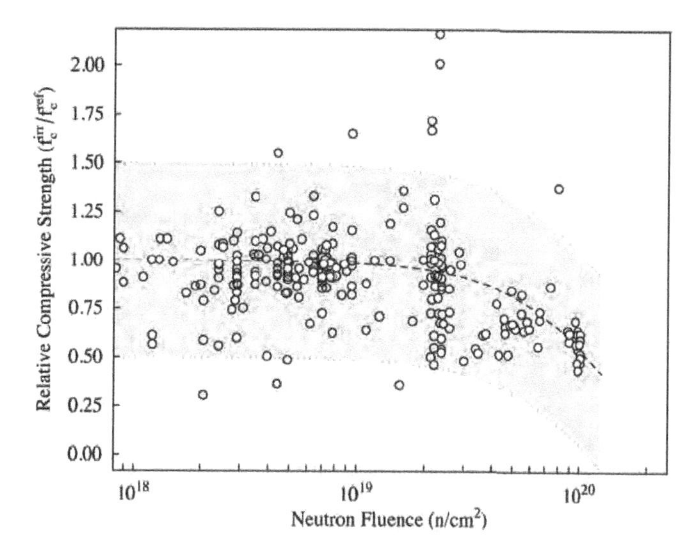

Figure 7.1. *Relative compressive strength of different concretes as a function of neutron fluence, from Field et al. (2015)*

7.2.2. Effects of irradiation on the cement matrix

In the case of Portland cement-based concretes, the cement matrix is essentially made up of hydrated calcium silicate (C–S–H) in a poorly crystallized state and calcium hydroxide (portlandite), with some secondary minerals (ettringite[2], hydrocalumite[3]). Due to the rather amorphous and non-stoichiometric nature of the main hydrate (a gel with its own microporosity), as well as the presence of a capillary porosity, the cement matrix appears as a disordered medium at different scales, which is likely to evolve more easily than the aggregates according to the thermo-hydro-mechanical stresses under irradiation The direct effect of gamma irradiation on the material does not seem to be significant in terms of modification of the Ca/Si ratio or the structural environment of the silicate tetrahedra (Tajuelo Rodriguez et al. 2020). However, it is possible to observe a more global effect under neutron and gamma irradiation. This is a creep variety, a phenomenon that is typically identified for mechanically loaded cement materials that are undergoing curing. The evolution of the pore size distribution in a Portland cement paste under

2. Ettringite: hydrated calcium aluminum trisulfate, $3CaO.Al_2O_3.3CaSO_4.32H_2O$.

3. Hydrocalumite: hydrated calcium aluminate, $4CaO.Al_2O_3.13H_2O$.

gamma irradiation (see Figure 7.2) is part of this phenomenon and illustrates a relative plasticity of the material when it receives energy, the total porosity remaining constant. This property is relatively "useful" in irradiated concrete, for accompanying and absorbing part of the stresses transmitted by aggregates that have a specific behavior under radiation.

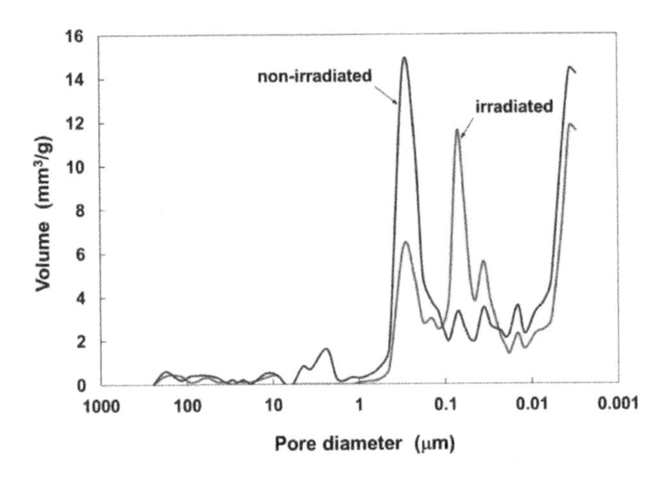

Figure 7.2. *Pore size distribution (mercury porosimetry) in a Portland cement paste with a water/cement mass ratio of 0.4 before and after gamma irradiation, for an integrated dose of 1.2 MGy (Bouniol 2004). For a color version of this figure, see www.iste.co.uk/bouffard/nuclear.zip*

If there is no mineralogical transformation of the cementitious solid phases under irradiation, the coexistence of portlandite and radiolytic peroxide within the pore solution can, depending on the conditions, give rise to the heterogeneous precipitation of calcium peroxide octahydrate at room temperature (Bouniol and Lapuerta-Cochet 2012): $Ca(OH)_2 + H_2O_2 + 6\ H_2O \rightarrow CaO_2.8\ H_2O$. However, the occurrence of this unstable compound is unlikely within concrete with a service temperature above 50°C.

7.2.3. *Effects of irradiation on aggregates*

It is known that neutron-irradiated siliceous aggregates are prone to swelling. For quartz, in particular, amorphization of the crystal lattice leads to a maximum volume increase of about 17% (see Figure 7.3). Many silicate minerals (nesosilicates, inosilicates, tectosilicates) display the same behavior (Krishnan et al. 2018), with the magnitude of the phenomenon depending on the degree of accommodation allowed by the initial crystal structure. There are therefore some exceptions (anorthite[4]), on the contrary, where densification is observed. In the case of quartz, and all the aggregates that contain it (quartzite, granite, etc.), we note that the fluences necessary to obtain a significant swelling are the same as those causing damage to concrete (10^{19} to 10^{20} n/cm^2).

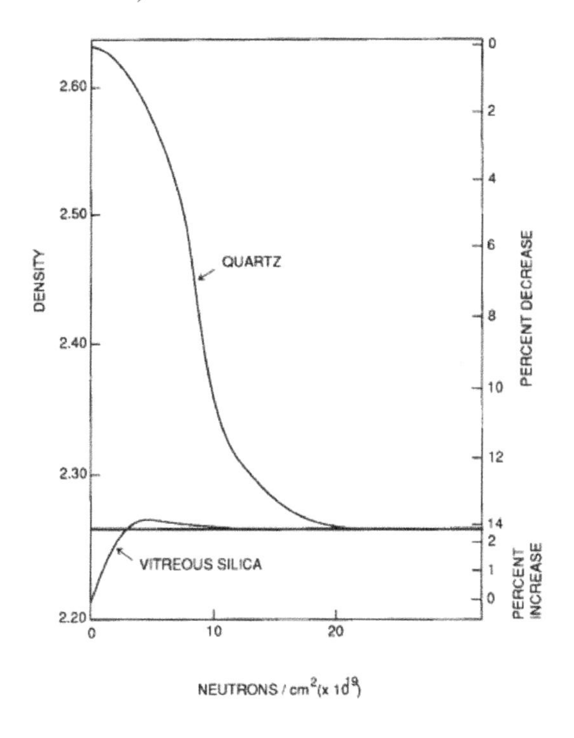

Figure 7.3. *Density variation of alpha quartz and silica glass under neutron irradiation, according to Lell et al. (1966)*

For ordinary concretes, the second source of aggregates are carbonate rocks. Contrary to siliceous aggregates, where the Si–O bond is of covalent nature, limestones (the very majority) and dolomites are characterized by crystalline structures, where the cation–CO_3 bond is of ionic nature. This results in a clear difference in behavior under irradiation, with relaxation and repair leading to an absence of significant swelling (Hsiao et al. 2019), even up to 10^{20} n/cm^2 ($E < 0.1$ MeV). For many minerals, the experimentally determined radiation-induced volumetric expansion (RIVE) appears to be a key input in neutron irradiation

4. Anorthite: formula tectosilicate, $CaAl_2Si_2O_8$.

damage models (Le Pape et al. 2018). For quartz, as for many minerals, it can be described by an empirical function with a sigmoidal shape:

$$\varepsilon(\phi) = \varepsilon_{max}^{*}\left(1 - exp\left(-\left(\frac{\phi}{\phi_{1/2(T)}}\right)^{d}\right)\right)$$

with ε_{max}^{*} being the maximum volume expansion (%), ϕ and $\phi_{1/2}(T)$ respectively being the nominal fast neutron fluence and the fluence corresponding to half of the maximum observed expansion at temperature *T*, *d*: exponent between 2 and 5. For carbonate materials, the sigmoid form is not observed and is replaced by a threshold and plateau function (see Figure 7.4).

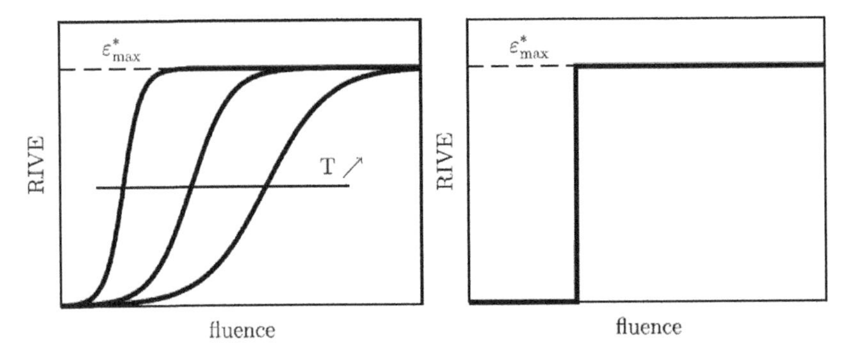

Figure 7.4. *Two empirical radiation-induced volumetric expansion (RIVE) models for quartz (left) and calcite (right), from Le Pape et al. (2018)*

From the point of view of the RIVE criterion, the gamma component remains negligible in the case of simultaneous gamma-neutron exposure. In the case of gamma irradiation alone, changes are nevertheless possible above 10^9 Gy (Denisov 2020).

7.2.4. *Prediction of concrete damage*

The macroscopic behavior of concretes that are subjected to a mixed irradiation of long duration is very complex, because it involves multi-scale and multi-phase processes. The existing models are based on a classical thermo-hydro-mechanical description, by integrating phenomenologies specific to irradiation: gradients of dose and temperature, specific creep, swelling of the aggregates and coupled modification of the parameters of elasticity and thermal conductivity. The difficulty is all the greater since the appearance of micro-cracks, which turn into fractures, is accompanied by a change in water transfer that is at the origin of the paste shrinkage, as well as additional creep. The models require the acquisition of very large amounts of experimental data (Maruyama et al. 2017), which are difficult to obtain from real materials whose history is partially or poorly known. Beyond the number of elementary data in the description of the behavior, the homogenization techniques used (mainly the Mori-Tanaka method) and the partial couplings (creep-damage, radiation transport-heat transfer) make it possible to account for some trends.

7.3. Waste conditioning matrices

7.3.1. *Overview*

Depending on the nature of the waste to be embedded, whether it is liquid (evaporator concentrates) or solid (metallic technological waste, ion exchange resins, incinerator ashes, etc.), the materials resulting from the mixture with cement are very varied and have a composite nature, like that of concretes. Compared to the latter, the service conditions and specifications are quite distinct, with little or no associated heat (ambient temperature) and an absence of a structural role (few requirements on mechanical performance). However, the integrity of the mix must be maintained over time, which is generally guaranteed by the absence of physico-chemical interactions between the waste and the cement matrix, as well as the absence of short-term hindrances in transport or storage situations. This last point particularly concerns the production of dihydrogen by radiolysis of the pore water (which is systematically present in any cement material), a phenomenon that appears as soon as the waste is brought into contact with the mix during its manufacture. Based on the mixes that are generally wrapped in "breathing" packages, that is, not gas-tight, the amplitude of H_2 gas emission depends on three parameters: 1) the dose rate associated with the activity and the nature of the radiation; 2) the chemical context of the cement (very high pH and reactive solutes); 3) the convective and diffusive transport of the gases (H_2 and O_2) permitted by the porosity and its liquid saturation rate. In practice, the estimation of the amount of H_2 produced is a coupled radiation chemistry-transport problem.

7.3.2. *Radiolysis of the cement matrix*

The presence of a residual liquid phase in the cement material after hardening makes the chemical aspect of the phenomenon predominant. The pH around 13 notably implies several changes, in relation to the description of the radiolysis of solutions at intermediate pH levels (4–10).

Due to the higher interaction of primary species with the OH^- ion in high concentration during the inhomogeneous stage (before 10^{-6} s), the primary yields for α and β γ radiation are characterized by a slight increase in the radical yield (see Table 7.1).

Radiation	H_2	e^-_{aq}	$H^\bullet$	H_2O	H_3O^+	$OH^\bullet$	H_2O_2	$HO_2^\bullet$
α	1.47	0.41	0	–3.87	0.41	0.32	1.065	0.30
βγ	0.421	2.755	0.55	–9.656	2.755	2.917	0.615	0

Table 7.1. *Primary yields (molecules/100 eV) used to simulate α radiolysis (LET = 130 keV/m) and β γ radiolysis in a cement medium at pH 13 and 25°C (Bouniol 2004)*

Due to the fact that all acid–base pairs of radiolytic species have a pK_a less than 13, the species present in the cement medium differ from those observed at acidic or intermediate pH levels. The $H^\bullet$, $OH^\bullet$, $HO_2^\bullet$ and H_2O_2 species are thus replaced by e^-_{aq}, $O^{\bullet-}$, $O_2^{\bullet-}$ and HO_2^- respectively, with distinct reactivities. As a result, among the 60 or so secondary reactions involving primary and newly derived species, the reaction chain responsible for regulating radiolysis at intermediate pH levels (Allen chain) is replaced by its equivalent at basic pH levels. Despite an additional intermediate stage, the basic form shows faster kinetics, leading in many cases to a more efficient recycling of H_2.

$$O^{\bullet-} + H_2 \xrightarrow{k_1} OH^- + H^\bullet \qquad k_1 = 1.3\ .10^8\ dm^3.mol^{-1}.s^{-1}\ (25°C)$$

$$OH^- + H^\bullet \xrightarrow{k_2} e^-_{aq} \qquad k_2 = 2.4\ .10^7\ dm^3.mol^{-1}.s^{-1}$$

$$e^-_{aq} + HO_2^- \xrightarrow{k_3} O^{\bullet-} + OH^- + H_2O \quad k_3 = 3.5\ .10^9\ dm^3.mol^{-1}.s^{-1}$$

This mechanism helps to understand why the production of H_2 from a cemented waste package, over time, is generally much lower than the theoretical primary production. However, its sensitivity to the chemical characteristics of the environment is comparable to that of Allen's chain, and the recycling of H_2 can be less efficient in the case of poisoning by species such as O_2, Fe^{3+} and SH^- (the most frequent solutes). Due to their higher kinetic constants, these solute reactions are in competition with those of the mechanism, which can even lead to the establishment of a parasitic chain, as is the case of iron with the simultaneous presence of Fe(+3) and Fe(+2):

$$O^{\bullet-} + Fe(+2) \rightarrow O^{2-} + Fe(+3) \qquad k = 3.8 \times 10^{9}\ dm^{3}\ mol^{-1}\ s^{-1}$$

$$e^{-}_{aq} + Fe(+3) \rightarrow H_2O + Fe(+2) \qquad k = 6.0 \times 10^{10}\ dm^{3}\ mol^{-1}\ s^{-1}$$

Experimentally, this results in a higher net production of radiolytic H_2 (see Figure 7.5), which actually means that for a given production, less is destroyed.

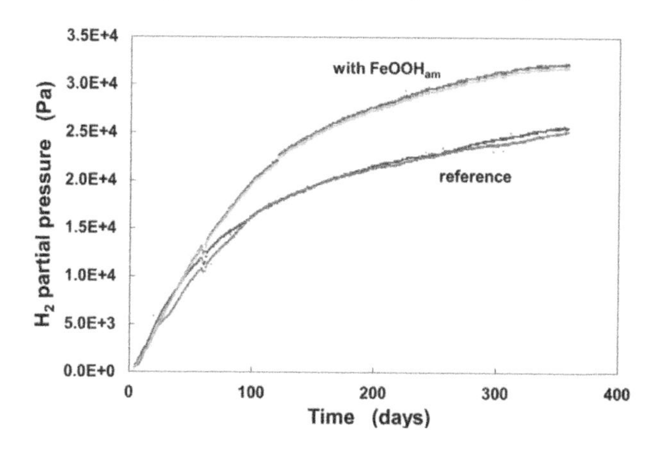

Figure 7.5. *Radiolytic H_2 production by cement pastes (pure Ca_3SiO_5) in a closed system, with and without $FeOOH_{am}$ with a gamma dose rate of 0.3 Gy/s in pore water (Bouniol et al. 2013). For a color version of this figure, see www.iste.co.uk/bouffard/nuclear.zip*

In the case of SH^-, the choice of cement is clearly involved. In a CEM III blended cement, the slag is an important source of sulfide, which gives the pore medium an immediate reducing character. Under γ irradiation, this configuration is characterized by a two-stage evolution of the system: 1) disruption of the recycling reaction chain and secondary formation of dihydrogen by the reaction $H^{\bullet} + H_2S \rightarrow H_2 + SH^{\circ}$ ($k = 9 \times 10^{9}\ M^{-1}.s^{-1}$); 2) conversion of the sulfide into polysulfides, and reactivation of the recycling chain with a strong regulating effect. As a result, the behavior of an ordinary Portland cement (CEM I) and a slag cement (CEM III) appear very different on the basis of the recycling rate R (see Figure 7.6). This criterion, which compares the effective and primary production rates, is useful when assessing the ability of a cement material to regulate radiolysis.

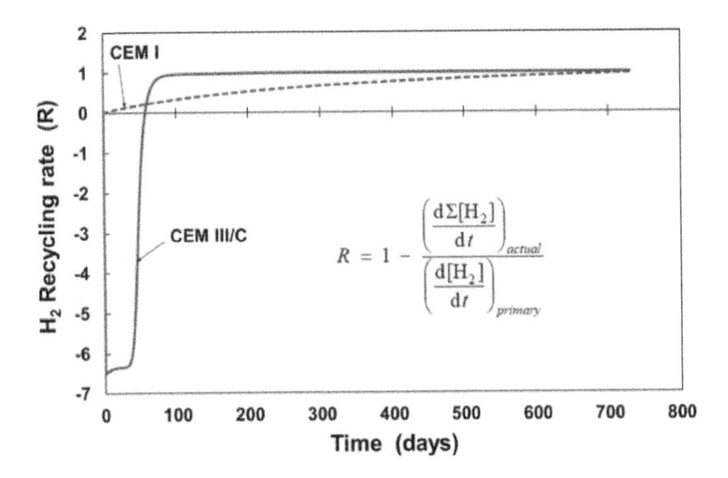

Figure 7.6. *Radiolytic H_2 recycle rate (R) for CEM I and CEM III cement pastes in a closed system, for a gamma dose rate of 0.1 Gy/s in pore water (Bouniol et al. 2018). For a color version of this figure, see www.iste.co.uk/bouffard/nuclear.zip*

7.3.3. *Phenomenological couplings*

Within a cemented waste mix, the description of radiolysis goes far beyond the sole aspect of alkaline water decomposition under radiation, which makes a global phenomenological approach necessary. Various *peripheral* phenomena, which are however strongly coupled with radiation chemistry, must be considered, as they contribute to the profound modification of the net radiolysis balance. Among the most determining couplings is the evolution of radiolytic chemistry with that of radiation types that naturally intervenes in the management of cemented waste, over the long term (a few hundred years). From a multi-emitter radiological inventory within which the long-lived α emitters remain present, the disappearance of the $\beta\gamma$ emitters first modifies the nature of the radiolysis. The alpha primary yield values in Table 7.1 not only predict a higher production of H_2, but a very low production of the $OH^{\bullet}$ radical supposed to attack it. The efficiency of the recycling reaction chain becomes very limited under these conditions. The coupling of the radiolytic chemistry with the gas transport is, however, the strongest. The passage of radiolytic H_2 into the gas phase within the porosity not occupied by the liquid phase is at the origin of this gas transfer (and of all the others that are present), according to two modes: gas diffusion, which is preponderant when the total pressure difference is low between the internal medium and the outside; and permeation in the opposite case. For very lightly saturated cement materials (the saturation rate of a sound material is around 90%), Figure 7.7 shows that H_2 production is proportional to the amount of water that can undergo radiolysis, with all of the gas being gradually expelled. Above 60% water occupancy in the porosity, the resistance to gas transport increases, resulting in increased H_2 residence time and its recombination in situ (Foct et al. 2013). In the vicinity of 100% saturation, radiolysis behaves the same as in a closed system. Consideration of this coupling is fundamental in assessing the amounts of H_2 produced by cemented waste packages which are affected by long-term drying, because the responses can be very different for comparable dose rates, depending on the nature of the mix.

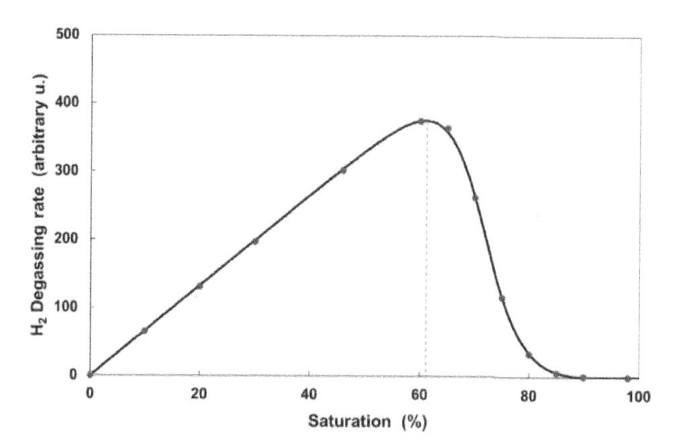

Figure 7.7. *Radiolytic H_2 degassing rate for a cemented waste package as a function of porosity saturation by liquid water, at a given integrated dose (simulations)*

7.4. Conclusion

Whether for reactor concretes or waste conditioning materials, obtaining operational models is essential in order to validate the industrial choices made with the safety authorities. As a result of the composite nature of cement materials, the development of these models must be based on the acquisition of numerous experimental data, on both the chemistry under radiation and the properties of the aggregates, as well as on a higher level of integration of the elementary mechanisms that have already been identified (couplings, homogenization methods). This does not exclude the consideration of other issues such as the corrosion of steel reinforcements, the role of additives inhibiting radiolysis, reactivity in porous media and the contribution of solid cement phases to the production of radiolytic H_2, to name but a few of the most prominent topics.

7.5. References

Bouniol, P. (2001). Bétons spéciaux de protection. *Techniques de l'ingénieur matériaux pour le nucléaire*, bn3740, 1–29.

Bouniol, P. (2004). État des connaissances sur la radiolyse de l'eau dans les colis de déchets cimentés et son approche par simulation. Report, CEA-R-6069.

Bouniol, P. and Lapuerta-Cochet, S. (2012). The solubility constant of calcium peroxide octahydrate in relation to temperature: Its influence on radiolysis in cement-based materials. *Journal of Nuclear Materials*, 420(1), 16–22.

Bouniol, P., Muzeau, B., Dauvois, V. (2013). Experimental evidence of the influence of iron on pore water radiolysis in cement-based materials. *Journal of Nuclear Materials*, 437(1), 208–215.

Bouniol, P., Guillot, W., Dauvois, V., Dridi, W., Le Caër, S. (2018). Original behavior of pore water radiolysis in cement-based materials containing sulfide: Coupling between experiments and simulations. *Radiation Physics and Chemistry*, 150(9), 172–181.

Denisov, A.V. (2020). Radiation changes of concrete aggregates under the influence of gamma radiation. *Magazine of Civil Engineering*, 96(4), 94–109.

Field, K.G., Remec, I., Pape, Y.L. (2015). Radiation effects in concrete for nuclear power plants – Part I: Quantification of radiation exposure and radiation effects. *Nuclear Engineering and Design*, 282, 126–143.

Foct, F., Di Giandomenico, M.-V., Bouniol, P. (2013). Modelling of hydrogen production from pore water radiolysis in cemented intermediate level waste. In *EPJ Web of Conferences*, 56, 05002.

Hsiao, Y.-H., Wang, B., La Plante, E.C., Pignatelli, I., Krishnan, N.M.A., Le Pape, Y., Neithalath, N., Bauchy, M., Sant, G. (2019). The effect of irradiation on the atomic structure and chemical durability of calcite and dolomite. *Npj Materials Degradation*, 3(1), 36.

Krishnan, N.M.A., Le Pape, Y., Sant, G., Bauchy, M. (2018). Effect of irradiation on silicate aggregates' density and stiffness. *Journal of Nuclear Materials*, 512, 126–136.

Le Pape, Y., Alsaid, M.H.F., Giorla, A.B. (2018). Rock-forming minerals radiation-induced volumetric expansion – Revisiting literature data. *Journal of Advanced Concrete Technology*, 16(5), 191–209.

Lell, E., Kreidl, N.J., Hensler, J.R. (1966). Radiation effects in quartz, silica and glasses. In *Progress in Ceramic Science*, Burke, J.E. (ed.). Pergamon Press, Oxford.

Maruyama, I., Kontani, O., Takizawa, M., Sawada, S., Ishikawao, S., Yasukouchi, J., Sato, O., Etoh, J., Igari, T. (2017). Development of soundness assessment procedure for concrete members affected by neutron and gamma-ray irradiation. *Journal of Advanced Concrete Technology*, 15(9), 440–523.

Rosseel, T.M., Field, K.G., Le Pape, Y., Naus, D.J., Remec, I., Busby, J.T., Wall, J.J., Bruck, P. (2015). Dommages d'irradiation dans les cavités en béton des réacteurs aux États-Unis. *Revue générale du nucléaire*, 1, 21–27.

Tajuelo Rodriguez, E., Hunnicutt, W.A., Mondal, P., Le Pape, Y. (2020). Examination of gamma-irradiated calcium silicate hydrates. Part I: Chemical-structural properties. *Journal of the American Ceramic Society*, 103(1), 558–568.

8

Organic Materials

Emmanuel BALANZAT[1] **and Muriel FERRY**[2]
[1] *CIMAP, CEA – CNRS – ENSICAEN, Université Caen Normandie, France*
[2] *CEA/Paris-Saclay, Service de Physico-Chimie, Gif-sur-Yvette, France*

8.1. Introduction

Organic materials, especially polymers, have a significant place in the nuclear power industry, and they are found everywhere in various forms. As such, their behavior under irradiation must be carefully studied. The fields concerned are: the safety of nuclear installations, the management of nuclear waste and the upstream cycle. Here, we will mostly deal with polymers, but not exclusively. Their structure presents several levels of organization at the molecular, macromolecular and supramacromolecular scales; they can be glassy or semi-crystalline. They have very diverse physical properties, for example, elastic, plastic or brittle. Faced with this enormous chemical and microstructural diversity, and as a consequence of a wide range of properties of use, it is impossible to individually deal with each polymer used in the nuclear industry. The aim of this chapter is to give the reader the most generic concepts possible, based on a few specific examples.

The vast majority of nuclear materials, metals and ionocovalent insulators have their structure modified by the effect of atomic displacement. On the contrary, in organic materials, damage is associated with the transfer of energy from the incident radiation to the electrons of the material, that is, the excitation of the target electrons and/or the ionization of the target atoms. The modification of organic materials is governed by the electron stopping power of the projectiles $(dE/dx)_e$, which we shall consider equivalent to the linear energy transfer (LET). These materials are said to be "sensitive to radiolysis", and the underlying concepts and mechanisms of damage

Nuclear Materials under Irradiation,
coordinated by Serge BOUFFARD and Nathalie MONCOFFRE.

have little in common with those that prevail, for example, in metals. The definition of radiolysis is the breaking of one or more chemical bonds by the action of ionizing radiation. Thus, in the nuclear sphere, in addition to polymers and organic materials, radiolysis effects are found for so-called "free" or "bound" water (see Chapter 7 and Chapter 8). In a gas-cooled reactor, GCR, we must also consider the radiolysis of CO_2 (see Chapter 4). For certain stages of nuclear industrial processes, lubricants may be necessary; their radiolysis is then inevitable. Lastly, to be complete, it should be mentioned that some insulators such as alkali halides are also sensitive to radiolysis, and should therefore be considered in the very specific case of a storage site construction for LL-HLW and LL-ILW[1] waste in a geological site where the barrier is salt.

As in the other chapters of this book that deal with materials that are sensitive to radiolysis, we will now present several important quantities. The quantity that will characterize irradiation above all others will be the dose (D). The dose is expressed in Gray (Gy); one Gray corresponds to one Joule of energy absorbed, in the form of ionization/excitation, by one kilogram of material. The response of the material to radiation is given by the radiochemical yield G (mol/J), which is the ratio between the molar quantity of modifications produced and the quantity of energy (ionization and excitation) absorbed by the material. The concentration of the induced modifications (mol/kg) is therefore given by the product of the radiochemical yield and the dose: $G \times D$. The radiochemical yields are experimental data, because they cannot currently be calculated. We will also see that the dose rate can come into play, in other words, that the result, at an equal dose, will depend on the exposure time. Controlling dose rate effects is essential when accelerated aging tests are conducted.

This dose rate effect almost always comes from the fact that polymers are relatively gas-permeable materials. This has two major consequences. The first is that in the presence of air, dioxygen enters the material and the unstable species created in the polymer will react with this gas, leading to a consumption of this dissolved oxygen and the formation of oxidized defects. This specific phenomenon is called radio-oxidation. However, in order for the presence of oxygen within the material to always be effective, it is necessary to have a sufficient permeation, in other words, a sufficiently fast diffusion of oxygen in the thickness of the polymer, so as to compensate for its consumption by radio-oxidation; in this case, a homogeneous radio-oxidation of the material will occur. The second consequence of

1. Long-lived high-level waste (LL-HLW) and long-lived intermediate-level waste (LL-ILW); deep geological disposal sites are being studied in several countries.

the permeability of polymers is that the gaseous molecules produced by radiolysis can escape from the material, finding themselves in the surrounding atmosphere. These gases rarely aggregate into bubbles, and only do so at high dose rates.

In the nuclear sphere, the radiolysis of polymers and organic materials unfortunately leads to a deterioration of the functionality and functional properties of the material. However, this should not obscure the fact that irradiation can be a remarkable tool to create new polymers, or to modify their properties in a beneficial way (Clough 2001). This is evidenced by the fact that more than 1,500 accelerators worldwide are dedicated to these commercial production activities.

Several books review the effects of irradiation in polymers; this chapter is not intended to deal with the subject in depth, but simply to introduce the technological context and to present the main effects of irradiation of organic materials on a slightly more fundamental level. For a more in-depth look at the subject, let us first mention the book edited by Dole (1973), *The Radiation Chemistry of Macromolecules*, which despite being published almost half a century ago, remains an essential reference. More recently, we recommend two books published in 1991, one by Clegg and Collyer (1991), and the other by Clough and Shalaby (1991). In addition to this, let us point out the very recent chapter "Polymers in the nuclear power industry" in the *Comprehensive Nuclear Materials* (Ferry et al. 2020) encyclopedia, which deals with the subject in more depth than in this chapter. Lastly, let us mention the chapter "Ionizing radiation effects in polymers" (Ferry et al. 2016), which gives the basics of the radiation/polymer interaction phenomena, before discussing some of the technological applications that can be envisaged in the industry, excluding nuclear power.

8.2. Technological context

8.2.1. *Organic materials of the nuclear industry*

In the nuclear industry, four major families of organic materials are used. The first and most abundant is the polymer family. The second one relates to ion exchange resins and reverse osmosis membranes. Although they are made of polymers, a focus is made on this type of material because of their specific use in the nuclear industry. The third category includes organic additives, which are traditionally used in the manufacture of oxide fuels by sintering powders for the implementation and control of the final porosity. The last category is that of bitumen, a blocking matrix for waste packages.

8.2.1.1. *Polymers*

Polymers are the most represented family of organic materials in the nuclear industry. These are mainly cables (electrical insulators and outer sheaths), elastomeric gaskets, adhesives, paints in the form of coatings, gloves and glove box panels, wipes, plastic boxes and vinyl films (which are used to bag and thus contain contamination). There is also a small number of silicon polymers, which formally speaking are not organic materials, but can be considered as such with respect to radiation effects. Figure 8.1 gives a schematic view of the distribution of different classes of polymers that are usually used in the nuclear industry.

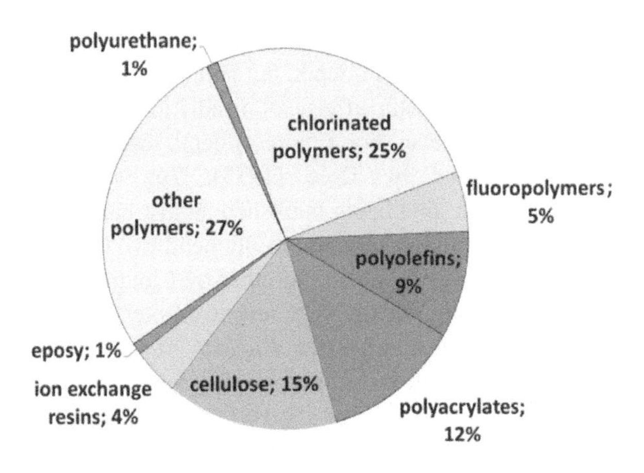

Figure 8.1. *Different classes of polymers used in the nuclear industry. Taken from Chantreux (2021). For a color version of this figure, see www.iste.co.uk/bouffard/nuclear.zip*

8.2.1.2. *Ion exchange resins and reverse osmosis membranes*

Ion exchange resins, or IERs, are widely used in the nuclear industry to purify water in the primary and secondary circuits of PWRs, water in storage pools and to decontaminate certain radioactive effluents. After use, the contaminated resins are generally mixed into cement materials. The effects of irradiation on polymer changes at the molecular level, as well as the functional properties of ion exchange resins, have been the subject of a relatively large body of work (Swyler et al. 1983; Pillay 1986; Rébufa et al. 2015). The dose range considered can exceed one MGy. Ion exchange resins are generally three-dimensional polymeric networks with functional polar groups. In the presence of oxygen, the formation of H_2 and CO_2 dominates gaseous emissions. One of the specificities of ion exchange resins comes from their hydration. In addition to the "direct" effect of radiation on the network, the surrounding water adds an "indirect" effect of attacking the polymer by the reactive species produced in the water. The effects of hydration on network modifications and gaseous emission were assessed for low LET irradiations (γ irradiation) and then for higher LET irradiations, which is the typical case of retention of α emitter actinides (Baidak LaVerne 2010; Boughattas et al. 2018).

The use of reverse osmosis membranes for the treatment of contaminated water is an emerging field, largely driven by the 2011 major accident at Fukushima-Daichi, where the operator TEPCO faced a compelling necessity to reprocess very large quantities of highly contaminated seawater (Combernoux et al. 2017). Reverse osmosis membranes are composites that are generally structured as a stack of three skin/porous sublayer/mechanical support layers, which can typically be polyamide that is heavily added with polyvinyl alcohol, polysulfone and polyester, respectively. It has recently been shown that the functional properties of these membranes are significantly degraded for doses that are typically higher than 500 kGy (Combernoux et al. 2016).

8.2.1.3. *Organic additives*

During the conditioning of UOX (for Uranium Oxide UO_2) and MOX (for Mixed OXides $(U, Pu)O_2$) nuclear fuel pellets, it is necessary to use organic or organometallic auxiliaries with lubricating, deagglomerating or pore-forming functions during the mixing, pressing and sintering stages (Gracia 2017).

These compounds, which are irradiated by the α particles of the fuel, must keep their functional lubricating properties during the process. The process is proven for UOX fuels. However, the integration of a high concentration of Pu in MOX, with a more unfavorable isotopic distribution, will induce an increase of the dose rate and thus of the dose deposited in the organic material. The radiation also induces an increase in the temperature of the additive. Studies on the behavior of additives under α irradiation and the search for more radioresistant compounds are still ongoing. The exposure times are short, so it is not necessary to use accelerated tests. As the radiation/matter interaction takes place at a temperature higher than room temperature, both effects must be considered simultaneously with a particular in-depth analysis of the influence of irradiation on thermal stability (Lebeau et al. 2017; Gracia et al. 2018).

8.2.1.4. *Bitumen*

The use of bitumen as a coating matrix for low- and intermediate-level waste is long-standing; it will continue until 2025 (at least), at a rate of a few dozen packages per year. In France, it began industrially in 1966 and has generated about 80,000 drums of bituminous mixes that are stored, pending permanent storage or reconditioning. The wastes conditioned in bitumen matrix are essentially sludges from the co-precipitation of radioactive effluents. These organic materials were initially chosen because they have good containment properties, high agglomeration power, high chemical and thermal inertia, good impermeability, low solubility in water and ease of processing. However, bitumen is an organic matrix, so under the effect of irradiation by the radioelements it contains, it produces gases – essentially dihydrogen – and its viscosity changes with time. Moreover, its gas impermeability properties lead to the retention of radiolysis gases, which causes the formation of bubbles and the swelling of said matrix. This is an effect that is obviously dose rate dependent (Phillips and Burnay 1991), as discussed in section 8.1.

8.2.2. *Polymers in the reactor building*

Polymers representing a class of radio-sensitive materials are banned from the reactor core. Nevertheless, they are found in the reactor building, near the core, for instance as electrical cables that are necessary for the control of essential equipment. As safety devices, their continued functionality must be ensured: their aging must therefore be understood to establish end-of-life criteria. To date, the end-of-life criterion chosen in the nuclear industry is the alteration of mechanical properties that indirectly leads to the loss of electrical functional properties. Generally speaking, it is the cracking of an electrical insulator that will cause a loss of electrical insulation, not the evolution of the resistivity of the material. Two very different cases are to be considered: normal operation, including the possible extension of the life of the plants, and accidental situations.

In normal operation and under the most severe conditions encountered in the reactor building, the dose rate is very low, of the order of 0.1 Gy/h, there is a long exposure time of 40 years (or even 60 years), and the temperature is about 50°C. In order to understand and validate the performance of cables and other safety devices under such conditions, accelerated aging tests must be performed, with the condition that the dose rate effect must be known. Moreover, since the temperature is around 50°C, thermal aging is not negligible; both types of aging (thermal and radiative) must be considered simultaneously. This complex problem has already been well addressed at technological and fundamental levels. Even though the predictions must be refined, no major problems are expected in terms of equipment performance, even if the lifetime of nuclear power plants is extended to 60 years.

In accidental conditions, the most discussed case and often defined as the reference accident case is the loss of coolant accident (LOCA). For a LOCA, depending on the scenarios considered, the inventory of radionuclides and their quantity may differ somewhat, but the following rules can generally be considered as true: α emitters do not have to be considered and the γ emission of radionuclides leads to low dose rates, when compared to β emitters. The materials in the reactor building would therefore essentially be irradiated by β radiation, which would generate a significant dose rate and cumulative dose. The radionuclides released have various lifetimes, but a significant fraction of them is short-lived. Thus, the dose rate at two hours would be about two orders of magnitude higher than that observed after about ten days, the period during which sensitive materials must retain their functional properties. In order to remain representative of a LOCA accident, it is therefore not essential to perform tests by accelerating the dose rate by several orders of magnitude. Actual irradiation conditions can be simulated using electron accelerators. β radiation is not very penetrating, less than a millimeter. For materials of one-millimeter thick or more, this leads to the existence of a marked dose profile and thus a marked damage profile. This is essentially where the difficulty lies when predicting the behavior of materials under conditions of this typical type of accident.

The case of a serious accident has been less studied in the case of organic materials, because it is possible to augur a loss of their functional properties, even without irradiation. In this case, the integrity of the fuel rods is lost, so in addition to the release of β/γ radioisotopes, α emitting radionuclides from the fuel are also dispersed in the reactor building and deposited on the materials. The α emitters have long half-lives, so there will only be a small decrease in dose rate with time. The α emitted by these isotopes have fairly close energies between 5 and 6 MeV. The physical nature and location of the deposits have a crucial impact on the dose rate. Indeed, as the ranges of the α are weak, this type of irradiation is very superficial and only the emitters in the immediate vicinity of the polymer will irradiate it. As a comparison, half of the dose is deposited by α on a thickness of about 10 μm and about 400 μm for β irradiation.

For a LOCA, the β dose at the surface of the material would be of the order of 2 MGy in the first few days, and 4 MGy after a month. For a serious accident, the α contribution is harsher in the long term: the surface dose range is of the order of 4 MGy in the first few days, and 30 MGy after a month. These dose values for accidental situations are orders of magnitude; it is unrealistic to try to accurately predict the conditions of irradiation of materials and the associated exact dose. Indeed, although the inventory of radionuclides present in the core is well known, estimates of the fraction released, of the physical form in which the radioelements would be found, and of their distribution in the reactor building are much more inaccurate. Hence, it is important to approach the problem, on the one hand in a rather fundamental way, for example, by analyzing how the relevant mechanical properties of a polymer can be affected by a very heterogeneous modification in depth (Corbin 2001), and on the other hand, by considering penalizing but nonetheless representative conditions. Lastly, in a general way, we must always consider that the accident could occur at the end of the reactor's life[2], thus on previously aged materials.

8.2.3. *Nuclear waste*

8.2.3.1. *Storage and disposal*

The nuclear power industry generates technological waste that is contaminated by radioelements. This waste must be disposed of in optimized safety conditions. In general, waste containing organic materials has a low or medium activity. Studies are currently focusing on long-lived intermediate-level waste (LL-ILW), since deep geological disposal is the solution currently being considered for this type of waste. The evolution of the waste package must be studied from its creation to a hundred thousand years: it is therefore necessary to use accelerated tests, and studying the effect of the dose rate is necessary.

Two classes of waste are to be considered: i) contaminated organic technological waste packaged for storage/disposal (all types of polymers that are usually used in the nuclear industry are to be considered), and ii) the case of an organic material used as an immobilization matrix (generally speaking, this is bitumen, even if polymers have sometimes been used) (Phillips and Burnay 1991; Özdemir 2014; Ojovan et al. 2019). When the dose is deposited by α actinide, which is primarily plutonium from MOX fuels, we find the same characteristics that have been mentioned for materials in power plants in the event of a serious accident, but over longer exposure times. As the α paths are short, we must expect a heterogeneous dose deposit at depth, as well as an impact of the granulometry of the MOX grains on the dose. The dose can be considerable, ranging from several dozen to a hundred MGy. In a case where the dose is deposited by β and γ emitters, the characteristics mentioned above for materials in power plants are found again, also over longer exposure times, and with a variable dose rate.

Packages for technological waste can be ventilated through a diffusive filter or sealed. For both types of packages, it is necessary to ensure the outcome of radiolysis gases according to their harmfulness. Thus, the emission of inflammable (H_2), combustible (CH_4), corrosive (HF, HCl) or toxic (CO) gases from the polymers must

2. Due to the possible extension of the life of the plants and to remain in a representative case, we consider that the materials have aged 60 years.

be accurately assessed, in order to properly size the disposal site to avoid an explosion or a fire by the excessive release of dihydrogen or methane, respectively. Corrosive gases can accelerate or at least initiate the degradation of the stainless-steel external container. CO_2 can alter the cement matrix. In the latter case, with the volume being constrained, it will be necessary to consider the pressure variation, due to the emission of radiolysis gases and to the consumption of oxygen induced by radio-oxidation. The resulting overpressure or depression will eventually impact the integrity of the package.

At the deep geological disposal site planned for France, the waste will be placed 500 meters below ground. Beyond the ventilated operating period, the site will be closed and water is expected to seep in through the multiple containment barriers until it reaches the core of the waste packages. It is estimated that the site will be saturated with water between 10,000 and 100,000 years after its closure (ANDRA 2005). During this stage, the radio-oxidized polymers will degrade by hydrolysis. This further degradation may lead to the formation of water-soluble organic compounds, potentially complexing radionuclides (Glaus and Van Loon 1998; Dannoux 2007; Fromentin et al. 2016; Fromentin 2017; Chantreux 2021). These compounds could promote the migration of radioelements outside the disposal site: it is therefore necessary to identify and quantify these degradation products.

8.2.3.2. *Transport*

Waste packages are transported from one site to another, for example, between conditioning, storage and/or disposal. For this purpose, the packages can be inerted in transport casks. Since most of the transport is by road, accident scenarios must be considered, with the worst case being a road accident with a fire. Under these conditions, it is estimated that the temperature will reach more than 800°C outside the waste package, and up to 150°C inside. In this case, the radiolysis gases can deform or even degrade the container, with the associated risk being the loss of integrity of the package and therefore the release of contamination into the environment.

Two cases can be considered. In the first case, the polymers contained in the package are not highly irradiated (recent waste package): the most representative accelerated aging conditions are irradiation in temperature. In the second case, the polymers are highly radioactive (old waste package): the representative experimental simulation consists of a sequential aging with a pre-irradiation of the polymers, followed by a thermolysis. The case of polyvinyl chloride (PVC) has been analyzed in detail (Boughattas et al. 2016a, 2016b). From the point of view of reaction mechanisms, this work clearly shows that in the case of sequential aging, defects induced by prior radio-oxidation influence thermal degradation by lowering the initial temperature of hydrochloric acid and benzene formation; these two molecules being indicators of material degradation.

8.3. Radiation exposure

We have seen in the previous section that depending on their use in the nuclear cycle, organic materials are irradiated by low LET particles, the γ and β radiations of radionuclides, which we will assimilate to electron irradiation, and by higher LET particles, the α emitted by actinides. In this section, we will look back on some practical and theoretical considerations on these two situations, and then briefly open the field to other irradiation conditions that deserve to be briefly mentioned, even if they do not directly involve "nuclear power energy".

8.3.1. *The LET effect*

The energies of the projectiles can range from a few keV to a few MeV (even GeV for fast heavy ions). They have no comparison to the two relevant energies in radiolysis phenomena, which are the band gap and the covalent bond energies, from a few eV to ten eV. The only relevant energy threshold is the first ionization threshold, hence the name "ionizing radiation", which distinguishes chemistry under radiation from photochemistry (aging under UV radiation). The projectile energy only influences the spatial distribution of ionization/excitation events. This is the reason for the so-called "LET effect".

In a very or even too schematic way, we could say that β/γ irradiations lead to a homogeneous distribution of excitation/ionization events, and α irradiations lead to a strongly heterogeneous distribution of these events, the latter being concentrated in a nanometric cylinder along the path of the ions. However, it is necessary to make a subtle distinction. β/γ irradiations never lead to a strictly homogeneous distribution at the nanometer level; it is customary to distinguish between spurs, blobs and short tracks, whose proportion depends on the energy of the primary electron (Mozumder and Magee 1966). These proportions are similar under β and γ radiation. There is no need to distinguish between them, except possibly for the very low β energy of tritium ($E < 18.6$ keV), which is not involved in our subject. For α (and other fast ions), the distribution of excitation/ionization events is very heterogeneous: some of the energy is transported away from the tracks by higher-energy secondary electrons, forming a less dense excitation/ionization halo around a dense core (Bouffard et al. 1995; Ngono-Ravache et al. 2016).

8.3.2. *β/γ irradiation*

Although, from a physical point of view, we will not distinguish irradiation with electrons from a γ radiation, we must not forget that from a more macroscopic and practical point of view, there may be a difference. The γ essentially interacts by the Compton effect, generating Compton electrons of high energy. As the emission of these Compton electrons are isotropic, a fraction of these electrons escape from the material, decreasing the dose near the surface; this loss of energy is not compensated by electrons from the environment. This is the difference between the dose, the energy actually received by the material, and the KERMA, the energy lost by the radiation. There is a dose deficit over an equilibrium distance, millimeter or submillimeter, which depends on the γ energy or the γ energy spectrum (Biggin 1991). During simulation irradiations performed with γ sources, the electronic equilibrium condition is usually ensured, so that the samples are irradiated homogeneously. This is not necessarily the case for γ-irradiated materials in real-life conditions. For irradiations with electrons, secondary electrons, which have a low-energy average, only induce an electronic imbalance in extreme surfaces, without practical consequence. It is, however, important to be aware of possible backscattering effects of the electrons from the beam when, for example, a film is placed on a substrate composed of heavier atoms (Fe, Cu, etc.).

8.3.3. *α irradiation*

For experimental irradiations simulating α irradiation, two strategies are used. The first is to perform experiments with actinides as an irradiation source by using, for example, stacks of PuO_2 pellets and polymer films, and possibly choosing a particular isotopy to increase the dose rate (Gracia et al. 2018; Venault et al. 2019). These are glove-box experiments, requiring a heavy nuclearized environment; online instrumentation is possible but complicated, and the experiments are long in duration, with no possibility to accelerate time by significantly increasing the dose rate. This is the closest approach to a real exposure situation, but it is barely used because of the inherent constraints. The second strategy is to irradiate with accelerated He ion beams, generally at energies higher than those of actinide α, so as to allow them to pass through a window and then the irradiation atmosphere. It is possible to perform irradiations with mono-energetic ions or to modulate the energies of the incident He ions, in order to reproduce the energy spectrum related to the slowing down of α in the fuel grains, as mentioned above.

Another possibility is to irradiate with heavier ions and at higher speeds, but whose LET in the organic material is equivalent to that of α. The French community has put forward this possibility and has used it extensively, because it allows irradiation with particles with a larger path length, and therefore to use irradiation conditions that are easier to implement and better defined (constant LET over a larger thickness). As an example, Figure 8.2 compares the electronic stopping powers as a function of the path for 4He of 1.25 MeV/A (5 MeV) – α equivalent – and ${}_{12}$C of 12 MeV/A (144 MeV), which is a beam available at GANIL in Caen, for example. In order to be thorough, it should be remembered that two ions of the same LET, but different speeds, do not deposit energy in a strictly identical way. The halo of the faster ion will be further extended, because the energy spectrum of the secondary electrons extends to higher energies (Bouffard et al. 1995; Ngono-Ravache et al. 2016). This is called the "velocity effect", which is an effect that can, as a first approximation, be neglected in our context.

We have already mentioned the real conditions of exposure to α particles and their complexity in section 8.2. The estimation of the deposited dose requires a simulation that takes the arrangement of actinides and polymers into account. The CEA teams have developed codes that consider typical cases that can occur: a flat layer of actinide oxide on a flat polymer surface; a spherical actinide oxide grain on a flat polymer surface; a spherical actinide oxide grain immersed in the polymer or surrounded by stearate; a homogeneous actinide oxide-polymer mixture (Gracia et al. 2018; Ferry et al. 2020; Esnouf et al. 2022a, 2022b).

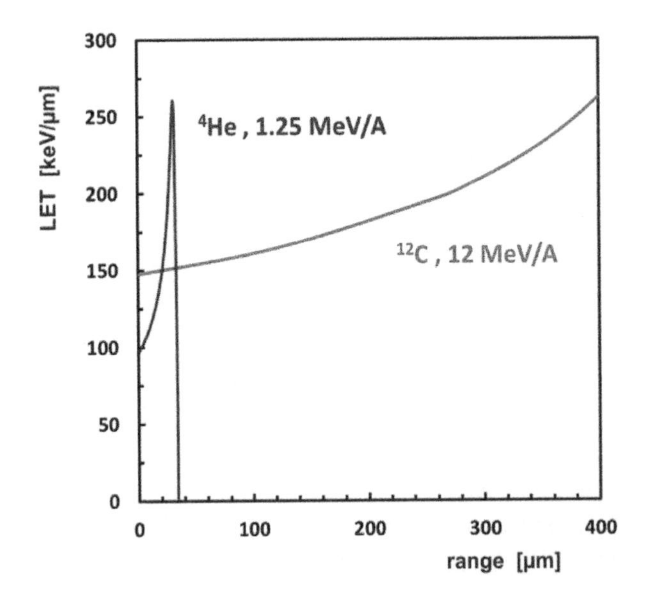

Figure 8.2. *Electronic stopping power as a function of the incident ion path in polyethylene. For a color version of this figure, see www.iste.co.uk/bouffard/nuclear.zip*

8.3.4. *Thermal neutrons*

The case of thermal neutrons is rarely mentioned explicitly, but deserves a mention. Thermal neutrons react with hydrogen in polymers by a capture reaction (n, γ) emitting a γ of 2.17 MeV. Thus, from the perspective of a material that is only made of carbon and hydrogen atoms, it is a γ irradiation that can be contemplated as the γ background of a mixed field of γ/thermal neutrons; we must only pay attention to the way the dosimetry has been performed. On the other hand, for chlorinated polymers (PVC, for example) or nitrogenous polymers (polyimides, polyurethanes, etc.), other reactions must be considered. Indeed, ^{14}N and ^{35}Cl have reactions (n,p) with large cross-sections, emitting, in the material, protons of energy in the range of 0.6 MeV, as well as a recoil nucleus of ^{14}C of 40 keV (^{14}N) and ^{35}S of 15 keV (^{35}Cl). Furthermore, ^{35}Cl has a reaction (n, γ) with an effective cross-section that is more than two orders of magnitude larger than that of hydrogen (45 10^{-24} cm^2 for ^{35}Cl versus 0.34 10^{-24} cm^2 for hydrogen). Polymers with Cl or N atoms therefore require careful dosimetry.

8.3.5. *Other projectiles*

Outside the nuclear sphere, there is an important scientific and technological activity around polymers that are irradiated by fast heavy ions, with LETs much higher than those of α. When the LET exceeds a critical threshold, strongly non-linear phenomena start to appear, leading to the formation of a specific damage known as latent track. This name is linked to the fact that these tracks can be revealed chemically: a nano-micrometric pore is created for each projectile ion. This specificity is used in very wide range of fields that are gathered under the acronym ITT for "ion track technologies" (Toulemonde et al. 2004; Ma et al. 2020). The threshold of track creation vastly depends on the material properties, and the analysis, as a function of the LET, of the gases emitted in different polyolefins (polyethylene, polypropylene, polybutene) makes it possible to determine the value. At low LET, the breaking of the polymer side group is the source of the gas emitted: dihydrogen for polyethylene, H_2 and methane for polypropylene, and H_2 as well as ethane and ethylene for polybutene. Thus, the emitted gases are characteristic of the initial structure of the polymer. However, above a LET threshold of about 500 keV/μm, the yields of low-molar-mass hydrocarbons increase significantly, indicating chain fragmentation and entry into a track creation regime (Picq et al. 1998; Picq 2000). For an α irradiation, the maximum LET, also known as the Bragg peak, is more or less equal to 250 keV/μm (Figure 8.2), and is thus significantly below the track creation threshold. At least for the aliphatic polymers presented in this paragraph, we can reasonably assume that we will remain in a "traditional" radiolysis regime during α irradiations.

We have spoken exclusively about the electronic energy deposit and neglected any effect of elastic collisions. Without wishing to treat this subject in an exhaustive way, we can affirm that this is absolutely true under the typical exposure conditions for nuclear organic materials. The question is somewhat more open for irradiations by heavy ions of a few tens of keV used, for example, to modify polymers on the surface (Marletta 1990). It is nevertheless important to remember that even if the nuclear stopping power of a given projectile is greater than its electronic stopping power, this does not necessarily mean that the energy received by the target in electronic form only comes from the electronic stopping power of the projectile. The ions ejected from their site by elastic collisions, the primaries (in general for light organic ions: hydrogen and carbon) lose part of their energy in electronic form. Part of the nuclear stopping power is thus "converted" through the primaries into electronic stopping power and, for materials sensitive to radiolysis, this must be taken into account (Gaté 1997).

8.4. Irradiated polymers: phenomenology

8.4.1. *Resistance of polymers to irradiation*

From a practical standpoint, the first piece of information needed is to know the dose range in which the polymers can be used. There are many published compilations that attempt to answer this question. Among these are those of Harwell (Phillips and Burnay 1991), Takasaki (Seguchi and Morita 2003), Oak Ridge (Bopp and Sisman 1953) and especially the exhaustive work performed at CERN (Van de Voorde 1970; Tavlet and Schönbacher 1989; Schönbacher et al. 1996; Tavlet et al. 1998; Guarino et al. 2001). Figure 8.3 shows the irradiation resistance of some thermoplastics.

Figure 8.3 shows a very large variability in the sensitivity of polymers to irradiation: nearly four orders of magnitude separate the least stable material from the most stable. The most resistant polymers are aromatic and can withstand up to ten MGy: polyimides (Kapton®), polyetheretherketone (PEEK) and polystyrene. Conversely, polytetrafluoroethylene (PTFE, better known by its commercial name Teflon®) is a poor candidate. For most polymers, a dose of a few MGy is the limit.

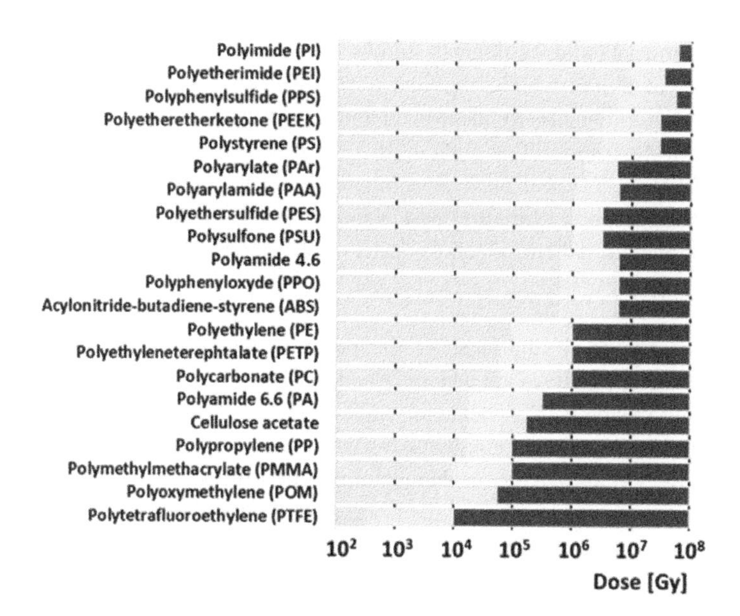

Figure 8.3. *Irradiation resistance of different thermoplastic polymers. Green: safe to use. Yellow: moderate damage, possible use. Red: severe damage, not recommended for use. For a color version of this figure, see www.iste.co.uk/bouffard/nuclear.zip*

All other things being equal, the sensitivity of materials depends on the irradiation conditions. Thus, an irradiation in air (in the presence of dioxygen) is much harsher than an irradiation in anoxic conditions, often reducing the limit dose of use by more than an order of magnitude. This difference is due to the phenomenon of radio-oxidation, which will be discussed in section 8.6. The dose rate is another parameter that can affect sensitivity. In general, the dose rate has little influence in anoxic conditions and a strong influence in the presence of oxygen; low-dose-rate irradiations are then harsh. In polymer comparison tables, "irradiation in anoxic conditions or in air at [a] very high dose rate" is sometimes indicated. This refers to another concept that we discuss in section 8.6, which is the limitation of oxidation by permeation. A thick material irradiated at a high dose rate will only be oxidized at the surface. This feature is used industrially, because the use of a high dose rate makes it possible to do without a vacuum or controlled atmosphere irradiation system, while keeping control of the defects formed in the polymer, and thus of the associated properties. These irradiation conditions are used, for example, to improve the temperature resistance of heating pipes or cable insulation.

The numerous phenomenological studies have made it possible to establish a series of rules of thumb which, although not perfectly rigorous, still contain some truth. In addition to the already mentioned and indisputable roles of aromaticity and the presence of dioxygen, let us mention that there is a greater fragility of branched polymers compared to linear ones. A comparison of the chemical groups gives the following classification: from the least sensitive to the most sensitive, we have aromatics, aliphatics, ethers, alcohols, esters and ketones (Kircher and Bowman 1964).

The above empirical studies show the performance of pure polymers. In the industry, fillers and additives are generally added, in order to improve the lifetime of materials (antioxidants), their behavior under irradiation (antirads), to light (UV stabilizers) and to fire (flame retardants), as well as plasticizers and/or fillers, which make it possible to improve the mechanical properties of materials. Each molecule is added for a specific purpose, to improve a property of use. However, under irradiation, they can also influence the aging mechanisms, for example, by energy transfer[3]. We will return to this phenomenon in section 8.5.2.

8.4.2. *Changes induced by irradiation*

Under irradiation, the chemical bonds of polymers are modified. At the macromolecular level, polymers can be classified into two main groups: those that cross-link and those that degrade. These two mechanisms are not mutually exclusive, they are classified by a dominant mechanism. The cross-linking consists of a covalent bridging between two chains, which increases the molar mass[4]. When all chains are bridged, a three-dimensional network (of "infinite" molar mass) is obtained and the polymer becomes insoluble. Degradation is the opposite effect: the irradiation leads to the chain breaking (scission) and the average molar mass decreases. As an example, in anoxic conditions, polyethylene is the archetype of the polymer that cross-links, polymethyl methacrylate (PMMA better known under its industrial name Plexiglas®) and polyisobutylene being the typical examples of polymers that degrade. Polypropylene has a balanced behavior, with approximately equal cross-linking and scission radiochemical yields. Comparison between polyethylene, polypropylene and polyisobutylene shows that scission is promoted by the presence of tertiary carbons, and even more so by quaternary carbons. This

3. The term *energy transfer* is a general term, which includes transport of mass or reactive species (radicals and ions), transfer of charges alone (electrons and holes), as well as transfer of electronic excitations.

4. It can also involve modifications that do not affect the molar mass, such as an intra-chain bridging.

mechanism of degradation perfectly explains the greater fragility of branched polymers compared to linear ones.

When the polymer chains are linear and stereoregular (tactical polymers), a crystalline order is established. By breaking the regularity of the chains, irradiation modifies this crystallinity (amorphization), often after an incubation dose necessary for the accumulation of a sufficient number of defects in the crystallites (Chailley et al. 1995). Sometimes, a slight increase in crystallinity is observed at low doses (chemi-crystallization). Thus, each modification will have an influence at several levels: chain scissions at the macromolecular level increase chain mobility, allowing recrystallization at the supramolecular level. Radiation-induced changes in tacticity (racemization) are likely to be infrequent and, in any case, are little studied.

The average molar mass, the distribution of molar masses, the presence of a crystalline fraction and the glass transition temperature largely determine the mechanical and rheological behavior of polymers. As previously indicated, irradiation primarily affects the molar mass, only affects the crystallinity a little (mostly at high dose) and affects the glass transition temperature even less. This explains the major influence of cross-linking and scission on the mechanical and functional properties under irradiation. Below a critical average molar mass that depends on the polymer, the functional properties of the material are lost. Ultimately, oxidation promotes degradation, hence its negative influence on the radioresistance of irradiated polymers.

Eventually, the aging of polymers leads to their yellowing. This is related to the formation of various chromophores, but this modification does not give much specific information on the chemical changes induced.

8.5. Radiolysis in anoxic polymers: fundamental effects

8.5.1. *Polymer radiolysis: introduction*

The initial process of radiolysis is the creation of excitations/ionizations. The radiochemical yield, G_i, associated with it depends very little on the projectile, and is only somewhat sensitive to the energy spectrum of primary and secondary electrons (Tessaro et al. 2019). The number of N_i events created per unit of deposited energy, E, can be estimated by: $N_i/E \approx 1/(2.5\,E_g)$, where E_g is the band gap width. This indicates that about 60% of the deposited energy is lost as heat.

With E_g = 8 eV (a typical value for a polymer), we get[5]: $G_i \approx 5.2\ 10^{-7}$ mol/J. For sensitive polymers such as polyolefins, the total radiochemical yield of induced modifications is of the same order of magnitude as G_i (Perera and Hill 2003), proving that a radiolysis mechanism can be very efficient. The presence of a chain reaction can result in a radiochemical yield much higher than G_i; this is often the case for oxygen consumption during radio-oxidation (see section 8.6).

Very briefly, the primary event creates excited molecules M* and cation–electron pairs M^+-e^- or, using a less molecular and more physical approach of the solid, electron–hole pairs e–h. In non-polar or low polar polymers such as polyolefins, the cation–electron recombination, which is often geminated, is very fast and leads to the formation of an excited state M*. This is an important difference with the case of water where the electron is strongly solvated. The atomic and electronic structure of these excited states (self-trapping excitons) in polyethylene has given rise to some theoretical studies (Ceresoli et al. 2005). The de-excitation of an M* will lead to a covalent bond breaking with the formation of two neutral radical species (two macro radicals, or one macro radical and one molecular radical). This is the starting point of a radical chemistry, which will lead to the formation of new stable species. This description is certainly simplifying (Kondoh et al. 2011), but it is widely used and sufficient to begin with.

As mentioned above, the modifications under irradiation occur at several scales: a) at the molecular level, which involves modifications of the initial monomers; b) at the macromolecular level, involving changes in the distribution of molar masses (of the length of the chains), cross-linking and degradations; c) at the supramolecular level, which involves modifications of crystallinity, crystallinity rate, size of the lamellae and even modifications in spherulites[6].

In the space dedicated to this chapter, it is not possible to discuss in detail the experimental approaches implemented to characterize modifications in irradiated polymers. The changes of crystallinity can obviously be followed by X-ray diffraction, even if it is not so frequent, or by infrared spectroscopy[7], but the technique of choice remains the calorimetric analysis, which allows the measurement of the temperature and the enthalpy of fusion of crystallites. The

5. $G_i[mol/J] = 1/(Avo\,[mol^{-1}]\,2.5\,E_g[J])$, with Avo being Avogadro's number, or: $G_i[mol/J] = 1{,}04\ 10^{-5}[mol\ eV\ J^{-1}]/2.5\ E_g\ [eV]$.

6. The lamellae are small crystallites, generally in the form of platelets, about 10 nanometers thick and with a dimension of about 100 nanometers. The lamellae are the bricks of larger-sized crystalline superstructures (100 μm), with the spherulites being constituted of the lamellae arranged in a radial and symmetrical manner.

7. FTIR for Fourier transform infrared spectroscopy.

identification and quantification, at the molecular level, of chain scissions and cross-linking bridges is complicated: these changes are usually estimated at the macromolecular level. If the polymers are quite "monodispersed"[8], steric exclusion chromatography gives accurate information (Bouffard et al. 1997).

As radicals are unstable species, their lifetime at room temperature is short. The mobility of radicals is facilitated by the various intra-chain molecular movements, and especially by the inter-chain molecular movements, which are triggered when the temperature exceeds the glass transition temperature T_g. Most often, the detection of radicals and their recombination is achieved by measurements at a low temperature (77 K or even lower), followed by a progressive annealing at room temperature (Rånby and Rabek 1977). Electron paramagnetic resonance (EPR) is the leading technique for the measurement of radicals, but FTIR can also be used. In general, when $T \gg T_g$, few radicals remain in the amorphous phase. The stability of radicals with temperature is much greater in the crystalline zones. However, it should be noted that a semi-crystalline polymer is not a strictly biphasic system: there is, indeed, some inter-phase mobility. Thus, a chain or a part of chain, which is integral part of a crystallite, can migrate in an amorphous phase and vice versa. The phenomenon can contribute to radical delocalization.

Most of the permanent modifications at the molecular level are studied by their mid-infrared signature (FTIR). When polymers remain soluble after irradiation, nuclear magnetic resonance (NMR) can be very useful. The NMR of the solid is sometimes used for insoluble polymers, but it is a much more complicated technique to implement. Lastly, it should not be forgotten that the gas emission is, to a large extent, the complementary aspect of modifications at the molecular level, the gases emitted can be measured by gas mass spectrometry or by FTIR.

8.5.2. *A textbook case: polyethylene*

As a result of the simplicity of its formula $(–CH_2–)_n$, polyethylene has often been considered as a model case. As we cannot discuss the different families of polymers, we will use this material to give examples of modifications induced by irradiation in anoxia.

In polyethylene, the primary act of radiolysis is the breaking of the C–H bond, giving an alkyl macroradical $P^•$ and an $H^•$ radical. The $H^•$ radical has never been

8. A monodispersed polymer has a very narrow molar mass distribution.

detected, even at temperatures below 4 K. This can be explained by a fast reaction of $H^{\bullet}$ by a tunnel effect with the surrounding polymer, which leads to the formation of dihydrogen and a new $P^{\bullet}$ (*irradiation* + $PH \rightarrow P^{\bullet}+H^{\bullet}$ then $H^{\bullet}+PH \rightarrow H_2+P^{\bullet}$). The primary act is thus written as *irradiation* + $2PH \rightarrow 2P^{\bullet}+H_2$. The radiochemical yield of the primary act is equal to $G(H_2)$; note that $G(P^{\bullet})$ is the double of $G(H_2)$. The value of $G(H_2)$, which is $\approx 3.5\ 10^{-7}$ mol/J, is similar to the estimated excitation/ionization radiochemical yield G_i. Thus, the process of converting excited states to defects is very efficient. Polyethylene does not undergo C–C scission of the main chain, so the alkyl radical is the only primary $P^{\bullet}$ radical. Upon radiolysis of polyethylene, the alkyl radicals will recombine to either give *trans*-vinylene (TV) unsaturation (CH_2–CH=CH–CH_2–) or a cross-link (X), in approximately equal proportions. An amount of *trans-trans*-diene (TTD) conjugated unsaturation approximately 10 times smaller is also formed. Laying out the atomic balance leads to the following equality:

$$\underset{\approx 3.5\ 10^{-7}\ \text{mol/J}}{G(H_2)} = \underset{\approx 1.6\ 10^{-7}\ \text{mol/J}}{G(TV)} + \underset{\approx 1.6\ 10^{-7}\ \text{mol/J}}{G(X)} + \underset{\approx 1.5\ 10^{-8}\ \text{mol/J}}{2\,G(TTD)}$$

For polyethylene, direct measurement of G(X) over a wide dose range is difficult at both the molecular and macromolecular levels, because polyethylenes are rarely monodisperse. Moreover, they become insoluble at low doses. However, the simultaneous measurement of $G(H_2)$, G(TV) and G(TTD) makes it possible to estimate G(X) using the balance equation above.

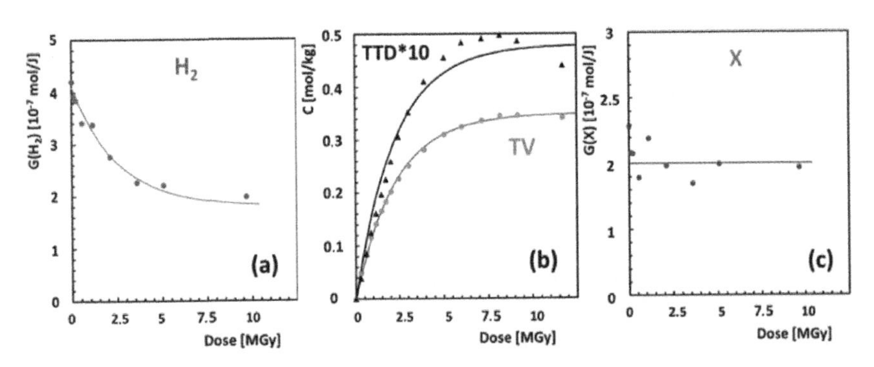

Figure 8.4. *Dose evolution of the different groups produced in polyethylene, by irradiation with 1 MeV electrons: (a) dihydrogen emission radiochemical yield, $G(H_2)$; (b) concentration of trans-vinylene, TV (dots), and trans-trans-diene, TTD (triangles), with the measured concentration of TTDs having been increased by a factor of 10; (c) cross-linking radiochemical yield, G(X). The lines in the figures are guides for the eye. Figures taken from Ventura (2013). For a color version of this figure, see www.iste.co.uk/bouffard/nuclear.zip*

What we have just described is the initial, so-called zero-dose behavior. As the dose increases, $G(H_2)$ decreases with the dose to tend towards a saturation value (Figure 8.4(a)). This decrease in $G(H_2)$ with dose is important from a technological point of view. For the management of the disposal of contaminated waste packages, the emission of H_2 must be predicted, so as to avoid any risk of fire or explosion. Knowing the decrease of $G(H_2)$ over time (with the dose) makes it possible to better adjust the predictive models by not having to take the initial $G(H_2)$ value, which would largely overestimate the emission of this gas in the long term. This decrease of $G(H_2)$ is, in fact, linked to the transfer of a fraction of the energy from the excited M* states to the radiation-induced defects (Ferry and Ngono 2021). This fraction does not contribute to the formation of $P^{\bullet}$, so it will not induce C–H bond breaking. It depends on the concentration of irradiation defects, as well as on their nature and distribution. The synthesis of "custom" polymers, in other words, specially designed to help shed some light on an irradiation mechanism, is a strategy that has been used to study energy transfer. Polyethylenes containing a defined quantity of *trans*-vinylenes groups have thus made it possible to quantify the effect of this group's presence in the energy transfers responsible for the decrease of $G(H_2)$, and then to estimate the contribution of the cross-links on this decrease. It is, for the same defect concentration, lower than that of the TVs. When the dose increases, the TV concentration reaches a saturation concentration (Figure 8.4(b)), which is explained by two effects that add up: the energy transfer to the radiation-induced defects and the radical attack by $P^{\bullet}$ that leads to the formation of a cross-link ($TV + P^{\bullet} \rightarrow X + P^{\bullet}$). The evolution of TTD with dose is similar (Figure 8.4(b)) and has the same origin. The cross-linking radiochemical yield remains stable with the dose (Figure 8.4(c)), therefore causing the cross-linking concentration to grow linearly, without any saturation. As a result, cross-linking bridges become the main defects at high doses (Ventura 2013; Ventura et al. 2016).

The addition of radio-resistant molecules such as aromatic molecules (sometimes called antirad) in mixture in a polymer, or the introduction of aromatic monomers in a covalent way in the chain, reinforce the polymer's resistance to the irradiation; it is the "protective" effect of aromatic molecules. It is believed that the radioresistance of aromatic molecules comes from the delocalization of electrons in the benzene ring because, on the one hand, this does not favor the localization of excited states and, on the other hand, the energy of the π^* excited states can be restored by luminescence, at least partly. These organic molecules have long been described as "energy sponges". It is the transfer of energy from the aliphatic phase to the aromatic molecules that is at the origin of this protective effect. A fairly comprehensive study has been carried out on random poly(ethylene-styrene) copolymers, as well as on cryogenic cyclohexane/benzene mixtures. It follows that, firstly, even at low temperature (15K) when the radical migration is frozen, the transfers remain efficient. Secondly, the efficiency of intermolecular transfers is of the same order of magnitude as that of intramolecular transfers. Lastly, the phenomenon of radioprotection of the donor group by the acceptor group is to the detriment of the latter (Ferry et al. 2012, 2013, 2019).

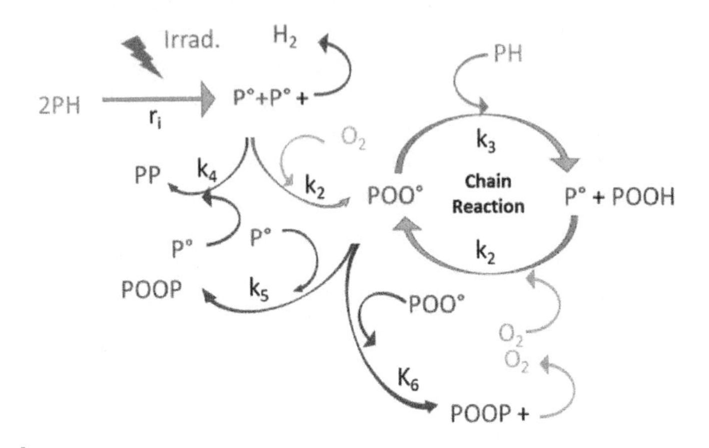

Figure 8.5. *Basic diagram of the radio-oxidation of a polymer. A model hydrocarbon polymer is considered here, of the polyethylene type, denoted as PH, and two macroradical species P• (the alkyl radical) and POO• (the hydroperoxyl radical). For a color version of this figure, see www.iste.co.uk/bouffard/nuclear.zip*

8.6. The radio-oxidation of polymers

Under the usual exposure conditions of organic nuclear materials, radio-oxidation governs the modifications under irradiation. It is therefore necessary to have a good understanding of the chemical reactions that cause the aging of polymers. Moreover, in many cases, the exposure times are very long, ranging from tens to a hundred years, which requires accelerated testing. Under these conditions, it is necessary to know the effect of the dose rate. These two points are discussed, then a specific focus is made on radio-oxidation coupled with α irradiation.

8.6.1. *Mechanism of radio-oxidation*

Bolland and Gee (1946) formalized the diagram that is briefly presented in Figure 8.5, in terms of chemical kinetics, by the following six reactions[9]:

9. These six reactions were originally proposed for rubber and related materials. This description is simplifying (Ahn et al. 2022), but it is widely used and sufficient to begin with.

$1)\ Irradiation + PH \rightarrow P^{\bullet}$ $\quad r_i \quad$ [8.1]

$2)\ P^{\bullet} + O_2 \rightarrow POO^{\bullet}$ $\quad k_2 \quad$ [8.2]

$3)\ POO^{\bullet} + PH \rightarrow POOH + P^{\bullet}$ $\quad k_3 \quad$ [8.3]

$4)\ P^{\bullet}+P^{\bullet} \rightarrow PP\ (inactive\ products)$ $\quad k_4 \quad$ [8.4]

$5)\ P^{\bullet}+POO^{\bullet} \rightarrow POOP\ (inactive\ products)$ $\quad k_5 \quad$ [8.5]

$6)\ POO^{\bullet} +POO^{\bullet} \rightarrow POOP\ (inactive\ products) + O_2$ $\quad k_6 \quad$ [8.6]

The quantities r_i and k_2 to k_6 are the rate constants of the reactions (kinetic constants). Equation [8.1] is the initiation: the ionizing radiation induces the formation of macroadicals, $P^{\bullet}$; equations [8.2] and [8.3] constitute propagation following the chain reaction: the macroadical $P^{\bullet}$ reacts with dissolved O_2 in the polymer to give a peroxide macroadical $POO^{\bullet}$, then the latter reacts with the polymer to form a hydroperoxide $POOH$ and a new macroadical $P^{\bullet}$. Lastly, equations [8.4] through [8.6] constitute the mutual recombination of the macroradicals $P^{\bullet}$ and $POO^{\bullet}$. It is the magnitude of these reactions, known as termination, that limits this free radical chain reaction mechanism.

The diagram in Figure 8.5 is a highly idealized transcription of reality. However, because it is generic and simple, it establishes the main concepts of radio-oxidation and retains a certain degree of predictiveness. The material is considered to be a homogeneous model solid, consisting of a single chemical group that will initially give rise to the formation of a single macroradical. A very linear polyethylene partially meets this ideal case, as the primary act of radiolysis almost exclusively leads to the creation of $P^{\bullet}$ alkyl radicals. However, it should be remembered that real polyethylene is far from homogeneous; it is a semi-crystalline polymer (even highly crystalline, if it is very linear) with crystallites immersed in an amorphous phase (at room temperature, it is a supercooled liquid, because the glass transition temperature is below room temperature). The solubility of dioxygen in the crystalline parts is very low, and they are, in fact, irradiated in anoxia. The role of crystallinity in radio-oxidation must be kept it mind, because it moves the study of the real polymer away from the diagram given in Figure 8.5. Our model material would thus be a theoretical amorphous polyethylene, denoted as *PH*. The diagram in Figure 8.5 is also simplified, although this is probably of little consequence to the kinetics. For initiation, equation [8.1] condenses the sequence $irradiation + PH \rightarrow P^{\bullet}+H^{\bullet}$ then $H^{\bullet}+PH \rightarrow H_2+P^{\bullet}$, as described in section 8.5.2. The recombination products of the macroradicals $P^{\bullet} + POO^{\bullet}$ and $POO^{\bullet} + POO^{\bullet}$ (equations [8.5] and [8.6]) are highly schematized here and denoted, on a simple atomic balance basis, as a *POOP* peroxide. In reality, the radio-oxidation of polyethylene leads to the formation of many chemical groups: peroxides, hydroperoxides, ketones, esters, alcohols and carboxylic acids. Any reader who wishes to go further can, for example, refer to the work on γ irradiation of a poly(ethylene-propylene-diene) terpolymer, carried out in Clermont Ferrand (Cambon 2001; Rivaton et al. 2005). The recombination products $P^{\bullet} + P^{\bullet}$ (equation [8.4]) are directly identifiable by the creation of cross-links and unsaturations.

Kinetic constants are rarely measured independently, and are done so with difficulty. This difficulty is a major weakness of kinetic models of radio-oxidation in polymers, and is inherent to the study of chemical reactions in solids, which contrasts with the situation that prevails in water radiolysis (Elliot and Bartels 2009) or in the gas phase in combustion (Westbrook and Dryer 1984). For some reactions, there is even doubt about the order of magnitude of the constant. The initiation rate can be estimated by the radiochemical yield of H_2 formation because $G(P^{\bullet})=2G(H_2)$, see section 8.5.2; the constant k_2 is clearly controlled by oxygen diffusion, which is well known, and there is consensus on the value of k_2. The constant k_6 can be estimated by measuring $POO^{\bullet}$ concentrations by EPR (Esnouf and Balanzat 2007) or by estimation of the chain mobility that controls this reaction (Gervais et al. 2021). The measurement of $G(-O_2)$ as a function of the dose rate I and the concentration of O_2 ($[O_2]$) in the polymer informs us about the k_3 propagation and the k_6/k_4 ratio, respectively (Dély 2005). $P^{\bullet}$ migration is faster than $POO^{\bullet}$ migration, because $P^{\bullet}$ migration is not only due to chain mobility. Hopping or jumping of a neighboring H of the alkyl radical also contributes to it; thus, k_4>>k_5>>k_6. The widely used kinetic model is the “homogeneous kinetic” version: the concentrations of the different species are average concentrations that do not depend on spatial coordinates. This is a priori not justified, given that, at the nanometric level, spatial heterogeneity is to be expected: we have indeed seen in section 8.3.1 that, even at low LET, the primary ionization/excitation events are not created homogeneously.

By transcribing the six Bolland and Gee equations into chemical reaction terms, it is easy to write the set of coupled differential equations that govern the evolution of each species with time (or dose). However, the system of differential equations is only numerically integrable. With the usual set of values of the kinetic constants, the temporal analysis of the evolution of the concentrations $[P^{\bullet}]$ and $[POO^{\bullet}]$ shows the establishment of a stationary regime with times of the order of 10 seconds. With this time interval having no common measure with the real durations of polymers used in the nuclear industry, we can assume that the stationary state reached, that is, $d[POO^{\bullet}]/dt = 0$ and $d[P^{\bullet}]/dt = 0$, with t being time. We are referring to stationary homogeneous chemical kinetics. The analysis as a function of $[O_2]$ shows three regimes: a) a regime of high oxygen partial pressure, where the preponderant recombination is $POO^{\bullet}$+ $POO^{\bullet}$; b) a regime of low O_2 partial pressure, where the preponderant recombination is $P^{\bullet} + P^{\bullet}$; c) an intermediate regime. For each regime, the system of equations can be simplified by retaining just one termination reaction: equations [8.6], [8.4] and [8.5], respectively. The system of differential equations can then be integrated analytically.

It is worth mentioning that the oxygen consumption radiochemical yield, $G(-O_2)$, can reach values that far exceed the macroradical formation radiochemical yield ($G(P^\bullet) \simeq 7.10^{-7}$ mol/J). The radio-oxidation of a polymer is a chain reaction for which it is usual to define the kinetic chain length by $l_{kc} = [G(-O_2) - G(P^\bullet)]/G(P^\bullet)$. This value quantifies the degree to which $P^\bullet$ radicals are used to consume oxygen. Intuitively, the greater the propagation and the lower the bimolecular recombination are, the greater the l_{kc} will be.

Several parameters modify the diagram in Figure 8.5. Elevated temperature adds a new initiation reaction by decomposing *POOH*, generating new $PO^\bullet$, $OH^\bullet$ and $POO^\bullet$ radicals, hence the importance of simultaneously and formally treating thermal and radio aging (Khelidj et al. 2005; Khelidj 2006). Moreover, additives including antioxidants are incorporated into the actual materials, which can be included in the kinetic diagram (Bannouf 2014). The antioxidants used were designed and chosen to be effective in thermo-oxidation, and in a temperature range that is not necessarily encountered under irradiation. Finding antioxidants that are more suitable for radio-oxidation and understanding their reaction mechanisms would be of great interest. Lastly, real polymers contain mineral fillers, which have a major role in terms of the mechanical properties, but whose influence on the kinetics of radio-oxidation is less well known.

8.6.2. *Chemical and physical influences of the dose rate*

Considering the mechanism of Bolland and Gee, it is easy to assume that increasing the dose rate I would lead to an increase in termination reactions and, thus, to an oxidation limitation. This is called the "chemical effect of the dose rate". The equation of this relationship was obtained experimentally (Decker and Mayo 1973): the radiochemical yields of the different reactions can be written as: $G = A + B/\sqrt{I}$. For each preponderant recombination regime (equation [8.4] to [8.6]) and for each product, it is possible to calculate A and B analytically (Dély 2005). For example, for the preponderant $POO^\bullet + POO^\bullet$ regime, which often corresponds to real conditions, we find $A = 0$ for hydroperoxides; therefore, G(POOH) is strictly proportional to the inverse of the square root of the dose rate. On the contrary, for *POOP*, B=0, so the radiochemical yield is independent of the dose rate. For $G(-O_2)$, A and B are non-zero. The chemical effect of the dose rate is specific to each created product or gas consumed. This has concrete consequences on the modeling of the long-term behavior of materials, as well as on the time acceleration factors. For example, in this oversimplified model, assuming that the harshest modification comes from $POO^{\bullet} + POO^{\bullet}$ recombinations, changing the dose rate will then have no consequence; only the dose counts.

We have already briefly mentioned that a "thick" sample is not oxidized in the core if the dose rate is too high. The oxygen supply from the surface, which is limited by diffusion, is not sufficient to compensate for the consumption of oxygen by irradiation. This is called the "physical effect of the dose rate". A profile of $[O_2]$ as a function of the distance x to the surface $[O_2](x,t)$ is then established. The stationary profile is calculated by solving the following differential equation:

$$0 = D \,.\, v_{-O_2}(x) \,.\, \partial([O_2](x))/\partial x^2 \text{ where } v_{-O_2}(x) = G(-O_2)\,(x).I$$

$G(-O_2)$ depends on x because this radiochemical yield depends on $[O_2]$, and $[O_2]$ depends on x (Gillen and Clough 1991; Corbin 2001; Dély 2005). Measuring $G(-O_2)$ as a function of O_2 partial pressure, $P_p(O_2)$, is necessary if we want to accurately calculate the oxidation profile in a sample. Usually, when $P_p(O_2)$ and thus $[O_2]$ increases from zero, $G(-O_2)$ initially increases linearly. Then, for low $P_p(O_2)$ values that are much lower than the ordinary pressure, the increase in $G(-O_2)$ saturates and a plateau is observed where the system is, in terms of chemical kinetics, supersaturated with O_2. From a practical point of view, it is customary to define a critical thickness, e_c, above which a sample is no longer homogeneously oxidized:

$$e_c = \sqrt{\frac{P_p(O_2)}{G(-O_2)\,.I}}\,.p\,.X$$

where p is the permeation coefficient and X is a parameter that depends on the evolution of $G(-O_2)$ as a function of $[O_2]$. If this is not known, we use X = 8.

Chemical kinetic models have been partly developed to provide a more rigorous basis for predicting the lifetime of nuclear materials, especially cable insulation. In parallel, more pragmatic approaches have been conducted (Plaček and Kohout 2010). For example, EDF proposed that the evolution over time of a characteristic property P, which is typically elongation at break, is given by $dP/dt = -K_t \cdot P^{\beta}$, with K_t being a pseudo-constant of a thermally activated reaction and dependent on the dose rate, and β being an apparent order of reaction (Pinel and Boutaud 1999; Monchy-Leroy and Therond 2011). More recently, the Sandia National Laboratory team (Celina et al. 2019) developed scaling laws for a time–temperature–dose rate (t–T–r) superposition, and gave a set of analytical formulas for the value known as

DED (Dose to Equivalent Damage). For a given sample, this DED indicates the dose that needs to be reached for an equivalent damage to be obtained, but under different temperature or dose rate conditions.

Let us mention that a polymer irradiated in anoxia, which is then released in the air, will most likely oxidize, even relatively long after the end of the irradiation: this is post-irradiation oxidation. This phenomenon is due to the persistence of radicals. For example, alkyl radicals initially created in the crystalline phase diffuse very slowly. They will, at a given time, find themselves in the amorphous phase, which is accessible to oxygen. The Bolland and Gee chain reaction will then be initiated. This has a practical consequence when a sample is returned to the air after an anoxic irradiation, so as to measure its modifications, which is very frequently the case. The oxidation can then somewhat distort the result.

8.6.3. *α irradiation*

The formalization of the equations was carried out through the exploitation of results obtained with low LET irradiations. When the radio-oxidation is produced by particles of higher LET, which is typical of α irradiations, the experiment reveals the overall expected effects: the polymer oxidizes less, as shown in Figure 8.6. As the primary species are locally more concentrated around the passage of the particles, this should promote the radical recombinations and thus limit the oxidation.

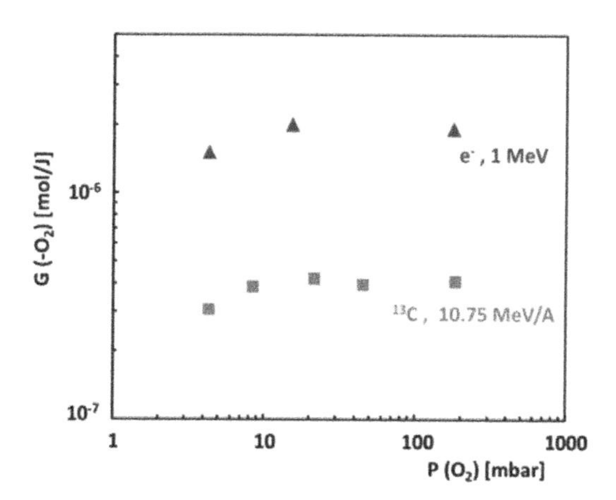

Figure 8.6. *Dioxygen consumption radiochemical yield G(–O_2) as a function of oxygen pressure and LET. From Dély (2005) and Dély et al. (2005). Copyright 5004621409076. For a color version of this figure, see www.iste.co.uk/bouffard/nuclear.zip*

We already know that for low LET irradiation, the homogeneity assumption made in chemical kinetics models is not perfectly valid; this assumption is quite unreasonable for α irradiation. Solving the kinetic diagram in Figure 8.5 by explicitly treating heterogeneity in the energy deposition is extremely difficult, because it is a diffusion/reaction problem with very different time and space scales. This has only very recently been examined using kinetic Monte Carlo simulations (Gervais et al. 2021). This study showed that at the very first instants of the irradiation, the concentration of macroradicals increased to tend progressively towards a stationary state, like in the homogeneous case. However, the concentrations never really homogenize. This is a stationary state in the sense that the average concentration and the average degree of heterogeneity[10] are constant, but it is still very far from spatial homogeneity. Figure 8.7 shows a "photo" of the spatial distribution of $POO^{\bullet}$ at a given moment and in the stationary state, the age of each $POO^{\bullet}$ being reported thanks to the color scale. This figure clearly shows that during irradiation with equivalent α ions, the distribution of $POO^{\bullet}$ is very heterogeneous: in blue, we can distinguish some areas of high concentration for the youngest radicals. An image at another time, but still in a stationary state, would have been visually different, but would have led to the same mean value of $POO^{\bullet}$ concentration and to the same radial distribution function.

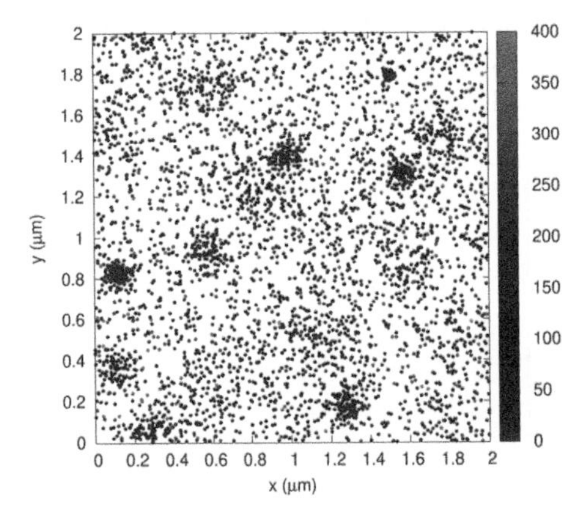

Figure 8.7. *Spatial distribution of $POO^{\bullet}$ macroradicals, in a stationary state, for irradiation with an LET that is equivalent to the LETs of α. The right scale gives the age of the radicals in seconds. Simulation conditions are given in the publication Gervais et al. (2021). Copyright 5004620957806. For a color version of this figure, see www.iste.co.uk/bouffard/nuclear.zip*

10. The average degree of heterogeneity is characterized by the radial distribution function, in other words, the probability for a given radical to find another radical at a given distance. It depends on the flux: the higher the flux, the less heterogeneous the system is.

8.7. Conclusion and perspectives

The behavior of organic materials under irradiation has certainly been less studied than that of nuclear fuel or of metallic alloys for vessels, cladding and internals. Nevertheless, these materials, which are sensitive to radiolysis, have a significant place in the nuclear power industry. All stages of the fuel cycle are concerned by their modifications under irradiation, from the front end of the nuclear cycle to the management of nuclear waste and the safety of nuclear installations. This is currently an active field of research. New topics have emerged, such as the study of radiolysis by actinide α. On the fundamental level, innovative experimental approaches (such as custom synthesized polymers or the contribution of low temperatures) have led to significant conceptual advances. However, it is possible to go further. The understanding of the radio-oxidation of polymers would certainly progress if time-resolved experiments and online monitoring of the dynamics of radical populations were developed (online EPR experiments, under irradiation). On a more technological level, there have also been significant advances. The experimental effort has, for example, made it possible to establish operational databases for gas emission. These fundamental and technological types of research feed one another, allowing better progress in both. Lastly, and we have not yet mentioned this here, all of the research has made it possible to establish changes in the choice of materials used. The most representative example is the replacement of polyvinyl chloride (PVC) films by polyurethane (PUR).

8.8. References

Ahn, Y., Roma, G., Colin, X. (2022). Elucidating the role of alkoxy radicals in polyethylene radio-oxidation kinetics. *Macromolecules*, 55, 8676–8684.

ANDRA (2005). Dossier 2005 argile – Synthèse : évaluation de la faisabilité du stockage géologique en formation argileuse [Online]. Available at: www.andra.fr/publications?f%5B0%5D=facet_doc_date%3A2005.

Baidak, A. and LaVerne, J.A. (2010). Radiation-induced decomposition of anion exchange resins. *Journal of Nuclear Materials*, 407(3), 211–219.

Bannouf, W. (2014). Analyse et modélisation cinétique de la perte physique et de la consommation chimique d'un mélange phénol/HALS au cours du vieillissement radio-thermique d'une matrice EPDM. PhD Thesis, École nationale supérieure d'arts et métiers – ENSAM Paris.

Biggin, H.C. (1991). An introduction to radiation units and measurerments. In *Irradiation Effects on Polymers*, Clegg, D.W. and Collyer, A.A. (eds). Springer, The Netherlands.

Bolland, J.L. and Gee, G. (1946). Kinetic studies in the chemistry of rubber and related materials II: The kinetics of oxidation of unconjugated olefins. *Transactions of the Faraday Society*, 42(0), 236–243.

Bopp, C.D. and Sisman, O. (1953). Radiation stability of plastics and elastomers (Supplement to ORNL-928). Report, Oak Ridge National Lab, ONRL-1373.

Bouffard, S., Gervais, B., Leroy, C. (1995). Basic phenomena induced by swift heavy ions in polymers. *Nuclear Instruments & Methods in Physics Research Section B: Beam Interactions with Materials and Atoms*, 105, 1–4.

Bouffard, S., Balanzat, E., Leroy, C., Busnel, J.P., Guevelou, G. (1997). Cross-links induced by swift heavy ion irradiation in polystyrene. *Nuclear Instruments & Methods in Physics Research Section B: Beam Interactions with Materials and Atoms*, 131(1–4), 79–84.

Boughattas, I., Ferry, M., Dauvois, V., Lamouroux, C., Dannoux-Papin, A., Leoni, E., Balanzat, E., Esnouf, S. (2016a). Thermal degradation of gamma-irradiated PVC: I-dynamical experiments. *Polymer Degradation and Stability*, 126, 219–226.

Boughattas, I., Pellizzi, E., Ferry, M., Dauvois, V., Lamouroux, C., Dannoux-Papin, A., Leoni, E., Balanzat, E., Esnouf, S. (2016b). Thermal degradation of gamma-irradiated PVC: II-Isothermal experiments. *Polymer Degradation and Stability*, 126, 209–218.

Boughattas, I., Labed, V., Gerenton, A., Ngono-Ravache, Y., Dannoux-Papin, A. (2018). Hydration effect on ion exchange resin irradiated by swift heavy ions and gamma rays. *Journal of Nuclear Materials*, 504, 68–78.

Cambon, S. (2001). Étude du mécanisme de dégradation radiochimique d'un élastomère de type EPDM. PhD Thesis, Université Blaise Pascal, Clermont-Ferrand.

Celina, M., Linde, E., Brunson, D., Quintana, A., Giron, N. (2019). Overview of accelerated aging and polymer degradation kinetics for combined radiation-thermal environments. *Polymer Degradation and Stability*, 166, 353–378.

Ceresoli, D., Righi, M.C., Tosatti, E., Scandolo, S., Santoro, G., Serra, S. (2005). Exciton self-trapping in bulk polyethylene. *Journal of Physics: Condensed Matter*, 17(29), 4621–4627.

Chailley, V., Balanzat, E., Dooryhee, E. (1995). Amorphization kinetics of poly(vinylidene fluoride) on high-energy ion irradiation. *Nuclear Instruments & Methods in Physics Research Section B: Beam Interactions with Materials and Atoms*, 105, 110–114.

Chantreux, M. (2021). Étude de la dégradation par radiolyse et/ou hydrolyse basique de PVC présents dans les stockages de déchets nucléaires. PhD Thesis, Université Aix-Marseille.

Clegg, D.W. and Collyer, A.A. (1991). *Irradiation Effects on Polymers*. Springer, The Netherlands.

Clough, R.L. (2001). High-energy radiation and polymers: A review of commercial processes and emerging applications. *Nuclear Instruments and Methods in Physics Research Section B: Beam Interactions with Materials and Atoms*, 185(1), 8–33.

Clough, R.L. and Shalaby, S.W. (1991). *Radiation Effects on Polymers*. American Chemical Society, Washington DC.

Combernoux, N., Labed, V., Schrive, L., Wyart, Y., Carretier, E., Moulin, P. (2016). Effect of gamma irradiation at intermediate doses on the performance of reverse osmosis membranes. *Radiation Physics and Chemistry*, 124, 241–245.

Combernoux, N., Schrive, L., Labed, V., Wyart, Y., Carretier, E., Moulin, P. (2017). Treatment of radioactive liquid effluents by reverse osmosis membranes: From lab-scale to pilot-scale. *Water Research*, 123, 311–320.

Corbin, D. (2001). Étude de l'oxydation et de la tenue d'élastomères irradiés : conséquences sur l'intégrité des câbles électriques lors d'une situation accidentelle d'un réacteur à eau pressurisée. PhD Thesis, Université de Caen Normandie.

Dannoux, A. (2007). Extrapolation dans les temps des cinétiques de production des produits de radiolyse : application à un polyurethanne. PhD Thesis, Université de Caen Normandie.

Decker, C. and Mayo, F.R. (1973). Aging and degradation of polyolefins II: γ-initiated oxidations of atactic polypropylene. *Journal of Polymer Science: Polymer Chemistry Edition*, 11(11), 2847–2877.

Dély, N. (2005). Radio-oxydation d'un élastomère de type EPDM lors d'irradiations faiblement ou fortement ionisantes : mesure et modélisation de la consommation de dioxygène. PhD Thesis, Université de Caen Normandie.

Dély, N., Ngono-Ravache, Y., Ramillon, J.M., Balanzat, E. (2005). Oxygen consumption in EPDM irradiated under different oxygen pressures and at different LET. *Nuclear Instruments & Methods in Physics Research Section B: Beam Interactions with Materials and Atoms*, 236, 145–152.

Dole, M. (1973). *The Radiation Chemistry of Macromolecules*. Academic Press, Cambridge, MA.

Elliot, A.J. and Bartels, D.M. (2009). The reaction set, rate constants and g-values for the simulation of the radiolysis of light water over the range 20 deg to 350 deg C based on information available in 2008. Report, IAEA, AECL --153-127160-450-001, Canada.

Esnouf, S. and Balanzat, E. (2007). Trapped and transient radicals observed in ethylene–propylene–diene terpolymers. *Polymer*, 48, 7531–7538.

Esnouf, S., Dannoux-Papin, A., Bossé, E., Roux-Serret, V., Chapuzet, C., Cochin, F., Blancher, J. (2022a). Hydrogen generation from α-radiolysis of organic materials in transuranic waste. Comparison between experimental data and STORAGE calculations. *Nuclear Technology*, 208(2), 347–356.

Esnouf, S., Dannoux-Papin, A., Chapuzet, C., Roux-Serret, V., Piovesan, V., Cochin, F. (2022b). STORAGE: A source term model for intermediate-level radioactive waste. *Nuclear Technology*, 208(12), 1806–1821.

Ferry, M. and Ngono, Y. (2021). Energy transfer in polymers submitted to ionizing radiation: A review. *Radiation Physics and Chemistry*, 180, 109320.

Ferry, M., Bessy, E., Harris, H., Lutz, P.J., Ramillon, J.M., Ngono-Ravache, Y., Balanzat, E. (2012). Irradiation of ethylene/styrene copolymers: Evidence of sensitization of the aromatic moiety as counterpart of the radiation protection effect. *Journal of Physical Chemistry B*, 116(6), 1772–1776.

Ferry, M., Bessy, E., Harris, H., Lutz, P.J., Ramillon, J.M., Ngono-Ravache, Y., Balanzat, E. (2013). Aliphatic/aromatic systems under irradiation: Fluence of the irradiation temperature and of the molecular organization. *Journal of Physical Chemistry B*, 117(46), 14497–14508.

Ferry, M., Ngono-Ravache, Y., Aymes-Chodur, C., Clochard, M.C., Coqueret, X., Cortella, L., Pellizzi, E., Rouif, S., Esnouf, S. (2016). Ionizing radiation effects in polymers. *Reference Module in Materials Science and Materials Engineering*. Elsevier, Amsterdam.

Ferry, M., Ramillon, J.M., Been, T., Lutz, P.J., Ngono-Ravache, Y., Balanzat, E. (2019). Energy migration effect on the formation mechanism of different unsaturations in ethylene/styrene random copolymers. *Polymer Degradation and Stability*, 160, 210–217.

Ferry, M., Roma, G., Cochin, F., Esnouf, S., Dauvois, V., Nizeyimana, F., Gervais, B., Ngono-Ravache, Y. (2020). Polymers in the nuclear power industry. In *Comprehensive Nuclear Materials*, 2nd edition. Konings, R.J.M. and Stoller, R.E. (eds). Elsevier, Amsterdam.

Fromentin, E. (2017). Lixiviation des polymères irradiés : caractérisation de la solution et complexation des actinides. PhD Thesis, Université Pierre et Marie Curie, Paris.

Fromentin, E., Pielawski, M., Lebeau, D., Esnouf, S., Cochin, F., Legand, S., Ferry, M. (2016). Leaching of radio-oxidized poly(ester urethane): Water-soluble molecules characterization. *Polymer Degradation and Stability*, 128, 172–181.

Gaté, C. (1997). Polymères aliphatiques sous irradiation par des ions lourds. PhD Thesis, Université de Caen Normandie.

Gervais, B., Ngono, Y., Balanzat, E. (2021). Kinetic Monte Carlo simulation of heterogeneous and homogeneous radio-oxidation of a polymer. *Polymer Degradation and Stability*, 185, 109493.

Gillen, M.A. and Clough, R.L. (1991). Accelerated aging methods for predicting long-term mechanical performance of polymers. In *Irradiation Effects on Polymers*, Clegg, D.W. and Collyer, A.A. (eds). Elsevier, Amsterdam.

Glaus, M.A. and Van Loon, L.R. (1998). Experimental and theoretical studies on alkaline degradation of cellulose and its impact on the sorption of radionuclides. Report, PSI Bericht 98–07, Paul Scherrer Institut, Villigen.

Gracia, J. (2017). Étude du comportement du stéarate du zinc en température et sous irradiation – Impact sur les propriétés de lubrification. PhD Thesis, École nationale supérieure d'arts et métiers (ENSAM), Paris.

Gracia, J., Vermeulen, J., Baux, D., Sauvage, T., Venault, L., Audubert, F., Colin, X. (2018). Stability of zinc stearate under alpha irradiation in the manufacturing process of SFR nuclear fuels. *Radiation Physics and Chemistry*, 144, 92–99.

Guarino, F., Hauviller, C., Tavlet, M. (2001). Compilation of radiation damage test data. Part 4: Adhesives for use in radiation areas. Report, CERN 2001–006, Geneva.

Khelidj, N. (2006). Vieillissement d'isolants de câbles en polyéthylène en ambiance nucléaire. PhD Thesis, École nationale supérieure d'arts et métiers (ENSAM), Paris.

Khelidj, N., Colin, X., Audouin, L., Verdu, J. (2005). A simplified approach for the lifetime prediction of PE in nuclear environments. *Nuclear Instruments and Methods in Physics Research Section B: Beam Interactions with Materials and Atoms*, 236(1), 88–94.

Kircher, J.F. and Bowman, R.E. (1964). *The Effects of Radiation on Materials and Components*. Reinhold, New York.

Kondoh, T., Yang, J., Norizawa, K., Kan, K., Yoshida, Y. (2011). Femtosecond pulse radiolysis study on geminate ion recombination in n-dodecane. *Radiation Physics and Chemistry*, 80(2), 286–290.

Lebeau, D., Esnouf, S., Gracia, J., Audubert, F., Ferry, M. (2017). New generation of nuclear fuels: Stability of different stearates under high doses gamma irradiation in the manufacturing process. *Journal of Nuclear Materials*, 490, 288–298.

Ma, T., Janot, J.-M., Balme, S. (2020). Track-etched nanopore/membrane: From fundamental to applications. *Small Methods*, 4(9), 2000366.

Marletta, G. (1990). Chemical reactions and physical property modifications induced by keV ion beams in polymers. *Nuclear Instruments and Methods in Physics Research Section B: Beam Interactions with Materials and Atoms*, 46(1), 295–305.

Monchy-Leroy, C. and Therond, P. (2011). Nuclear cables: Lifetime simulation and new approach for the study of polymer ageing. In *8th Internatiional Conference on Insulated Power Cables (Jicable'11)*, Versailles [Online]. Available at: www.jicable.org.

Mozumder, A. and Magee, J.L. (1966). Model of tracks of ionizing radiations for radical reaction mechanisms. *Radiation Research*, 28(2), 203–214.

Ngono-Ravache, Y., Ferry, M., Esnouf, S., Balanzat, E. (2016). Polymers under ionizing radiations: The specificity of swift heavy ions. *EPJ Web of Conferences*, 115, 02003.

Ojovan, I., Lee, W., Kalmikov, S. (2019). *An Introduction to Nuclear Waste Immobilisation*, 3rd edition. Elsevier, Amsterdam.

Özdemir, T. (2014). Monte Carlo simulations of radioactive waste embedded into EPDM and effect of lead filler. *Radiation Physics and Chemistry*, 98, 150–154.

Perera, M.C.S. and Hill, D.J.T. (2003). Radiation chemical yields: G values. In *Polymer Handbook*, 4th edition, Brandup, J., Immergut, E.H., Grulke E.A. (eds). John Wiley & Sons, New York.

Phillips, D.C. and Burnay, S.G. (1991). Polymers in the nuclear power industry. In *Irradiation Effects on Polymers*, Clegg, D.W. and Collyer, A.A. (eds). Elsevier, Amsterdam.

Picq, V. (2000). L'émission gazeuse des polymères aliphatiques sous irradiation : effet du pouvoir d'arrêt électronique. PhD Thesis, Université de Caen Normandie.

Picq, V., Ramillon, J.M., Balanzat, E. (1998). Swift heavy ions on polymers: Hydrocarbon gas release. *Nuclear Instruments and Methods in Physics Research Section B: Beam Interactions with Materials and Atoms*, 146, 496–503.

Pillay, K. (1986). A review of the radiation stability of ion exchange materials. *Journal of Radioanalytical and Nuclear Chemistry*, 102(1), 247.

Pinel, B. and Boutaud, F. (1999). A methodology to predict the life duration of polymers used in nuclear power stations. Industrial needs and their approach. *Nuclear Instruments and Methods in Physics Research Section B: Beam Interactions with Materials and Atoms*, 151(1), 471–476.

Plaček, V. and Kohout, T. (2010). Comparison of cable ageing. *Radiation Physics and Chemistry*, 79(3), 371–374.

Rånby, B. and Rabek, J.F. (1977). *ESR Spectroscopy in Polymer Research*. Springer, Berlin, Heidelberg.

Rébufa, C., Traboulsi, A., Labed, V., Dupuy, N., Sergent, M. (2015). Experimental design approach for identification of the factors influencing the γ-radiolysis of ion exchange resins. *Radiation Physics and Chemistry*, 106, 223–234.

Rivaton, A., Cambon, S., Gardette, J.L. (2005). Radiochemical ageing of EPDM elastomers 3. Mechanism of radiooxidation. *Nuclear Instruments and Methods in Physics Research Section B: Beam Interactions with Materials and Atoms*, 227(3), 357–368.

Schönbacher, H., Szeless, B., Tavlet, M., Humer, K., Weber, H.W. (1996). Results of radiation tests at cryogenic temperature on some selected organic materials for the LHC. Report, CERN 96–05, Geneva.

Seguchi, T. and Morita, Y. (2003). Radiation resistance of plastics and elastomers. In *The Wiley Database of Polymer Properties*. John Wiley & Sons, New York.

Swyler, K.J., Dodge, C.J., Dayal, R. (1983). Irradiation effects on the storage and disposal of radwaste containing organic ion-exchange media. Brookhaven National Lab report, NUREG/CR--3383, Upton, New York.

Tavlet, M. and Schönbacher, H. (1989). Compilation of radiation damage test data, Part I: Halogen-free cable-insulating materials, 2nd edition. Report, CERN-89-12, Geneva.

Tavlet, M., Fontaine, A., Schönbacher, H. (1998). Compilation of radiation damage test data, Part II: Thermoset and thermoplastic resins, composite materials, 2nd edition. Report, CERN 98–01, Geneva.

Tessaro, V.B., Poignant, F., Gervais, B., Beuve, M., Galassi, M.E. (2019). Theoretical study of W-values for particle impact on water. *Nuclear Instruments and Methods in Physics Research Section B: Beam Interactions with Materials and Atoms*, 460, 259–265.

Toulemonde, M., Trautmann, C., Balanzat, E., Hjort, K., Weidinger, A. (2004). Track formation and fabrication of nanostructures with MeV-ion beams. *Nuclear Instruments & Methods in Physics Research Section B: Beam Interactions with Materials and Atoms*, 216, 1–8.

Van de Voorde, M.H. (1970). Effects of radiation on materials and components. Report, CERN 70-05, Geneva.

Venault, L., Deroche, A., Gaillard, J., Lemaire, O., Budanova, N., Vermeulen, J., Maurin, J., Vigier, N., Moisy, P. (2019). Dihydrogen H_2 steady state in α-radiolysis of water adsorbed on PuO_2 surface. *Radiation Physics and Chemistry*, 162, 136–145.

Ventura, A. (2013). Polymères sous rayonnements ionisants : étude des transferts d'énergie vers les défauts d'irradiation. PhD Thesis, Université de Caen Normandie.

Ventura, A., Ngono-Ravache, Y., Marie, H., Levavasseur-Marie, D., Legay, R., Dauvois, V., Chenal, T., Visseaux, M., Balanzat, E. (2016). Hydrogen emission and macromolecular radiation-induced defects in Polyethylene irradiated under an Iiert amosphere: The role of energy transfers toward trans-Vinylene unsaturations. *Journal of Physical Chemistry B*, 120(39), 10367–10380.

Westbrook, C.K. and Dryer, F.L. (1984). Chemical kinetic modeling of hydrocarbon combustion. *Progress in Energy and Combustion Science*, 10(1), 1–57.

9

Irradiation Tools

Serge BOUFFARD[1] and Nathalie MONCOFFRE[2]

[1] *CIMAP, CEA – CNRS – ENSICAEN, Université Caen Normandie, France*

[2] *Institut de Physique des 2 Infinis de Lyon, Université Claude Bernard Lyon 1, CNRS/IN2P3, Villeurbanne, France*

9.1. Why experiment with accelerators?

Studying the aging of materials under real conditions is essential for the safety of nuclear installations and important for obtaining feedback to base predictive approaches on. However, this approach faces two hurdles: on the one hand, the duration, which can amount to decades in the case of nuclear reactor components, or even hundreds of thousands of years for waste disposal, and on the other hand, industrial materials and their environments are so complex that it is often difficult to extract the physico-chemical phenomena governing their behavior under irradiation. The experimental simulations will have objectives to accelerate the time and carry out parametric studies with separate variables on model materials, in order to understand the fundamental processes of material aging under irradiation.

There are many benefits of experimenting with accelerators:

– the precise monitoring of the irradiation parameters;

– the possibility of varying these parameters in a wide range (energy and mass of the projectile, flux, maximum dose, temperature and environment of the samples, etc.);

– the possibility of installing high-performance characterization tools online;

– the absence or low induced radioactivity;

– the low cost compared to an irradiation in a reactor;

– etc.

Nuclear Materials under Irradiation,
coordinated by Serge BOUFFARD and Nathalie MONCOFFRE.

However, these experiments have limitations that should not be overlooked:

– changing the particle implies that the primary spectrum and the distribution of the initial defects will be different;

– the presence of the surface near the irradiated zone can induce biases, as well as the size of the grains in relation to the projectile range;

– an increase in the damage rate can modify the conditions of nucleation and growth of the extended defects, and even make the material bifurcate towards a different structural phase;

– the simultaneous production of gas by nuclear reaction and the production of defects can only be reproduced by multi-beam techniques;

– etc.

9.2. Irradiation conditions in nuclear energy

In nuclear energy, there are numerous conditions of irradiation that cover extremely broad fields, and this is true for particles (type, energy, flux, fluence), the physico-chemical conditions (temperature, presence of water, CO_2, etc.) and the materials concerned (metals, iono-covalent ceramics, concretes, clays, organic materials).

Table 9.1 summarizes the particles that can damage materials in the nuclear power industry.

9.2.1. *Characteristics of these particles*

From the perspective of material irradiation, each type of particle is characterized by its nuclear and electronic stopping powers, flux and fluence.

Particles	Energy	Comments
Neutrons	0.1–14 MeV	One fission: 2 fission fragments and 2.8 neutrons
Fission fragments	60–100 MeV	~10^{19} neutrons/m²/s in the core of a 1,300 MWe PWR
Alphas	3–8 MeV	~ 4 10^9 α/g/s, in the fuel, four years after its release
Alpha recoil	80–90 keV	from the reactor, main emitter: ^{241}Pu
Gammas	Wide range	Their number is difficult to assess; participate in heating, radiolytic processes
Betas	Wide range from keV to MeV	~ 10^{10} β/g/s, in the fuel, four years after its release from the reactor, main emitter: ^{147}Pm, ^{137}Cs, ^{90}Sr, etc.

Table 9.1. *Particles that can damage materials in the nuclear power industry*

9.2.2. *How is irradiation simulated in a nuclear environment?*

9.2.2.1. *Simulation of neutron irradiation*

The simulation of neutron irradiation is often performed by ion irradiation, which makes it possible to adjust irradiation parameters such as the damage rate, the dose rate, and the irradiation temperature, with great flexibility. Moreover, the absence or the low activation of the samples gives these parametric studies access to all the techniques of characterization of the materials, some of which can be implemented online. However, there are differences between ions and neutrons, the main one being their range. While ions deposit their energy inhomogeneously on thicknesses between 0.1 and 100 μm depending on their energy, neutrons create a uniform damage profile on centimetric thicknesses. Moreover, the energy spectra of the first atoms that are knocked on (PKA: primary knock-on atom), which directly depend on the interaction potentials of the two types of particles, are fundamentally different. In Figure 9.1, a plot is shown for the fraction of displacements created by all primaries of energy lower than T, for T varying from the threshold energy of displacement to the maximum energy that can be transmitted. These curves unambiguously show that large collision cascades are favored in the case of neutrons. In reality, neutrons in a reactor are not monokinetic and, if we consider the energy spectrum of fast neutrons, the difference between ions and neutrons is not very large. From the point of view of the creation of defects, the choice of using a heavy ion of a hundred keV to simulate the effects of neutron irradiation seems wise, but there is a real penetration depth problem (< 100 nm), as well as a damage rate that is orders of magnitude faster (typically a year for neutrons and a day for ions).

By interacting with the nucleus of the target atoms directly, neutrons cause numerous nuclear reactions that change the chemical composition of the material, introducing gaseous hydrogen and helium atoms. Two approaches are used to simulate these nuclear reactions: multi-beam irradiation or irradiation by high-energy light ions. In the latter method, it is mainly spallation reactions that produce the gas in the material, but at a ratio of ~50 appm He/dpa, which is much higher than that created in the core of a PWR (~0.1 appm He/dpa^{-1})[1].

The use of multi-beams is certainly the most representative method to simulate neutron irradiation: a heavy ion beam creates defects with a size distribution of collision cascades close to that of the fast neutron spectrum, and a helium beam implants the gas. Ideally, these operations should be performed in parallel, but it can

1. This ratio is close to that produced by the 14 MeV neutrons of a fusion reactor.

be done sequentially, provided that the state of the material does not change too much during each sequence.

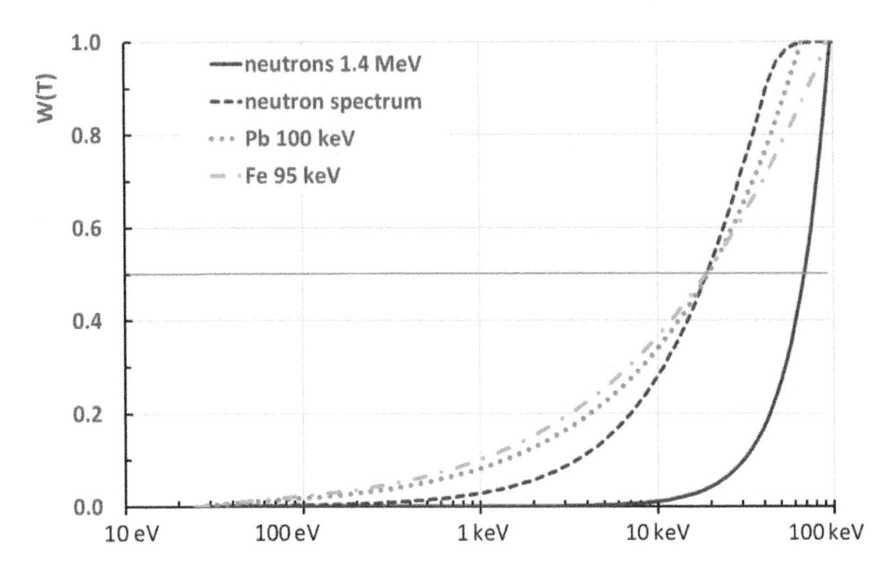

Figure 9.1. *Fraction of the displacements W(T) created by primaries of energy lower than T (e.g. 20% of the displacements are produced by primaries of energy lower than 4 keV, in the case of 100 keV Pb). Calculation of the number of displacements following the NRT formalism (see Chapter 1). For a color version of this figure, see www.iste.co.uk/bouffard/nuclear.zip*

The main advantage of ion irradiation is obviously the speed with which the final dose is reached. While 3 to 5 dpa/year are created in an experimental reactor, and more than 10 in a fast reactor, 1 dpa/day is easily reached with an accelerator. Is the ion-neutron equivalence maintained when the damage rate is multiplied by two orders of magnitude? In order to ensure that the deduced values are representative, temperature shift proposals are often made (Was 2015).

9.2.2.2. *Simulation of alphas and alpha recoils*

The emission of an alpha by an actinide is accompanied by the recoil of the daughter nucleus (conservation of momentum). Thus, the radioactive decay of ^{239}Pu consists of the emission of an alpha of 5.2 MeV and the recoil of a ^{235}U nucleus of 89 keV, two particles that can be delivered by accelerators. The study of their separate effects does not pose any problem, except for the short range of the recoil nucleus (of the order of 70 nm in nuclear glass). The role in the damage of the α recoil depends on the way the actinides are incorporated into the material. When they are isolated or in small clusters, the α recoil fully participates in the creation of defects. On the other hand, if the α emitters are included in grains of micrometric size, the matrix will only be irradiated by the alphas (except at the periphery of the grain). On the other hand, the study of the behavior of a material under the coupled effect of alphas and alpha recoils is only possible with external beams, by modifying the energy of one or both particles (see Figure 9.2).

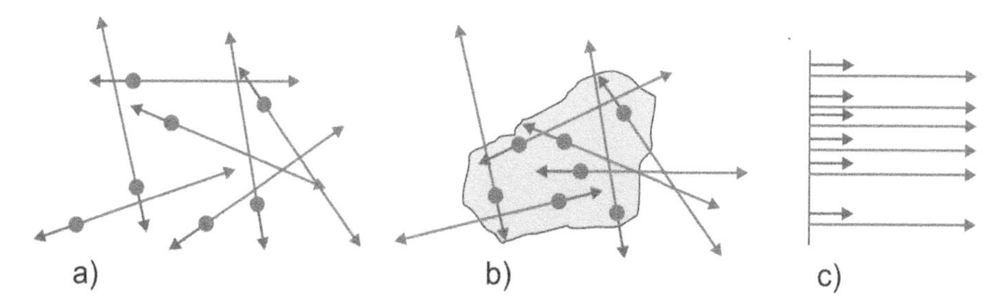

Figure 9.2. *The blue arrows symbolize the path of the alphas, and the red arrows show that of the α recoil. To make the drawing readable, the lengths of the red arrows have been deliberately increased, they should be 300 times shorter than the blue ones. Three cases are considered: a) irradiation of a material during the α decay of an actinide dispersed in the host matrix; b) included in grains and c) simulation by two external beams. For a color version of this figure, see www.iste.co.uk/bouffard/nuclear.zip*

9.2.2.3. *Fission fragments*

The fission fragments can be accurately simulated by accelerated ions. Indeed, they are a light fragment A = ~95, ~98 MeV and a heavy fragment A = ~138, ~68 MeV, so typically an Mo ion and an Xe ion with a respective range of 7.5 μm and 5.5 μm in UO_2.

9.2.2.4. *Gamma rays*

In nuclear energy, many nuclei de-excite by emitting a gamma ray. The materials that are subjected to gamma irradiation are in a large energy range. In order to simulate these irradiations, we have access to sources of ^{60}Co and ^{137}Cs. The first one emits two γ rays of 1.1732 and 1.3325 MeV corresponding to the de-excitation of ^{60}Ni, and the second one emits just one γ ray at 661.7 keV (de-excitation of ^{137m}Ba). In this energy range, gammas essentially interact by Compton effect with matter and thus by emission of high-energy electrons, which are at the origin of the majority of the defects (see Figure 9.3). It is therefore possible to simulate gamma irradiations by electron irradiations, but the dose distribution in a thick sample is the main difference.

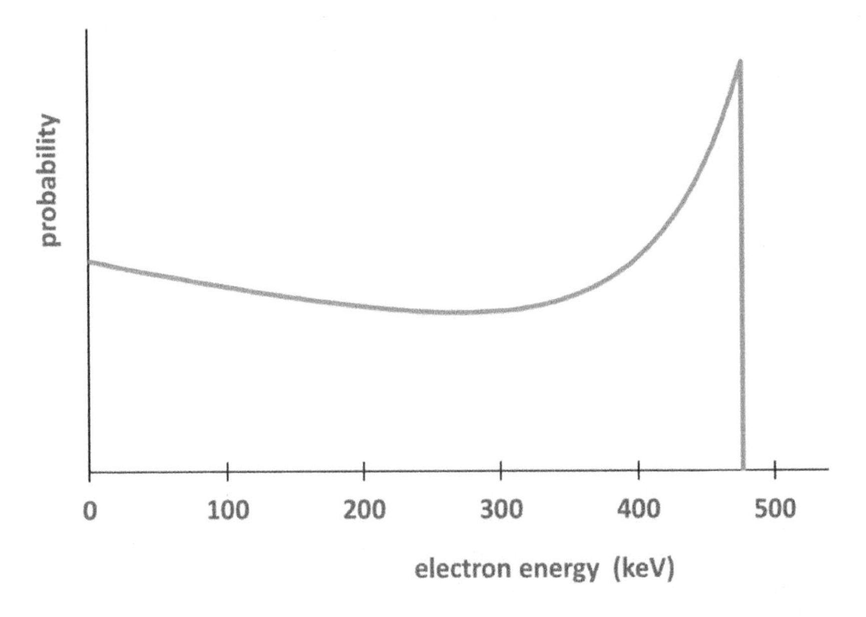

Figure 9.3. *Energy distribution of Compton electrons emitted by the 661 keV gammas of ^{137}Cs. For a color version of this figure, see www.iste.co.uk/bouffard/nuclear.zip*

9.3. Tools for simulation

Four types of tools are available for the studies of nuclear material aging under irradiation: experimental reactors, accelerators, gamma sources and doping with radioactive elements.

9.3.1. *Research reactors*

According to the IAEA[2], in 2017, there were 15 research reactors in the world with the ability to irradiate materials, and five are reportedly under construction,

2. IAEA: International Atomic Energy Agency, see: https://nucleus.iaea.org/RRDB/RR/ReactorSearch.aspx.

including the Jules Horowitz Reactor (JHR) in France. Most of these reactors have been in operation for more than 40 years, as shown in the list of the eight major reactors (see Table 9.2).

Operation	Country	Reactor	Power (MW)
1961	Belgium	BR-2	100
1961	Russia	SM-3	100
1963	Netherlands	HFR	45
1965	USA	HFIR	85
1967	USA	ATR	250
1967	Russia	MIR.M1	100
1968	Japan	JMTR	50
1968	Russia	BOR-60	60

Table 9.2. *List of major reactors that allow testing of materials under irradiation (IAEA 2018)*

These reactors, which allow technological irradiations to be carried out, have fast neutron fluxes that are higher than those of a PWR, hence the possibility of carrying out complete irradiation aging studies by accelerating the time, while remaining representative of the conditions of use of the material tested. In BOR-60, which is a sodium-cooled fast reactor with a harder fast neutron spectrum, it is possible to carry out 25 dpa/year. The planned HFR will allow damage of the order of 15 dpa/year. These tests in the reactor have the disadvantages of their advantages. They are essential to validate a material. However, the activation of the samples requires access to a hot laboratory (see Figure 9.4), delays their study and limits the number of accessible characterizations. The difficulty of access is a significant point. The time needed to set up the partnership with the national agency that manages the reactor and to design the irradiation capsule, as well as the cumbersome safety documents to be validated before each experiment, mean that a decade has passed between the design of the experiment and the first results being obtained. The issue of the cost of the experiment is also a limiting factor.

Figure 9.4. *C19 shielded cells of the "High Activity Waste" unit of the ATALANTE installation at the CEA Marcoule site (photo credit: CEA). For a color version of this figure, see www.iste.co.uk/bouffard/nuclear.zip*

9.3.2. *Accelerators*

Figure 9.5 shows the ions emitted by the irradiation sources in nuclear energy. In order to reproduce all of these particles, accelerators in three energy ranges must be implemented.

Low-energy accelerators. Commonly known as implanters, low-energy accelerators are generally made of an ion source mounted on a platform, carried at a high voltage of a few dozen to a few hundred kilovolts. These machines are used to introduce foreign atoms into materials by perfectly controlling their concentration profiles at depth. They were originally developed for the doping of semiconductors, before being used to improve the surface resistance of materials (wear, corrosion) with numerous applications in the field of metallurgy, for example. In this energy range, the ions create high damage rates by elastic collisions. Their short path in the material restricts the studies to layers that are a few tenths to a few micrometers thick, depending on the ions and their energy, which is the main drawback. With a 200 kV implanter and with Bi^{2+} ions, it is possible to implant about 10^{11} ions.cm^{-2}.s^{-1} at 120 ± 44 nm depth, while producing about 5,000 displacements per incident ion. It can be seen in Figure 9.6 that the implantation and damage profiles are shifted in depth. The presence of a defect gradient that is different from the implanted atom gradient complicates the interpretation of the implanted atom kinetics.

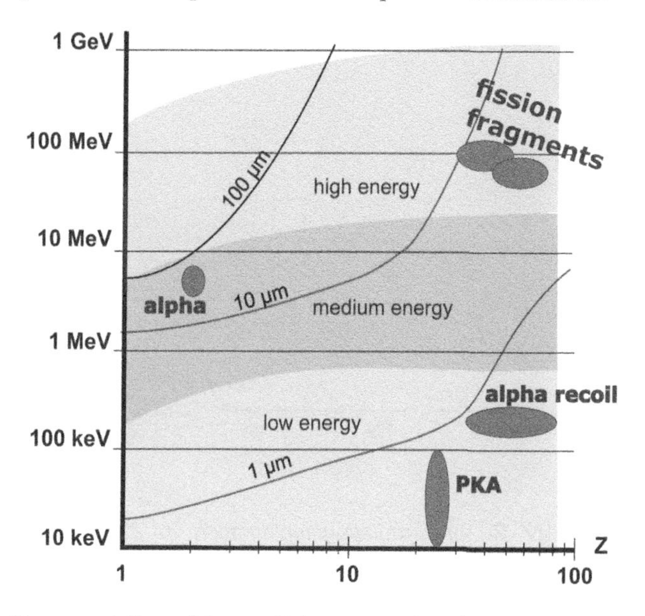

Figure 9.5. *Representation of the particles encountered in nuclear energy using an atomic number–energy reference (PKA, for primary knock-out atom, indicates the energy range of primaries created in iron by fast neutrons). The colored areas arbitrarily represent the energy ranges of the accelerators. For a color version of this figure, see www.iste.co.uk/bouffard/nuclear.zip*

Medium-energy accelerators. At the beginning of 2021, the International Atomic Energy Agency (IAEA) database listed 241 electrostatic accelerators in this energy range (1 MV ≤ V < 10 MV), 74% of which were in the tandem configuration (IAEA 2021). In the tandem configuration, the high-voltage terminal is connected to the ground by two insulating columns. Negative ions are injected at one end with an energy E_i to be accelerated to the terminal, where by collision with a gas or a very thin sheet, their charge will change from -1 to q+. They are then accelerated in the second column. Their final energy is equal to $E_i + (1 + q)V$, V being the high voltage of the terminal. As a result of the characteristics of their ions and their ease of adjustment, electrostatic accelerators are the most used machines for material irradiation, especially given that the same machines are used for ion beam analysis (RBS, ERDA, NRA, PIXE, etc., see Chapter 10).

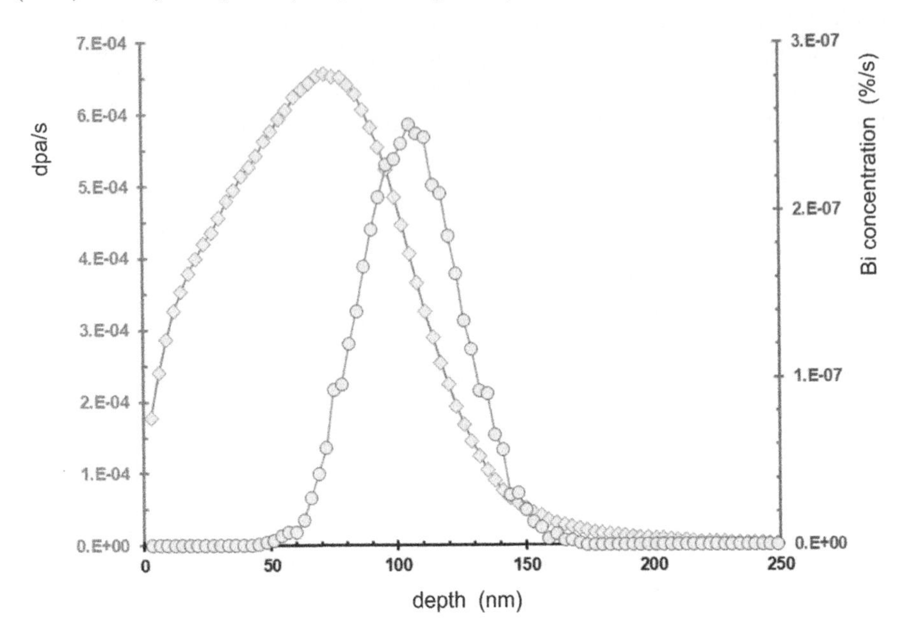

Figure 9.6. *400 keV Bi ion irradiation/implantation in a glass with a flux of 10^{15} ions/m²/s. Distribution as a function of depth, damage rate (red) and concentration (%/s) of implanted Bi ions (blue). Calculation performed with SRIM-2008 (www.srim.org). For a color version of this figure, see www.iste.co.uk/bouffard/nuclear.zip*

This is the energy range in which we find electron accelerators dedicated to the study of irradiation defects. As a result of their low mass, electrons only transmit low kinetic energy to the target atoms. An electron of 1 MeV transmits a maximum of 78 eV to an iron atom (see equation [9.1]).

$$T_{max} = 2\frac{E_1(E_1+2m_ec^2)}{M_2c^2} \qquad [9.1]$$

with $m_ec^2 = 511$ keV and $M_2c^2 = 931.6\ M_2$ (MeV).

With a 2.5 MV accelerator, the electrons only create point defects, allowing us to study their dynamics. Furthermore, by varying the energy of the electrons, the measurement of a property that is sensitive to the presence of defects makes it possible to determine the displacement thresholds. In addition, electron accelerators are particularly useful for radiolysis studies, for either continuous or pulse radiolysis. For the latter case, it is easier to use a machine like the linac, which accelerates electrons in bunches. The duration of these bunches is of the order of a nanosecond with a recurrence rate of about 10 megahertz. It is therefore necessary to suppress a certain number of pulses, so as to adapt the time between two pulses to the characteristic times of the radiation chemistry considered. It is possible to obtain shorter pulses by injecting electrons for a sub-nanosecond duration using, for example, a photo-cathode triggered by a femtosecond laser. Based on this principle, the ELYSE accelerator at Orsay delivers pulses of a few picoseconds at a frequency of the order of 50 Hz (Belloni et al. 2005).

The principle is extremely simple (its implementation, however, a little less so): the accelerator is made up of an insulating column that carries a source of particles at one end and a grounded electrode at the other. The potential between the two ends produces the field that accelerates the particles. Thanks to its electrodes, the column also has a focusing action. The inside of the column is under vacuum, and the outside is under high pressure of an insulating gas (usually SF_6). A DC voltage greater than 1 million volts can be obtained either by rectifying an AC voltage (Cockcroft and Walton 1930) or by transporting charges using a moving belt and depositing them at the terminal (Van de Graaff et al. 1933). In modern machines, the charge-carrying belt is replaced by a chain of "pellets" with insulated links. The largest machines have reached voltages of the order of 30 MV, but hardly exceed 20 MV in operation (24 MV for the Oak Ridge National Lab.). For more information, see Hellborg (2005).

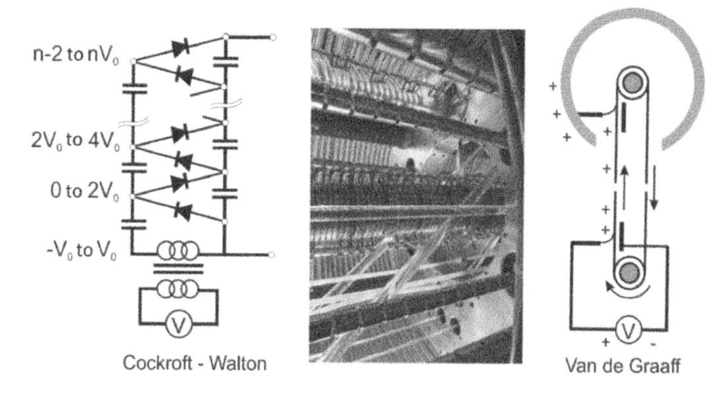

Figure 9.7. *On the left is the operating principle of a Cockroft–Walton machine, and on the right is the operating diagram of a Van de Graaff accelerator. In the center is a photograph of a Van de Graaff column. Photo credit: L. Riux (CEA/SRMP/JANNUS). For a color version of this figure, see www.iste.co.uk/bouffard/nuclear.zip*

Box 9.1. *How does an electrostatic accelerator work?*

The IAEA accelerator database lists 311 electrostatic machines, 197 of which are tandems (IAEA 2021). The distribution by country is given in Figure 9.8.

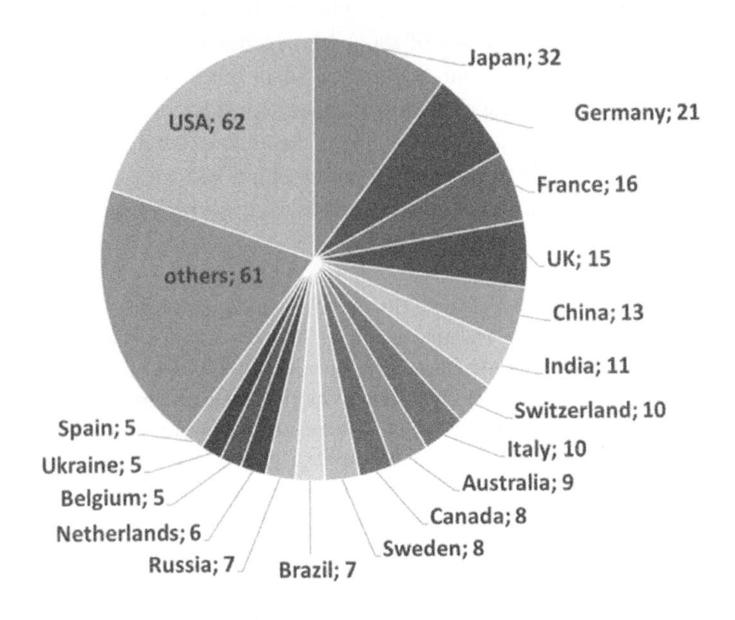

Figure 9.8. *Distribution by country of electrostatic accelerators according to the IAEA (2021). For a color version of this figure, see www.iste.co.uk/bouffard/nuclear.zip*

High-energy accelerators. These high-energy accelerators are generally machines designed for nuclear physics that dedicate all or part of their time to irradiation experiments. Three types of machines coexist: Van de Graaff-type tandems, cyclotrons and linacs.

In the 1980s, two very large nuclear physics research facilities (GSI in Darmstadt and GANIL in Caen) enabled the development of research on the effects of high electronic excitations in materials, thanks to Program Advisory Committees and the presence of material physics teams that were working on these accelerators, respectively, *Materialforschung* and *CIRIL* (Interdisciplinary Center for Research with Heavy Ions).

– GSI is a linear accelerator which, for material physics, delivers ions at 11.4 MeV/A in a room called M-branch or acts as an injector to an SIS-18 synchrotron which accelerates ions to 1 GeV/A. In M-branch, the following in-situ experiments are available: SEM, X-ray diffraction, Raman and infrared spectroscopy, AFM/STM, etc., and at very high energy, irradiations under very high pressure.

– At GANIL, the ions, from the source to the high-energy experiments, undergo four successive accelerations. The irradiation experiments that are available online are: X-ray diffraction, infrared spectroscopy, time-resolved ionoluminescence, gas analysis, sputtering, etc.

– In China, the Institute of Modern Physics has a set of accelerators, with which it is possible to perform irradiation experiments at energies as high as 4 GeV/A.

There are about 20 tandems in the world with a high voltage that is equal to or greater than 10 MV (a dozen ≥14 MV), and about half of them are equipped with an irradiation station. Some are boosted by the addition of a radiofrequency chamber or a linac.

Figure 9.9. *The left side of the photo shows an irradiation chamber installed on the GANIL IRRSUD line; the right side shows a device for in situ monitoring of the evolution of the infrared spectrum at low temperature (8 K), as a function of the irradiation dose (photo credit: S. Bouffard). For a color version of this figure, see www.iste.co.uk/bouffard/nuclear.zip*

Regardless of the energy of the accelerated beams, they can only be used if the accelerator has an irradiation chamber with the following main functions:

– measurement of the particle flux (preferably continuous);

– control of the irradiated surface and homogeneity of the fluence on this surface;

– fixing of samples and possibly automatic holding of samples.

9.3.3. *Use of radioactive elements*

Without going into detail, it is possible to carry out irradiations by inserting radioactive atoms into the materials, so as to create an internal source of irradiation, or by depositing a more or less thick layer of radioactive material on the surface. The flux or dose rate can be adapted to a certain extent, for example, by adjusting the isotopy of the plutonium: from 0.15 to 633 $Bq.kg^{-1}$ for ^{242}Pu and ^{238}Pu, respectively.

It is evident that this type of experiment requires an adapted environment, in particular for nuclear materials which, like plutonium, fall under the defense code (authorization to hold and use sensitive radioactive materials[3], glove box or even shielded cell, etc.). This approach is only accessible to a few research centers in the world.

9.4. Some major irradiation research centers

Table 9.3 lists the research centers that have significant irradiation resources. It should be noted that this is a subjective vision of the authors of this chapter; therefore, this list is far from being exhaustive.

Sites	Accelerators	Online features
Australia Australian Nuclear Sciences and Technology Organization	- Tandem10 MV - Tandem 6 MV - Tandem 2 MV - Tandem 1 MV	- IBA - Mass spectroscopy
China Institute of Modern Physics	- Linac 15 MV - Cyclotron GeV - ECR platform 320 kV	

3. For ^{239}Pu, the reporting threshold is 1g of Pu or 10^4 Bq/g.

Sites	Accelerators	Online features
France EMIR&A French network of accelerators for irradiation and analysis of molecules and materials	- Cyclotron 13 MeV/A - Cyclotron 1 MeV/A - VdG 3.5 MV - VdG (x2)3 MV - VdG (x2)2.5 MV - Tandem (x2)2 MV - Implanter 190 kV - VdG e^- 2.5 MV - Linac e^- (x2)10 MeV - HVTEM 1 MeV	- Triple beam - Double beam - Micro-beam - Cluster beam - Irradiation - TEM - Raman, XRD - UV-visible-IR Spectro. - RBS-C – NRA - Gas analysis
Germany Helmholz Zentrum Dresden Rossendorf – Ion Beam Center	- Tandem 6 MV - Tandem 3 MV - VdG 2 MV - Implanter 500 kV - Implanter 40 kV	- Double beam - RBS-C – ERDA – NRA
India Bhabha Atomic Research Centre, Mumbai	- Tandem 6 MV - Implanter 500 keV	- ERDA - Irradiation
India Inter-University Accelerator Centre, Delhi	- Tandem15 MV + booster 5 MeV/A - VdG 1.7 MV - Implanter 200 keV - Implanter 1 MeV	- RBS-C – ERDA - Irradiation
Japan Takasaki Advanced Radiation Research Institute	- Cyclotron500 MeV - Tandem 3 MV - VdG 3 MV - Implanter 400 kV - Implanter40 kV	- Triple beam - Large area irradiation - Single ion - micro-PIXE - Cluster beam - In-line TEM (400 kV)
Portugal Campus Tecnológico e Nuclear	- Tandem 3 MV - VdG 2.5 MV - Implanter 210 kV	- Micro-beam - RBS-C – NRA – PIXE - Ionoluminescence - AMS
Spain Centro Nacional de Aceleradores, Seville	- Cyclotron 18 MeV - Tandem 3 MV	- RBS – ERDA – NRA - PIXE - Ionoluminescence
United Kingdom University of Surrey Ion Beam Centre	- Tandem 2 MV - Implanter2 MV - Implanter 200 kV	- RBS-C – NRA – PIXE - Micro-beam

Sites	Accelerators	Online features
USA Argonne National Lab.	- linac 1.5 MeV/A - Tandem 2 MV - Implanter 650 kV	- On-line TEM - Irradiation
USA Michigan Ion Beam Laboratory	- Tandem 3 MV - Tandem1.7 MV - Implanter 400 kV	- Triple beam - On-line TEM - IBA - Irradiation
USA Sandia National Lab. Ion Beam Centre	- RFQ linac 1.9 MeV/A - Tandem 6 MV - Tandem 1 MV - VdG 3 MV - Implanter 350 kV - Implanter 100 kV	- On-line TEM - RBS-C – ERDA – ... - Irradiation
USA Texas A&M Univ. Dept. of Nuclear Engineering	- Tandem1.7 MV - Tandem 1 MV - Implanter 200 kV - Implanter 150 kV - Implanter10 kV	- RBS – ERDA – NRA - PIXE - Irradiation - Electrical measurements

Table 9.3. *Non-exhaustive list of centers of research on irradiated materials*

The EMIR&A research federation is a network of ion and electron accelerators for the irradiation and analysis of materials and molecules, which brings together 15 accelerators distributed across six sites in France (CEA-Saclay, CEMHTI-Orléans, CIRIL@GANIL-Caen, ICP and IJCLab-Orsay, INSP-Paris and LSI-Palaiseau). Created in 2014, this network aims to provide France with the equivalent of a large accelerator research center, like those that exist around the world (see Table 9.3). To this end, EMIR&A provides a forum for dialogue between the platforms and, above all, offers any researcher working in France or elsewhere a unified procedure for accessing the Federation facilities. The EMIR&A facilities practically cover all the fields of ion beam analysis (RBS-C, ERDA, NRA, PIXE, etc.). The same applies to irradiations with continuous and pulse ns and ps electron beams, and with ions from a few dozen keV to several hundred MeV. Most of these facilities are equipped with instrumentation (TEM, Raman, XRD, etc.), making it possible to follow in situ and even operando the evolution of the properties of the materials under irradiation.

The scientific topics are related to very fundamental aspects, such as the role of defects on the properties of solid-state physics (high-temperature superconductivity, topological insulators, etc.) or the studies of fast radiolysis kinetics. However, most of the experiments have an applied aspect in the fields of energy, electronics, space, etc., as well as in nanotechnologies.

For more information, see: https://emira.in2p3.fr

Box 9.2. *The EMIR&A French National Research Federation*

9.5. Conclusion

Irradiation facilities exist in all fields that are required for the study of nuclear materials for the purpose of energy. Many of these machines host external research projects by dedicating up to 100% of their beam time. The terms of this hosting range from simple collaboration to the assessment of the project by a scientific committee. The beam time is most often charged at a marginal cost, but it can be free.

Irradiation is important, but equally important is characterizing the induced modifications in the material. The characterization tools are therefore the subject of the next chapter.

9.6. References

Belloni, J., Monard, H., Gobert, F., Larbre, J.P., Demarque, A., De Waele, V., Lampre, I., Marignier, J.L., Mostafavi, M., Bourdon, J.C. et al. (2005). ELYSE – A picosecond electron accelerator for pulse radiolysis research. *Nuclear Instruments and Methods in Physics Research Section A: Accelerators, Spectrometers, Detectors and Associated Equipment*, 539(3), 527–539.

Cockcroft, J.D. and Walton, E.T.S. (1930). Experiments with high velocity positive ions. *Proceedings of the Royal Society of London, Series A: Mathematical and Physical Sciences*, 129(811), 477–489.

Hellborg, R. (2005). *Electrostatic Accelerators, Fundamentals and Applications*. Springer, Berlin, Heidelberg.

IAEA (2018). *Research Reactors for the Development of Materials and Fuels for Innovative Nuclear Energy Systems*. IAEA Nuclear Energy Series, N° NP-T-5.8, Vienna.

IAEA (2021). Accelerator knowledge portal [Online]. Available at: https://www.iaea.org/resources/databases/accelerator-knowledge-portal [Accessed 22 August 2021].

Van de Graaff, R.J., Compton, K.T., Van Atta, L.C. (1933). The electrostatic production of high voltage for nuclear investigations. *Physical Review*, 43, 149–164.

Was, G.S. (2015). Challenges to the use of ion irradiation for emulating reactor irradiation. *Journal of Materials Research*, 30(9), 1158–1182.

10

Characterization of Irradiation Damage

Aurélie GENTILS[1], Stéphanie JUBLOT-LECLERC[1] and Patrick SIMON[2]

[1] *Laboratoire de Physique des 2 Infinis Irène Joliot-Curie (IJCLab), CNRS, Université Paris-Saclay, Orsay, France*

[2] *Conditions Extrêmes et Matériaux: Haute Température et Irradiation (CEMHTI), CNRS, Orléans, France*

10.1. Introduction

The irradiation of a material, whether it be internal or induced by incident particles (ions, electrons, neutrons), causes microstructural changes that can range from the simple formation of point defects to more extensive damage, including the formation of extended defects (such as dislocations loops), cavities and even phase transformations such as amorphization in crystalline materials (see Chapter 1).

The choice of techniques to characterize this irradiation damage is related to several aspects:

– the nature and level of damage;

– the nature of the material;

– the desired scale of characterization (from the atom to the micrometer).

The techniques presented in this chapter are those most commonly used for studying the microstructures of nuclear materials damaged by irradiation, with the exception of surface analysis techniques, such as optical microscopy, scanning electron microscopy (SEM) and atomic force microscopy (AFM), which will not be discussed here. Studies of changes in physical and mechanical properties are not covered in this chapter either. Most of the experimental approaches are performed after irradiation of the samples. However, some of the methods can be implemented in situ, or even operando[1], that is, the measurement is performed during irradiation, in an accelerator or on active samples that are subjected to internal irradiation (self-irradiation) in a specifically protected environment.

10.2. Characterization of point defects

There are only a few experimental techniques that can characterize point defects: vacancies, interstitials and their associations (Frenkel pairs, Schottky defects, etc.). These techniques will essentially probe the electronic configuration of these point defects.

10.2.1. *Positron annihilation spectroscopy*

Positron annihilation spectroscopy is used to study native or irradiation-induced vacancy defects in any type of material (Barthe et al. 2003). The positron, which is positively charged, is the antiparticle of the electron: these two particles annihilate one another by emitting two gamma photons of 511 keV. In practice, positrons (most often emitted by a radioactive source) penetrate the material to a depth defined by their incident energy (see Figure 10.1). The positrons are rapidly thermalized by inelastic collisions. They are then annihilated preferentially by the conduction electrons in metals or by the external electrons of negative ions in ionic crystals. Repelled by the ions, they are annihilated in areas of low ionic density: they are therefore sensitive to the presence of vacancies and vacancy complexes, for which they constitute a non-destructive means of study. However, only negatively charged or neutral vacancies and vacancy complexes can be detected. The characteristics of the annihilation gamma photons (number, energy, angle between their directions of emission), as well as the positron lifetime, give information on the annihilation site, in particular the electron density and the distribution of the momentum of the electrons that annihilate with the positrons. Two main techniques are used to study irradiated materials, the principle of which is illustrated in Figure 10.1: (i) Doppler broadening spectroscopy (DBS), which measures the energy difference ΔE, that is, the Doppler broadening, between the two photons that are

---

1. For example, during an irradiation that is coupled with a chemical reaction with an external atmosphere (gas or liquid), even without radiolysis.

emitted at about 180° during the annihilation of an electron-positron pair[2]; (ii) positron annihilation lifetime spectroscopy (PALS), which measures the lifetime τ of the positron in the material, between the detection of its annihilation and the detection of a 1.27 MeV gamma emitted during positron production. The first technique gives access to the momentum distribution of the electrons involved in the annihilation, and the second to the types of defects that are present (mono-, di- and tri-vacancies, and vacancy clusters).

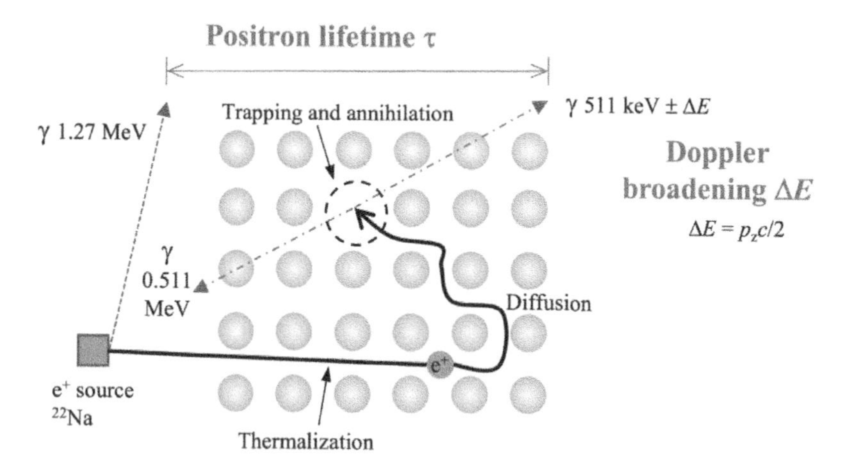

Figure 10.1. *Schematic diagram of the two main positron annihilation spectroscopy techniques: measurement of Doppler broadening ΔE and positron lifetime τ. For a color version of this figure, see www.iste.co.uk/bouffard/nuclear.zip*

Other experimental methods are very sensitive to the presence of point defects, but do not "see" these defects directly. This is particularly the case for Raman scattering spectroscopy, which we will look at in more detail.

10.2.2. *Raman scattering*

Raman scattering spectroscopy is based on inelastic light scattering, in which a monochromatic light beam is scattered at slightly different frequencies than the incident wave. The difference in frequency between the incident and scattered beams is equal to the vibrational frequency of the molecular, radical or crystalline entities within the material being analyzed. These vibration frequencies are

2. From the angle θ between the two photons, the angular correlation technique gives the same type of information.

governed by three types of factors: 1) the mass of the ions and thus the chemical composition; 2) the molecular or crystalline arrangement and the potential disorder, and 3) the macroscopic mechanical constraints. We can immediately see that an irradiation can have an effect on these three factors, particularly the last two. Raman scattering can therefore provide relevant information on the damage induced in the structure.

A more exhaustive presentation of Raman scattering (fundamentals, applications, instrumental aspects) can be found in the recent book *Spectroscopies vibrationnelles* (Simon 2020). The wavelengths used to probe matter are in the visible, near-IR or near-UV range, which are wavelengths that are not noticeably absorbed by silica. Thus, a simple silica optical window can be used to enclose an irradiation or confinement system for radioactive species, as the optical signals can be easily transported by optical fibers. The Raman method is therefore well adapted to the characterization of irradiated samples (*post-mortem*), as well as to an in situ approach giving access to damage kinetics or to possible intermediate states. By adjusting the focus of the optical beam in transparent samples, the evolution of the physical parameters studied can be followed as a function of depth. On the other hand, opaque systems can only be characterized at the surface.

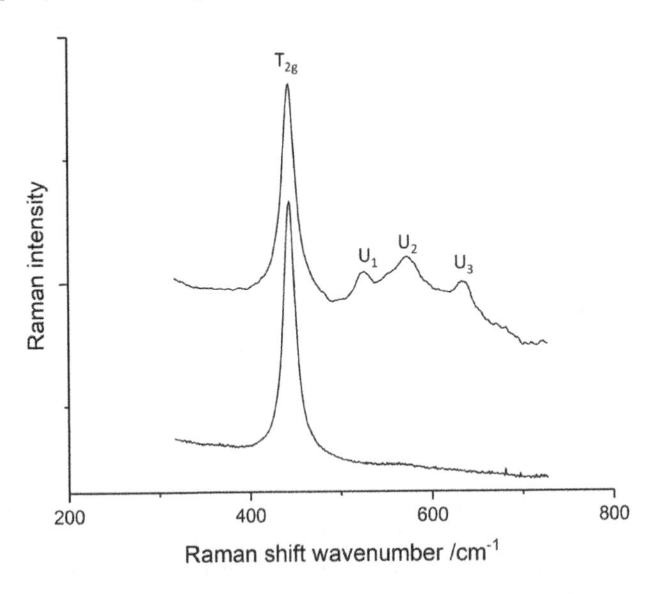

Figure 10.2. *Raman spectra of UO_2, non-irradiated (bottom) and irradiated (top) by 21 MeV He^{2+} ions (2 h, 50 nA). On non-irradiated UO_2, only the 445 cm^{-1} line is visible*

In practice nowadays, a Raman system is most often composed of a laser and a spectrometer coupled with an optical microscope. The analysis is done through microscope optics, using backscattering. The analyzed areas are of the order of μm^2, which gives a spatial resolution on the scale of the distances covered by ions of energy higher than a few MeV.

In this section, we will consider an example of Raman spectra modification by irradiation, where the source of the spectrum modification is the creation of point defects. Uranium oxide UO_2 is a model case: a single Raman line allowed by the symmetry of the fluorite structure (the so-called T_{2g} symmetry line)[3], and damage under irradiation that does not lead to amorphization. Figure 10.2 represents the spectra of virgin and irradiated ceramic UO_2 (He^{2+} 21 MeV beam). Whatever the irradiation (ion or electron), when considering energies of a few MeV, three additional lines (denoted here as U1, U2 and U3) will always appear on the spectrum. This is also the case for the isomorph ThO_2, as well as for a MOX fuel under conditions of self-irradiation and neutron irradiation, after having undergone a four-year cycle in a reactor (Jégou et al. 2015). The three lines do not depend on the type of irradiation, and therefore do not depend on the organization of the defects. In fact, they are characteristic of the fluorite structure. Although pristine UO_2 only has one first-order vibration mode that is visible in Raman, other vibration modes do exist, but are inactive in Raman for symmetry reasons. The point defects created through irradiation by locally lowering the symmetry make these modes possible in Raman scattering. We therefore do not see the vibration of the defect itself, but that of the matrix, a vibration that is made active in Raman by symmetry lowering. There can be two kinds of symmetry lowering: it is either a modification of the local symmetry (loss of a center of inversion, for example) or linked to the loss of the translational symmetry. Here, it is a lowering of the local symmetry. The U_2 mode, for example, is a T_{1u} symmetry mode (more specifically, its LO – Longitudinal Optical component), which is a mode that can normally only be detected by infrared reflectivity spectroscopy. The point defects will also contribute to a broadening of the allowed T_{2g} line, as well as to a slight frequency shift of this line (local stress effects – swelling of the crystal network). In the case of UO_2, through a joint Raman – positron annihilation study (Mohun et al. 2019), these point defects at the origin of the Raman lines have been identified as Frenkel pairs on U sites.

3. Symmetries T_{1u}, T_{2g}: a more complete discussion of symmetries is beyond the scope of this chapter. Let us just specify that T = triply degenerate mode (cube symmetry), u = ungerade (odd) and g = gerade (even), for modes that are inverted, or unchanged, by an inversion center. In a point group with an inversion center, Raman modes can only be "g" and infrared modes can only be "u" (see Rousseau, D.L. et al. (1981). Normal mode determination in crystals. *Journal of Raman Spectroscopy*, 10, 253–290).

10.2.3. *Other techniques*

Other techniques directly analyze the electronic configuration of defects such as electron paramagnetic resonance (EPR), which determines the fundamental electronic level of ions or radicals with partially filled electronic layers (Lund and Shiotani 2014), or photoluminescence (PL), making it possible to study the excited electronic states (Defresne 2016; Plantevin et al. 2016). Several decades ago, the EPR of irradiated glasses was not actually used to understand the damage phenomenon created by irradiation, but the actual structure of the glass (Griscom 1985).

Studies of point defects in metals and metal alloys have heavily relied on electrical resistivity measurements at low temperatures. Indeed, in this type of material, electrical resistivity follows Matthiessen's rule (Matthiessen and Holzmann 1860)[4]:

$$\rho = \frac{m}{ne^2}\frac{1}{\tau} \text{ with } \frac{1}{\tau} = \frac{1}{\tau_{phonons}} + \frac{1}{\tau_{defects}}$$

The contribution of phonons to resistivity decreases with temperature, so it is therefore at low temperature that resistivity will be the most sensitive to the presence of point defects. A typical order of magnitude of the resistivity of defects is a few μΩ.cm for a concentration of Frenkel pairs of 1% (1.3 μΩ.cm/at.% for Cu and 12.5 for Fe) (Lucasson and Walker 1962). The monitoring of electrical resistivity provides a lot of information on point defects in metals: displacement threshold, recombination volume, annealing temperatures, etc. Aside from being a very powerful technique to characterize point defects in pure metals and alloys, electrical resistivity is much less interesting in the case of industrial materials, even simplified ones, hence its current lack of appeal.

Infrared spectroscopy is widely used for the characterization of defects in polymers. For other materials – ceramics and metals – it is generally not applicable in transmission mode, as these materials are too absorbent. For polymers, however, this method is very sensitive to the appearance of electric charge defects, which leads to changes in electric dipoles.

---

4. In reality, Augustus Matthiessen was never interested in irradiation defects, and for good reason. He died in 1870 long before the discovery of irradiation effects. He explained the different electrical conductivity values of copper through the contribution of impurities.

10.3. Characterization of the global disorder and elastic strain

10.3.1. *Raman spectroscopy*

In the case of the creation of a very large number of point defects, this time the translational symmetry will be affected: the crystal network is no longer invariant by the translation of a lattice parameter. This results in a major modification of the spectra, with distorted lines that can be extremely broad. At the extreme, we can completely lose the translational symmetry: this is amorphization. In order to understand these evolutions, we must go back to the phonon dispersion curves, which govern all the vibrational dynamics in a crystal structure. In a periodic structure, only the so-called "Brillouin zone center" modes, at $q = 0$, are active in Raman or infrared. As soon as the disorder is such that the translational symmetry is not respected, all the vibrational modes, at any point of the Brillouin zone, can contribute to the Raman spectrum. The result is a "density of states" spectrum, often known as VDOS for "vibrational density of states", resulting from the projection of dispersion curves on the vertical axis of vibration frequencies. The spectrum then presents maxima at frequencies where dispersion curves are horizontal, thus contributing to the scattered intensity in a preponderant way. This is directly illustrated in Figure 10.3, where the Raman spectra of 6H-SiC single crystals that are irradiated with 20 MeV Au ions are shown. The spectra were acquired on a cross-section (cleavage plane) of the sample made after irradiation. The black dotted spectrum was acquired in the blank zone, where we can find the characteristic fine and intense lines of an ordered crystal. The three other spectra, whose very broad shapes are characteristic of a VDOS spectrum, were acquired at different depths in the irradiated area, and thus for different damage rates.

Similar results have been obtained on other structures, under electronic and/or nuclear stopping power conditions, with high-energy heavy ions (U, Xe on GaN, for example; see Moisy et al. (2016)).

As Raman spectroscopy analyses areas of the order of μm^2, it is possible to constitute a hyperspectrum, a matrix that gathers the spectral data of all the points. The latter is achieved by making measurements on a large number of points (a few tens or hundreds of thousands of points are technically possible). This of course makes it possible to probe the homogeneity of the surface, especially given that the large number of spectral data allows the use of extremely powerful methods of multivariate analysis (such as principal component analysis), which can highlight minimal differences between spectra. Today, most commercial Raman spectroscopy microscopes are equipped with such an imaging system. There is also at least one Raman imaging system in a glove box, making it possible to work on plutonium-based samples and, thus, in self-irradiation conditions (MOX fuel imaging (Medyk 2021)). One of the focuses of Raman scattering spectroscopy is, as previously highlighted, to be relatively adaptable to in situ measurement conditions under irradiation.

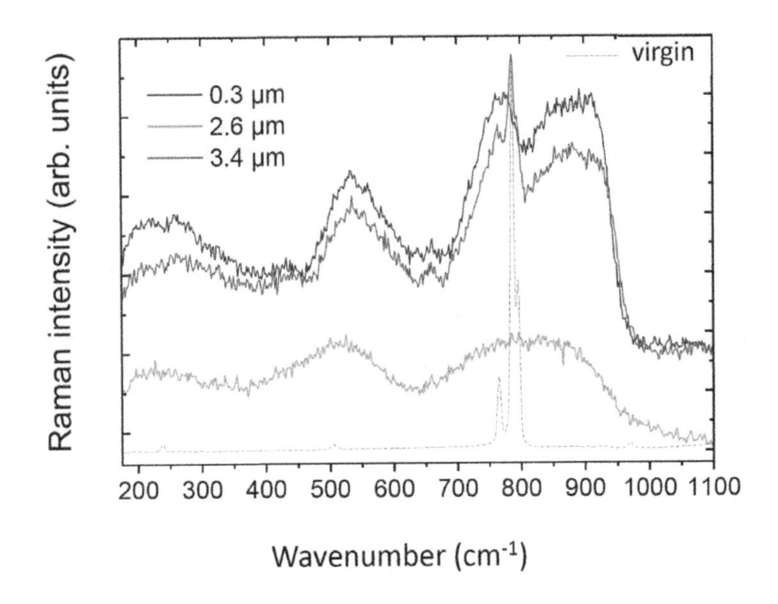

Figure 10.3. *Raman spectra of 6H-SiC, irradiated with 20 MeV Au ions at a fluence of 10^{14} cm^{-2}. Black dotted line: virgin crystal. Blue/green/red line: increasing depths from the surface (ion shut-off at around 3.5–4 µm) (from Linez et al. (2012)). For a color version of this figure, see www.iste.co.uk/bouffard/nuclear.zip*

The resulting images can be one-dimensional (line imaging) or two-dimensional (surfaces). Figure 10.4 repeats the example of irradiated SiC in the previous figure, but this time several dozen spectra (every 100 nm) were acquired from the surface to the underlying virgin material. In the damaged area, three different damage regimes can be distinguished, corresponding to the three curves in Figure 10.3. Depths (1) and (3), containing red areas, are weakly damaged, and their spectra are purely VDOS (deduced directly from the dispersion curves). The intermediate zone (2) in green, from 1 to 3 µm, presents larger forms, showing a more significant disorder: here, we not only have translational disorder, but a disorder between nearest neighbor ions that is notably modifying the dispersion curves. In this zone (2), we have created vibrational frequency states (> 970 cm^{-1}), which do not exist for an ordered crystal, on the whole Brillouin zone. In fact, it was in area (3) that the energy deposit was at its maximum during irradiation. It is likely that the local temperatures reached at this location led to local annealing of the defects and a final appearance of a less significant disorder (Linez et al. 2012).

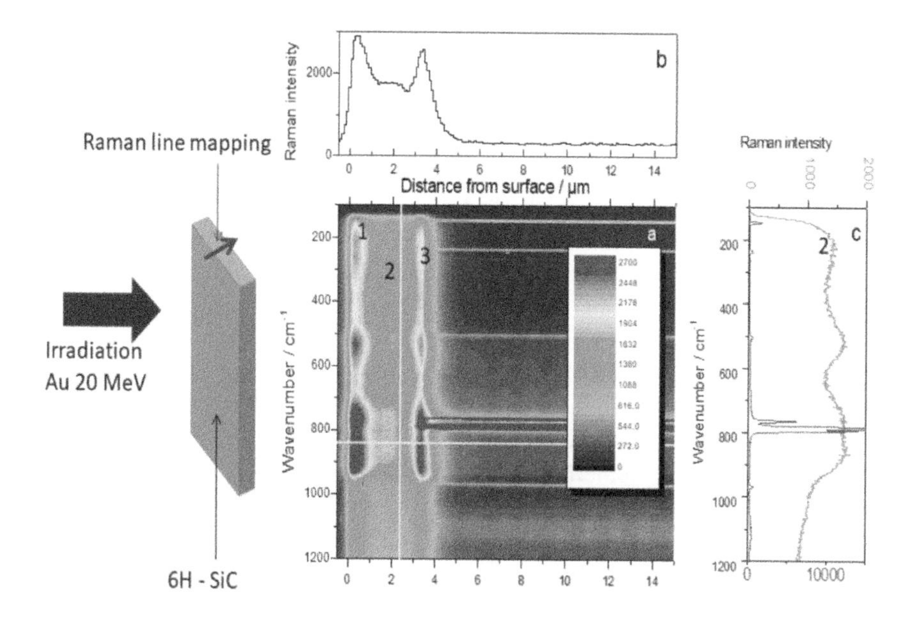

Figure 10.4. *"Line" images of Raman spectra of 6H-SiC, irradiated with 20 MeV Au ions, using a color map (red: high intensity, blue: low). The scattering geometry is the "side-view" configuration, shown on the left. The insets at the top and on the right give the profile of the curves on the yellow reticle (the spectrum of the pristine region is added in red in the right inset). On the map, we can see the pristine region (in blue, at more than 4 µm depth), as well as three different damage zones (1 red – 2 green – 3 red) between the surface at 0 µm and the stop zone at 4 µm. For a color version of this figure, see www.iste.co.uk/bouffard/nuclear.zip*

10.3.2. *Ion beam analysis*

Material damage can be measured experimentally by ion beam analysis techniques (Trocellier and Trouslard 2002), and in particular by Rutherford Backscattering Spectrometry (RBS), in channeling geometry (RBS-C). These techniques require the use of ion beams delivered by particle accelerators, such as those presented in Chapter 9. The principle of RBS is based on the fact that during an elastic collision at a given incident energy and deflection angle, there is a one-to-one relationship between the energy loss of the projectile and the mass of the target atom. For backscattering to occur, the projectile must be lighter than the nucleus of an atom of the material to be characterized. This is the reason why RBS experiments are most often performed with helium ions. Moreover, the projectile ion loses energy by electronic excitation during its path to and from the material, and this additional energy loss gives information on the depth at which the collision took place. For a given backscattering angle, the energy of the collected ions allows us to determine the nature of the elements that are present in the material, as well as the evolution of their concentration under the surface: each element is characterized by the presence of a front, at an energy that is defined by the laws of conservation of energy and momentum (kinematic factor). The heaviest elements appear at high energy, and the lightest at low energy. A typical spectrum recorded on a $MgAl_2O_4$ spinel crystal can be seen in Figure 10.5(a) (closed symbols): the signals (backscattering efficiency) arising from the different elements that are present in the target, which are Al, Mg and O, by decreasing mass (and thus by decreasing energy front), are visible. The chemical composition of the material can be determined in absolute terms (if the geometry of the experiment and the number of projectiles are known), thanks to the Rutherford backscattering cross-section. Ion beam analysis techniques can therefore also determine the chemical composition of a target as a function of depth.

In the case of a single crystal, the direction of the incident ion beam can be chosen to coincide with the direction of a crystal plane of the target, thanks to an accurate goniometer that can manipulate the crystal. A large number of incident ions then enter the target without encountering any atoms, traveling between the atomic planes, and the elastic scattering efficiency is therefore very low in this case: this phenomenon is called channeling. In the example given in Figure 10.5(a), cesium was incorporated by ion implantation into a single crystal of $MgAl_2O_4$ at different concentrations (indicated here in fluence, a number of ions per unit area). RBS-C spectra recorded with a helium ion beam, aligned with the <100> crystal axis (open symbols), show bumps for each sublattice of the crystal, relative to the existing disorder. Analysis of these signals makes it possible to obtain the fraction of atoms that are displaced in each sublattice, taken as a measure of the disorder rate. Several stages of damage can be seen in Figure 10.5(b), where the maximum disorder rate is plotted for different values of fluence and converted here into a unit of measurement of damage, which is the number of displacements per atom (dpa). For a value of disorder rate f_D equal to 1, the material is completely amorphous: the atoms no longer adhere to a periodic arrangement in space.

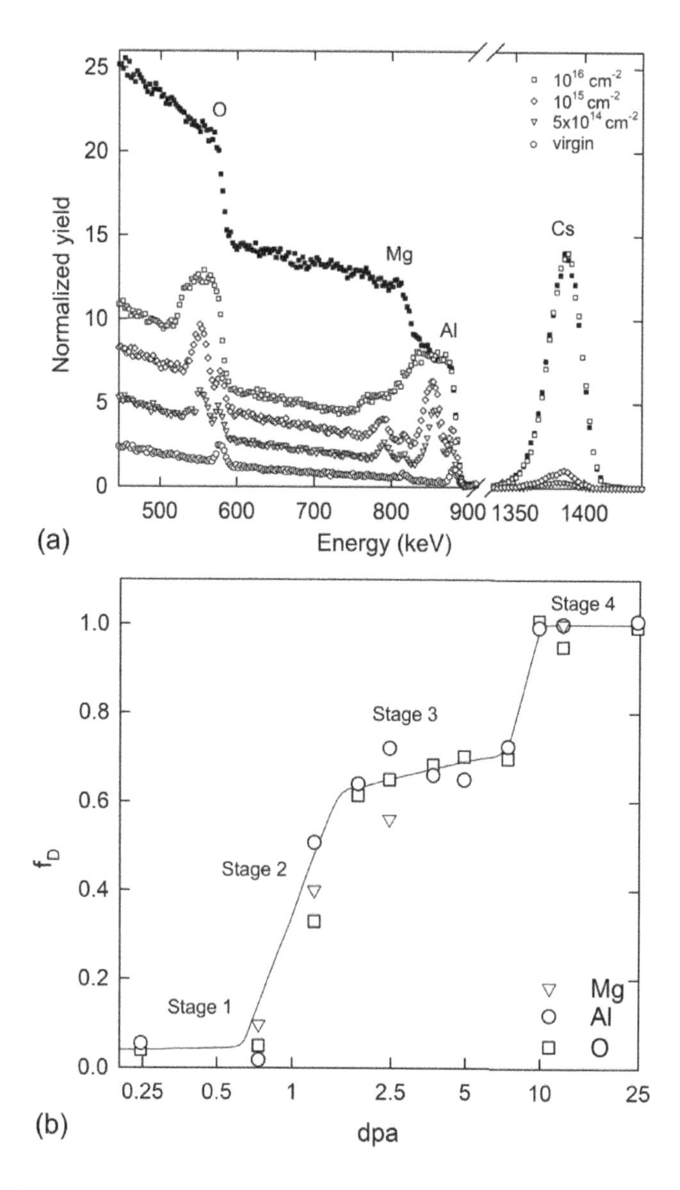

Figure 10.5. *(a) RBS spectra obtained with He ions of 1.6 MeV, recorded in random (closed symbols) and channeling directions along the <100> axis (open symbols) on a virgin $MgAl_2O_4$ spinel single crystal (circles), and irradiated with different fluences of 150 keV Cs ions – the standardized elastic scattering yield is shown here as a function of the incident ion kinetic energy. (b) Disorder rate f_D measured by RBS-C in the different sublattices of a $MgAl_2O_4$ single crystal, irradiated with Cs ions as a function of the number of displacements per atom (dpa) (from Gentils (2003))*

10.3.3. *X-ray diffraction*

X-ray diffraction is a non-destructive technique that is complementary to the widely used techniques of Rutherford backscattering spectrometry in channeling geometry, and transmission electron microscopy, for the characterization of irradiation damage in crystalline samples. While RBS-C gives a measure of the disorder, X-ray diffraction allows the determination of elastic strains, or variations in lattice parameters, induced in single-crystal layers by the presence of irradiation-induced defects. The presence of point defects, in particular interstitial defects, can generate significant elastic strains. The technique is therefore sensitive to relatively low levels of damage, generally before the detection of a signal in RBS-C and before the observation of microstructural defects in TEM.

The wavelength of X-rays is of the order of 10^{-11} to 10^{-8} m, which includes the order of magnitude of inter-reticular distances in crystalline materials. In the laboratory, X-rays are most often generated by a copper anode and have a wavelength of about 1.54 Å (main copper line $CuK_{\alpha 1}$). For this wavelength, the penetration depth of X-rays varies between a few microns and a few dozen microns, depending on the material and the studied diffraction spots or reflections. Each diffraction spot corresponds to the diffraction of a family of reticular planes, or atomic planes. For single-crystalline samples implanted with particles of a few keV to a few MeV, it is therefore possible to probe both the layer of material damaged by the implantation and a part of the non-implanted material, which serves as a reference. Reciprocal space maps acquired on such samples show that the resulting elastic strain is always normal to the sample surface. In practice, ω-2θ scans, or θ-2θ if the planes under study are perfectly parallel to the surface, are performed on single-crystalline samples in the vicinity of symmetric diffraction spots, in other words, induced by diffraction of planes that are parallel or near-parallel to the sample surface. ω refers to a rotation of the sample, while 2θ refers to the simultaneous rotation of the X-ray detector. An example of the curves obtained is plotted in Figure 10.6 (Leclerc et al. 2008). The material studied is 4H-SiC, implanted at room temperature with 160 keV He ions. In general, the (hkl) planes of the material layer that are not disturbed by the implantation diffract at an angle θ_{Bragg}, defined by Bragg's law $n\lambda = 2d_{hkl} \sin\theta_{Bragg}$, where n is the order of the diffraction and d_{hkl} is the inter-reticular distance of the (hkl) planes. In Figure 10.6, the corresponding intense diffraction peak, or Bragg peak, is visible at the abscissa $(\Delta d/d)_N = 0$, with $(\Delta d/d)_N$ denoting the value of the strain that is normal to the sample surface. The planes that are elastically strained by the ion implantation diffract at an angle that is slightly different from θ_{Bragg}. By derivation of Bragg's law, it is possible to associate this angle with the elastic strain thus plotted on the x-axis of Figure 10.6. To the left of the Bragg peak, diffracted intensity is visible for positive strain values (lattice expansion). The most intense peak immediately to the left of the Bragg peak is attributed to the near-surface of the material, which is little damaged compared to the deeper layers in the vicinity of the projected path of the ions, where the elastic strain is generally at its maximum. In the example shown, the implanted sample has a near-surface strain of 0.7% and a maximum strain of 4.3%. The recombination of point defects during thermal annealing causes the elastic strain of the material to relax (horizontal arrow in Figure 10.6). For an annealing temperature greater than or equal to 700°C, the decrease of the diffracted intensity in the zone of maximum strain (vertical arrow) is due to an increasing structural disorder (growth of He bubbles that expel dislocation loops). Note that the intensity diffracted by the damaged layers is very low compared to the intensity diffracted by the undisturbed substrate, which, in this case, represents several dozen microns. The intensity is thus plotted in logarithmic scale to highlight the distribution of the intensity diffracted by the implanted layer. From these curves, and when oscillations are visible as in Figure 10.6, it is possible to obtain a strain profile as a function of depth by simulating the XRD curve (Souilah et al. 2016; Stepanov 2020). Oscillations are indeed the result of interference between two areas of the same strain, on either side of an area of maximum strain, and the width of the oscillations therefore depends on the width of the strain profile.

X-ray diffraction can also be used to study irradiation-induced phase transformations in polycrystalline materials (powder diffraction). For example, the ALIX device allows in situ studies under irradiation with fast heavy ions on the IRRSUD beam line of GANIL (Grygiel et al. 2012). The MARS beam line of the SOLEIL synchrotron also allows high-resolution X-ray diffraction studies of radioactive samples (Schlutig et al. 2010). For the first time in the world in 2018, experiments were thus performed on a massive sample of UO_2 fuel irradiated in a reactor.

Besides these conventional X-ray diffraction techniques, which are limited to the study of the long-range periodic structure in single crystals and polycrystals, other diffraction techniques make it possible to obtain information on the organization of matter at short and medium range. For example, the analysis of the pair distribution function (PDF), which is experimentally obtained from the Fourier transform of a powder diffraction pattern collected on a synchrotron, provides the distribution of interatomic distances in a material, and can in particular be used to characterize the local order in amorphous or nanocrystalline materials (Bordet 2008).

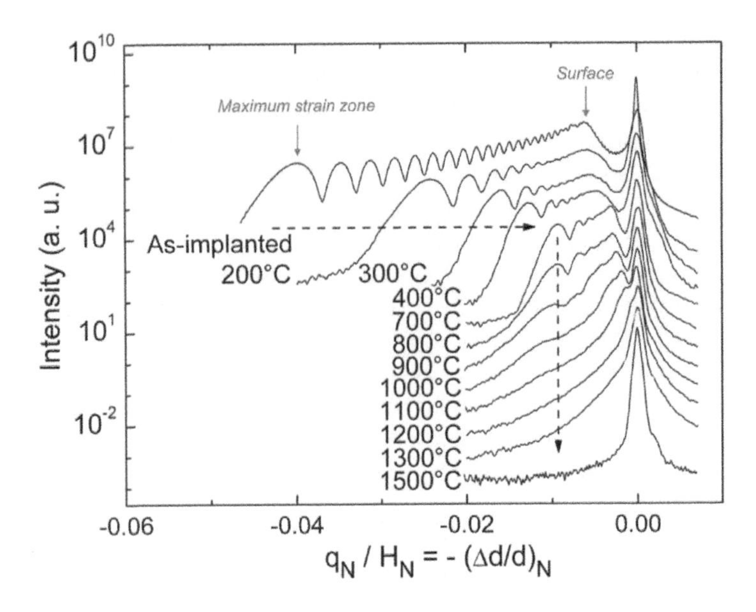

Figure 10.6. *X-ray scattered intensity distribution along the direction normal to the sample surface in the vicinity of the (0004) reflection of 4H-SiC, implanted at room temperature with 160 keV He ions at a fluence of 10^{16} ions.cm^{-2} (as-implanted) and annealed at different temperatures. $(\Delta d/d)_N$ denotes the value of the strain that is normal to the surface. The curves are plotted in logarithmic scale in arbitrary units and are vertically offset from each other for readability (after Leclerc et al. (2008))*

10.4. Imaging of extended defects and cavities

Cavities and extended defects, such as dislocations, dislocation loops and stacking faults, are most often characterized by conventional transmission electron microscopy (TEM). Passing an electron beam through the material to be analyzed makes it possible to obtain local structural and chemical information on a nanometric scale or even on an atomic scale in the case of the latest generation of microscopes. Such a microscope consists of a high-voltage electron source (typically 200 kV), a system of electromagnetic lenses used to focus the electrons on the sample and a system to form a contrasted image, as described in Louchet et al. (1988).

This technique can be used to characterize a wide range of materials, whether they are metallic or insulating. However, the study of some electron-sensitive materials (such as oxides) is delicate, and certain precautions must be taken during the observation. In addition, in order to transmit enough electrons, the samples must be transparent to electrons and therefore be prepared in the form of thin foils, with a thickness of 10 to 200 nm, depending on the material, for an acceleration voltage of 200 kV. This is a crucial step for the subsequent observation of microstructural defects. This meticulous preparation is generally carried out by successive cutting and polishing (mechanical, ionic, electrolytic and/or chemical), depending on the type of material (insulating or conducting).

While passing through the material, the electrons undergo different types of elastic and inelastic interactions. In particular, one part of the electron beam is transmitted directly without deflection, and another part is diffracted by the crystal planes of the material (elastic scattering). For a given orientation of the crystal, the small wavelength of the electrons (of the order of 10^{-12} m, which is much smaller than that of X-rays) and the nanometric thickness of the sample allow a number of planes to diffract simultaneously. Electron diffraction images can thus contain several diffraction spots, each corresponding to a family of reticular planes. The observation of extended defects by TEM is made possible by the diffraction (or amplitude) contrast: the image is formed by selecting the directly transmitted electron beam, or one of the diffracted beams, with an objective aperture. In the first case, it is called a bright-field image: the defects are observed in dark contrast on a bright background (see the example of dislocation loops, Figure 10.7(a)). In the second case, it is a dark-field image, in which the diffracting defects appear in light contrast on a dark background (see Figure 10.7(b)). In order to image these extended defects, the crystal is oriented so as to have just one family of reticular planes (hkl) that diffract, that is, in a Bragg position: here, we are talking about the "two-beam" condition. The technique of weak beam dark field is frequently used, because it makes it possible to get a finer image of the defects.

Another important aspect of the characterization of nuclear materials is the study of cavities induced by irradiation. The observation of these cavities by TEM is usually done by defocusing the electron beam. The cavities appear bright and surrounded by a dark fringe when under-focused, or appear dark and surrounded by a bright fringe when over-focused (see Figure 10.7(c)).

A more complete characterization of extended defects may require the use of TEM in high-resolution mode, which involves the interference of several diffracted beams. In this case, the sample requires a specific orientation, known as "in zone axis", where a maximum number of planes diffract. A large objective diaphragm is used to simultaneously select several diffracted beams, that is, diffraction spots corresponding to different atomic planes. The image obtained makes it possible to visualize the crystalline planes whose inter-reticular distance is superior to the resolution of the microscope.

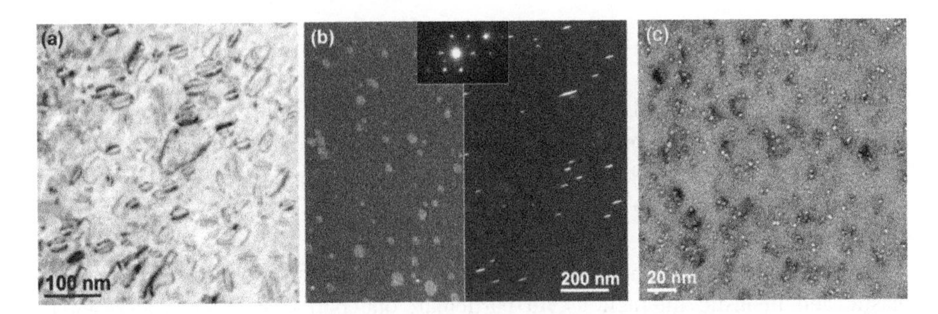

Figure 10.7. *Conventional TEM images of austenitic stainless steel, irradiated with 4 MeV Au ions (a and b) or implanted with He ions (c) at 550°C. (a) Bright-field image of perfect and faulted dislocation loops, (b) dark-field image of faulted Frank loops, (c) bright-field under-focused image of bubbles – the bubbles are gas-filled cavities, which in this case is helium (from Jublot-Leclerc et al. (2015, 2016))*

The disadvantage of the techniques associated with TEM is that the information obtained is very local, unlike the techniques described above, such as RBS-C or XRD, which provide statistical information.

10.5. Elemental analysis

TEM also allows the analysis of chemical elements in a sample at the nanometric scale, thanks to several analytical techniques such as EELS (Electron Energy Loss Spectroscopy) and EDX (Energy-Dispersive X-ray Spectroscopy), coupled to the STEM (Scanning TEM) mode of the microscope. In the case of EELS, the electrons, having crossed the sample and undergone energy losses, are sorted according to their energy loss using a magnetic prism. The inelastic interaction of the incident electrons with the core electron layers of the material's atoms results in characteristic energy losses that induce ionization peaks on the EELS spectra, from which an elemental identification can be made. In EFTEM (Energy Filtered TEM) mode, slits that select an energy window of the EELS spectrum allow the imaging of a chosen element. Light elements such as gases (He, H) are difficult to detect by EELS, but it is possible with adequate equipment. In the case of EDX, it is the X-rays generated by the de-excitation of atoms, which are ionized by the incident electron beam, that are detected/used for the identification and imaging of elements with atomic numbers greater than beryllium.

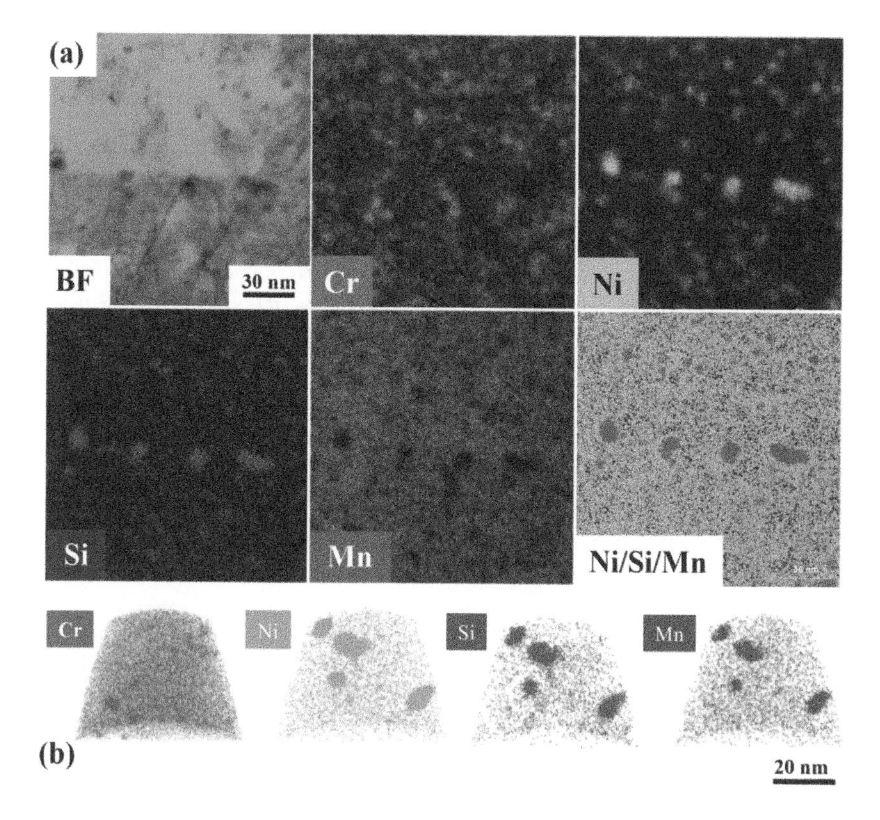

Figure 10.8. *Cr and Mn/Ni/Si-enriched precipitates, induced in HT9 ferritic-martensitic steel by neutron irradiation at 35 dpa in the BOR-60 reactor at 650 K, imaged by (a) STEM-EDX and (b) APT (from Zheng et al. (2019)). For a color version of this figure, see www.iste.co.uk/bouffard/nuclear.zip*

An example of chemical analysis is shown in Figure 10.8. Precipitates formed by segregation during neutron irradiation of a steel are imaged by STEM-EDX (see Figure 10.8(a)) and atom probe tomography (APT). APT is another technique for the characterization of materials under irradiation, and is very useful for observing chemical heterogeneities in a material in three dimensions (such as nanoprecipitates, grain boundaries or dislocations that are depleted or enriched in elements). Its principle is based on the evaporation of surface atoms under the effect of an electric field from a sample in the form of a sharp tip, which are collected by a locating detector that is also acting as a time-of-flight mass spectrometer. The localization makes it possible to calculate the initial position of the atom, and the time of flight allows its mass to be determined. The result is a three-dimensional map of the distribution of atoms in the sample, as shown in Figure 10.8(b).

Other techniques are also used to obtain the depth distribution of chemical elements such as fission products. This is the case for secondary ion mass spectrometry (SIMS) in a dynamic mode, which has excellent sensitivity (down to 10 ppb) with a depth resolution that can reach one nanometer (Darque-Ceretti et al. 2014; Peres et al. 2018). NanoSIMS provides access to a lateral resolution better than 100 nm. Ion beam analysis techniques (such as RBS-C, NRA or ERDA (Trocellier and Trouslard 2002)) are also widely used to obtain the profile of an element that is below the surface of a sample, with a wide range of analyzable depth ranging from a few nanometers to several dozen microns, and with good resolution (of about 10 nm). They can detect light elements, as well as their isotopes, and one of their specificities is that they can perform absolute quantitative analyses without a standard. These measurements are generally averaged over a surface of the order of a millimeter, but are also possible on micrometer surfaces with the help of microprobes.

10.6. In situ microstructural characterization of materials subjected to irradiation

The possibility of in situ measurements, during an irradiation, was mentioned in the preceding sections. These in situ measurements make it possible to access damage kinetics with much higher accuracy, and to highlight the existence of intermediate states, which would have disappeared during an analysis that was time-delayed. Indeed, the analyses can sometimes be carried out several weeks or months after the irradiation, which is mandatory when the sample becomes radioactive following the irradiation.

These in situ characterizations considerably reduce the duration of the experiments, in particular when studying the evolution of the microstructure as a function of the irradiation fluence. Temperature studies can also be facilitated. Moreover, working on a single sample makes it possible to get rid of a possible inhomogeneity of the material and limit the cost of the experiment in a case where the material is expensive and/or with very peculiar specificities. For local characterization techniques such as transmission electron microscopy, in situ studies make it possible to observe, for example, the evolution of characteristic values (size, shape, composition, density) of a given specific object under irradiation (e.g. gas bubbles or precipitates), which is very difficult or even impossible to achieve in ex situ microscopy.

Several irradiation facilities with charged particles are now coupled with in situ characterization equipment. Some examples of experiments and in situ devices are presented below, in addition to those already mentioned in this chapter.

In France, the imaging of extended defects and cavities, as well as the analysis of chemical elements in a sample at the nanometric scale, can be performed by in situ transmission electron microscopy during ion irradiation, with single or double ion beams, in the JANNuS-Orsay experimental hall of IJCLab. The coupling of two ion accelerators within the same facility also allows in situ RBS-C characterizations of material damage induced by irradiation.

Techniques for the characterization of organic materials are also available on site at GANIL in Caen. Infrared (IR) or UV–visible spectroscopy experiments are, for example, performed in situ with high-energy heavy ions. A photoluminescence device, such as the one coupled with an electron accelerator on the SIRIUS platform at LSI, allows the characterization of irradiation-induced modifications in glass. The optical time-resolved system (picosecond resolution) coupled to an electron accelerator, ELYSE, which is available at the ICP in Orsay, can probe the first stages of damage to the material during irradiation (for these very short times, the applications are more related to radiolysis). Several accelerators in France are now equipped with in situ Raman devices (CEMHTI Orléans, Arronax Nantes, JANNuS-Saclay). Here, we will give an example (see Figure 10.9) of the contribution of these in situ systems, by comparing the irradiation of surfaces of UO_2 samples that are in contact with water or with a neutral gas (argon). The strictly identical irradiation conditions make it possible to compare the damage kinetics, estimated by the intensity of the previously mentioned U_1 U_2 U_3 lines. It appears that damage reaches saturation faster under water, probably due to the recombinations of radiolytic species (oxygen) that are migrating from the surface (Mohun et al. 2016).

Lastly, these in situ Raman measurements also make it possible to quantify an important parameter that is difficult to approach under irradiation conditions: the local temperature of the sample in the ion impact zone. If we know the temperature behavior of a solid line through preliminary measurements (the vibration frequency directly depends on the temperature via the anharmonicity (Simon 2020)), then we can estimate the local heating from a frequency shift measured under irradiation. In an experiment done at room temperature on UO_2 polycrystals irradiated by He^{2+}

with an energy of 45 MeV (with a flux of 5.10[11] He^{2+} $cm^{-2}.s^{-1}$), the local temperature on the output side of the ions (with a residual energy of 5 MeV), where the nuclear stopping power is at its maximum, could be estimated at 170°C (Guimbretiere et al. 2016).

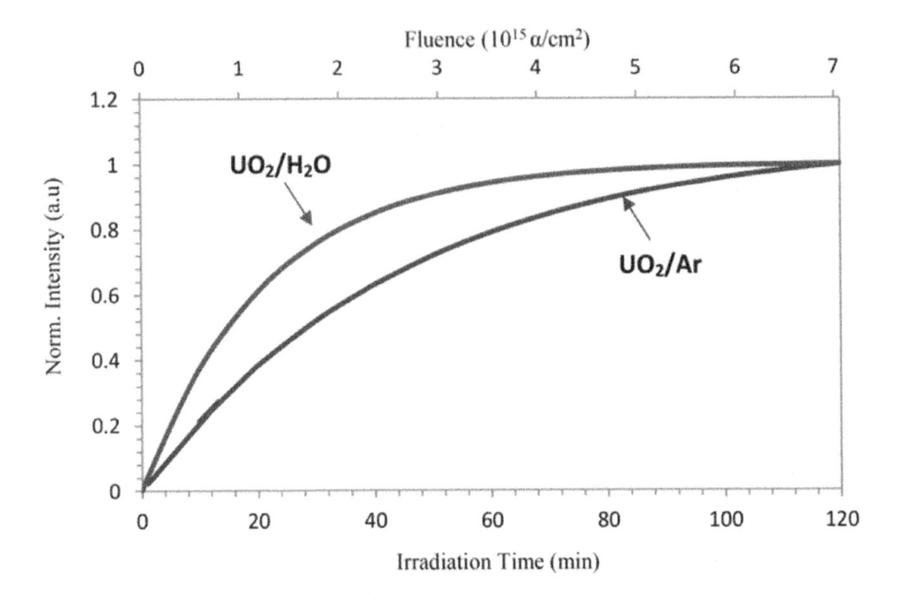

Figure 10.9. *Comparison of the irradiation of two interfaces UO_2/argon and UO_2/water (the irradiation beam crosses UO_2 to reach the studied interface). The irradiations are carried out with He^{2+} ions of 45 MeV, under strictly identical conditions. The intensity measured on site is that of the three Raman lines, U_1 U_2 U_3. The initial damage is faster in the presence of water (effect of radicals coming from the radiolytic decomposition of water), according to Mohun (2017). For a color version of this figure, see www.iste.co.uk/bouffard/nuclear.zip*

10.7. Conclusion and perspectives

This chapter has discussed various common techniques for characterizing irradiated materials, in a non-exhaustive way. Some of these techniques give access to very local information, such as TEM, while others allow a more statistical characterization, in other words, averaged over a larger volume of material. The studies of the modifications induced by irradiation in a material very often require coupling various techniques of characterization, bringing information at various scales. This complementarity of characterization techniques is an important point to obtain a global understanding of the phenomena involved.

The understanding of the phenomena of modification of materials under irradiation may also require the use of in situ characterization techniques during irradiation, with one or more beams of particles. The platforms of in situ characterization have been strongly developed over the last few years. They give access to information that is inaccessible to *post-mortem* characterizations such as fast transients, or complete damage kinetics at high or low temperature.

The development of increasingly efficient tools, in terms of sensitivity and resolution, is leading to a rapid increase in the amount of experimental data to be analyzed. The use of artificial intelligence should be an invaluable aid with significant time savings in extracting relevant data from the sum of acquired results (Li et al. 2018). This should be especially true for high-resolution TEM images, for which atomic-scale crystallographic structure information is difficult for the human eye to extract.

Lastly, in the near future, it would be highly beneficial to have data formats that are compatible with one another, so as to be able to compare data from experiments carried out on different equipment, especially images, and on a larger scale to be able to process results from completely different experiments together (several types of imaging, for example). This ongoing movement towards open data (Srivastava et al. 2020) is rapidly becoming more widespread today.

This open data could facilitate the use of new methods of processing complex data sets, based on algorithms and artificial intelligence learning, as mentioned above, with a significant gain in accuracy in spatial resolution. This could be particularly relevant in the case of complex, multi-parameter systems, such as those generated by irradiation in a material.

10.8. References

Barthe, M.F., Corbel, C., Blondiaux, G. (2003). Caractérisation de défauts lacunaires par annihilation de positons. *Techniques de l'Ingénieur*, 2610.

Bordet, P. (2008). Étude de la structure locale par la fonction de distribution de paires. *Collection SFN*, 9, 139–147.

Darque-Ceretti, E., Aucouturier, M., Lehuédé, P. (2014). Spectrométrie de masse d'ions secondaires : SIMS et ToF-SIMS – Principes et appareillages. *Techniques de l'Ingénieur*, 2618.

Defresne, A. (2016). Amélioration de la passivation de cellules solaires de silicium à hétérojonction grâce à l'implantation ionique et aux recuits thermiques. PhD Thesis, Université Paris-Saclay.

Gentils, A. (2003). Irradiation effects and behaviour of fission products in zirconia and spinel. PhD Thesis, Université Paris Sud, Paris XI.

Griscom, D.L. (1985). Defect structure of glasses – Some outstanding questions in regard to vitreous silica. *Journal of Non-Crystalline Solids*, 73(1–3), 51–77.

Grygiel, C., Lebius, H., Bouffard, S., Quentin, A., Ramillon, J.M., Madi, T., Guillous, S., Been, T., Guinement, P., Lelievre, D. et al. (2012). Online in situ x-ray diffraction setup for structural modification studies during swift heavy ion irradiation. *Review of Scientific Instruments*, 83(1), 013902.

Guimbretiere, G., Canizares, A., Desgranges, L., Caraballo, R., Duval, F., Jégou, C., Magnin, M., Simon, P. (2016). In situ Raman estimation of irradiation-induced heating of UO2. *Journal of Nuclear Materials*, 478, 172–175.

Jégou, C., Gennisson, M., Peuget, S., Desgranges, L., Guimbretière, G., Magnin, M., Talip, Z., Simon, P. (2015). Raman micro-spectroscopy of UOX and MOX spent nuclear fuel characterization and oxidation resistance of the high burn-up structure. *Journal of Nuclear Materials*, 458, 343–349.

Jublot-Leclerc, S., Lescoat, M.L., Fortuna, F., Legras, L., Li, X., Gentils, A. (2015). TEM study of the nucleation of bubbles induced by He implantation in 316L industrial austenitic stainless steel. *Journal of Nuclear Materials*, 466, 646–652.

Jublot-Leclerc, S., Li, X., Legras, L., Lescoat, M.L., Fortuna, F., Gentils, A. (2016). Microstructure of Au-ion irradiated 316L and FeNiCr austenitic stainless steels. *Journal of Nuclear Materials*, 480, 436–446.

Leclerc, S., Beaufort, M.F., Declemy, A., Barbot, J.F. (2008). Evolution of defects upon annealing in He-implanted 4H-SiC. *Applied Physics Letters*, 93(12), 122101.

Li, W., Field, K.G., Morgan, D. (2018). Automated defect analysis in electron microscopic images. *npj Computational Materials*, 4.

Linez, F., Canizares, A., Gentils, A., Guimbretiere, G., Simon, P., Barthe, M.F. (2012). Determination of the disorder profile in an ion-implanted silicon carbide single crystal by Raman spectroscopy. *Journal of Raman Spectroscopy*, 43(7), 939–944.

Louchet, F., Verger-Gaudry, J.-L., Thibault-Desseaux, J., Guyot, P. (1988). Microscopie électronique en transmission conventionnelle et balayage en transmission. *Techniques de l'Ingénieur*, 875.

Lucasson, P.G. and Walker, R.M. (1962). Production and recovery of electron-induced radiation damage in a number of metals. *Physical Review*, 127(2), 485–500.

Lund, A. and Shiotani, M. (eds) (2014). *Applications of EPR in Radiation Research*. Springer, Cham.

Matthiessen, A. and Holzmann, M.V. (1860). On the effect of the presence of metals and metalloids upon the electric conducting power of pure copper. *Philosophical Transactions of the Royal Society of London*, 150, 85–92.

Medyk, L. (2021). L'enjeu de la maîtrise des propriétés locales (stoechiométrie, répartition cationique) lors de la fabrication des combustibles (U,Pu)O2-x : potentialités de la microscopie Raman. PhD Thesis, Université d'Orléans.

Mohun, R. (2017). Raman spectroscopy for the characterization of defective spent nuclear fuels during interim storage in pools. PhD Thesis, Université Aix-Marseille.

Mohun, R., Desgranges, L., Simon, P., Canizarès, A., Raimboux, N., Omnee, R., Jégou, C., Miro, S. (2016). Irradiation defects in UO2 leached in oxidizing water: An in-situ Raman study. *Procedia Chemistry*, 21, 326–333.

Mohun, R., Desgranges, L., Jégou, C., Boizot, B., Cavani, O., Canizares, A., Duval, F., He, C., Desgardin, P., Barthe, M.F. et al. (2019). Quantification of irradiation-induced defects in UO2 using Raman and positron annihilation spectroscopies. *Acta Materialia*, 164, 512–519.

Moisy, F., Sall, M., Grygiel, C., Balanzat, E., Boisserie, M., Lacroix, B., Simon, P., Monnet, I. (2016). Effects of electronic and nuclear stopping power on disorder induced in GaN under swift heavy ion irradiation. *Nuclear Instruments & Methods in Physics Research Section B: Beam Interactions with Materials and Atoms*, 381, 39–44.

Nordlund, K., Zinkle, S.J., Sand, A.E., Granberg, F., Averback, R.S., Stoller, R.E., Suzudo, T., Malerba, L., Banhart, F., Weber, W.J. et al. (2018). Primary radiation damage: A review of current understanding and models. *Journal of Nuclear Materials*, 512, 450–479.

Peres, P., Choi, S.-Y., Desse, F., Bienvenu, P., Roure, I., Pipon, Y., Gaillard, C., Moncoffre, N., Sarrasin, L., Mangin, D. (2018). Dynamic SIMS for materials analysis in nuclear science. *Journal of Vacuum Science & Technology B*, 36(3).

Plantevin, O., Defresne, A., Roca i Cabarrocas, P. (2016). Suppression of the thermal quenching of photoluminescence in irradiated silicon heterojunction solar cells. *Physica Status Solidi (a) Applications and Materials Science*, 213(7), 1964–1968.

Schlutig, S., Solari, P.L., Hermange, H., Sitaud, B. (2010). MARS, a new facility for X-ray diffraction and X-ray absorption for radioactive matter studies. *MRS Online Proceedings Library*, 1264(1), 103.

Simon, G. (ed.) (2020). *Spectroscopies vibrationnelles. Théorie, aspects pratiques et applications.* Editions des Archives Contemporaines.

Souilah, M., Boulle, A., Debelle, A. (2016). RaDMaX: A graphical program for the determination of strain and damage profiles in irradiated crystals. *Journal of Applied Crystallography*, 49, 311–316.

Srivastava, D.J., Vosegaard, T., Massiot, D., Grandinetti, P.J. (2020). Core scientific dataset model: A lightweight and portable model and file format for multi-dimensional scientific data. *PLoS One*, 15(1), e0225953.

Stepanov, S. (2020). GID_SL on the Web. Dynamical X-ray diffraction from strained crystals, multilayers and superlattices at usual and grazing incidence angles [Online]. Available at: https://x-server.gmca.aps.anl.gov/GID_sl.html.

Trocellier, P. and Trouslard, P. (2002). Spectrométrie de collisions élastiques et de réactions nucléaires. Théorie. *Techniques de l'Ingénieur*, 2560.

Zheng, C., Reese, E.R., Field, K.G., Marquis, E., Maloy, S.A., Kaoumi, D. (2019). Microstructure response of ferritic/martensitic steel HT9 after neutron irradiation: Effect of dose. *Journal of Nuclear Materials*, 523, 421–433.

List of Authors

Emmanuel BALANZAT
CIMAP
CEA – CNRS – ENSICAEN
Université Caen Normandie
France

Nicolas BÉRERD
Institut de Physique des 2 Infinis de Lyon
Université Claude Bernard Lyon 1
IUT Lyon 1
CNRS/IN2P3
Villeurbanne
France

Serge BOUFFARD
CIMAP
CEA – CNRS – ENSICAEN
Université Caen Normandie
France

Pascal BOUNIOL
CEA/Paris-Saclay
Service d'Étude du Comportement des Radionucléides
Gif-sur-Yvette
France

Christine DELAFOY
Fuel Design
Framatome
Lyon
France

Christophe DOMAIN
EDF R&D
Département Matériaux et Mécanique des Composants (MMC)
Les Renardières
Moret-sur-Loing
France

Muriel FERRY
CEA/Paris-Saclay
Service de Physico-Chimie
Gif-sur-Yvette
France

Frederico GARRIDO
Laboratoire de Physique des 2 Infinis Irène Joliot-Curie (IJCLab)
CNRS
Université Paris-Saclay
Orsay
France

Aurélie GENTILS
Laboratoire de Physique des 2
Infinis Irène Joliot-Curie (IJCLab)
CNRS
Université Paris-Saclay
Orsay
France

Stéphanie JUBLOT-LECLERC
Laboratoire de Physique des 2
Infinis Irène Joliot-Curie (IJCLab)
CNRS
Université Paris-Saclay
Orsay
France

Sophie LE CAËR
CEA/Saclay
DRF/IRAMIS/NIMBE
UMR 3685
Gif-sur-Yvette
France

Nathalie MONCOFFRE
Institut de Physique des 2 Infinis
de Lyon
Université Claude Bernard Lyon 1
CNRS/IN2P3
Villeurbanne
France

Philippe PAREIGE
Groupe de Physique des Matériaux
Université Rouen Normandie
INSA Rouen Normandie
CNRS
France

Laurent PETIT
EDF
Laboratoire Les Renardières
Moret-Loing-et-Orvanne
France

Yves PIPON
Institut de Physique des 2 Infinis
de Lyon
Université Claude Bernard Lyon 1
CNRS/N2P3
Villeurbanne
France

Jean-Philippe RENAULT
CEA/Saclay
DRF/IRAMIS/NIMBE
UMR 3685
Gif-sur-Yvette
France

David SIMÉONE
CEA/Paris-Saclay
Service de Recherches
Métallurgiques Appliquées
Gif-sur-Yvette
France

Patrick SIMON
Conditions Extrêmes et Matériaux: Haute
Température et Irradiation (CEMHTI)
CNRS
Orléans
France

Magaly TRIBET
CEA
Centre de Marcoule
DES/ISEC/DE2D/SEVT/LMPA
Bagnols-sur-Cèze
France

Index

E, F

G, H

I, M

N, O

P, R

S, T

U, V

W, Y, Z

Printed and bound by CPI Group (UK) Ltd, Croydon, CR0 4YY

20/12/2023

08212804-0005